CHALLENGING THE MODERN SYNTHESIS

CHALLENGING THE MODERN SYNTHESIS

Adaptation, Development, and Inheritance

Edited by

Philippe Huneman and Denis M. Walsh

OXFORD
UNIVERSITY PRESS

OXFORD
UNIVERSITY PRESS

Oxford University Press is a department of the University of Oxford. It furthers
the University's objective of excellence in research, scholarship, and education
by publishing worldwide. Oxford is a registered trade mark of Oxford University
Press in the UK and certain other countries.

Published in the United States of America by Oxford University Press
198 Madison Avenue, New York, NY 10016, United States of America.

CIP data is on file at the Library of Congress
ISBN 978–0–19–937717–6

9 8 7 6 5 4 2 1

Printed by Sheridan Books, Inc., United States of America

CONTENTS

Contributors   vii

Introduction: Challenging the Modern Synthesis—DENIS M. WALSH
and PHILIPPE HUNEMAN   1

PART I: *Adaptation and Selection*

1. Natural Selection, Adaptation, and the Recovery of Development—
   DAVID J. DEPEW   37

2. Why Would We Call for a New Evolutionary Synthesis? The Variation
   Issue and the Explanatory Alternatives—PHILIPPE HUNEMAN   68

3. Genetic Assimilation and the Paradox of Blind Variation—
   ARNAUD POCHEVILLE and ÉTIENNE DANCHIN   111

4. Evolutionary Theory Evolving—PATRICK BATESON   137

PART II: *Development*

5. Evo-Devo and the Structure(s) of Evolutionary Theory: A Different
   Kind of Challenge—ALAN C. LOVE   159

6. Toward a Nonidealist Evolutionary Synthesis—STUART A. NEWMAN 188

7. Evolvability and Its Evolvability—ALESSANDRO MINELLI 211

8. "Chance Caught on the Wing": Metaphysical Commitment or Methodological Artifact?—DENIS M. WALSH 239

PART III: *Inheritance*

9. Limited Extended Inheritance—FRANCESCA MERLIN 263

10. Heredity and Evolutionary Theory—TOBIAS ULLER and HEIKKI HELANTERÄ 280

11. Serial Homology as a Challenge to Evolutionary Theory: The Repeated Parts of Organisms from Idealistic Morphology to Evo-Devo—STÉPHANE SCHMITT 317

Index 349

CONTRIBUTORS

Patrick Bateson is emeritus professor of ethology at the University of Cambridge. Much of his scientific career has been concerned with bridging the gap between the studies of behavior and those of underlying mechanisms, focusing on the process of behavioral imprinting in birds. Another aspect of his work was on the development of play and the induction of alternative pathways in cats depending on their early experience. Subsequently he has published extensively on development and evolution, including the modern study of epigenetics.

Étienne Danchin is director of research at the French CNRS at the Evolution & Diversité Biologique (EDB) laboratoire in Toulouse gathering 60 academics, plus about 50 PhD and postdoc students. He coleads the Laboratoire of Excellence TULIP that regroups 5 laboratories totaling 400 staff. He is an expert in behavioral ecology, on which he wrote a textbook for Oxford University Press. His fields of interest range from social information use in habitat and mate choices to the evolution of coloniality in birds, and nongenetic inheritance (epigenetic and cultural), as well as information flows as determinants of ecological and evolutionary dynamics. He advocates the necessity to generalize the modern synthesis of evolution into an inclusive evolutionary synthesis incorporating all dimensions of nongenetic inheritance.

David J. Depew is professor emeritus of rhetoric of inquiry at the University of Iowa. His work is in the history, philosophy, and rhetoric of evolutionary biology, with special interest in the history and conceptual foundations of the modern evolutionary synthesis. He is the coauthor, with Bruce H. Weber, of *Darwinism Evolving: Systems Dynamics and the Genealogy of Natural Selection* (1995); with the late Marjorie Grene of *Philosophy of Biology: An Episodic History* (2005); and most recently with John P. Jackson of an extended study of interactions between anthropologists and evolutionary biologists in 20th-century America.

Heikki Helanterä is an evolutionary biologist with his main focus in social evolution. His research tackles the causes and consequences of variation in social traits at both individual and group levels, and employs a wide range of methods from theoretical and conceptual work to genomic, population genetic, comparative, and behavioral studies. He has published over 50 papers and is currently a Kone Foundation Research Fellow at the University of Helsinki, Finland.

Philippe Huneman is a research director at the Institut d'Histoire et de Philosophie des Sciences et des Techniques (CNRS/Université Paris I Sorbonne). A philosopher of biology, he works and publishes papers on issues related to evolutionary biology, on the concept of organism and its relations to Kant's metaphysics, and on kinds of explanation in biology and ecology (with a focus on structural/topological explanations). He has edited several books, including *Understanding Purpose: Essays on Kant's Philosophy of Biology* (2007), *Functions: Selection and Mechanisms* (2013); *From Groups to Individuals* (2013, with F. Bouchard); and *Handbook of Evolutionary Thinking in the Sciences* (2015, with T. Heams, G. Lecointre, M. Silberstein). He is coeditor of the book series History, Philosophy, and Theory in Life Sciences.

Alan C. Love is Associate Professor of Philosophy at the University of Minnesota and Director of the Minnesota Center for Philosophy of Science. His areas of specialization are philosophy of biology and philosophy of science. Alan's research focuses on a variety of conceptual issues in evolutionary and developmental biology with an emphasis on philosophical questions about conceptual change, explanatory pluralism, the structure of evolutionary theory, reductionism, the nature of historical science, and interdisciplinary epistemology.

Francesca Merlin is researcher in philosophy of biology at CNRS (IHPST, UMR 8590, Paris, France). She obtained a PhD in philosophy at Université Paris 1—Panthéon Sorbonne in 2009. Her research focuses on central concepts in biology such as chance and probability, inheritance and epigenetics, in particular in the context of evolutionary theory. She currently works on the extension of biological inheritance in the light of the fact that organisms inherit much more than DNA. She is a recipient of the 2010 Young Researcher Prize from the Société de Philosophie des Sciences in France. She published *Mutations et aléas: Le hasard dans la théorie de l'évolution* (2013). She has also coedited, with Thierry Hoquet, *Précis de philosophie de la biologie* (2014).

Alessandro Minelli was full professor of zoology at the University of Padua, Italy, until his retirement in 2011; he has been vice president (1997–1999) of the European Society for Evolutionary Biology. For several years his research focus was biological systematics—the subject of his book *Biological Systematics: The State of the Art* (1993)—but since the mid-1990s his research interests have turned toward evolutionary developmental biology, the subject of his monographs *The*

Development of Animal Form (2003) and *Perspectives in Animal Phylogeny and Evolution* (2009). He is currently serving as specialty editor-in-chief for evolutionary developmental biology of *Frontiers in Ecology and Evolutionary Biology*.

Stuart A. Newman is professor of cell biology and anatomy at New York Medical College, where he directs a research program in developmental biology. He has contributed to several scientific fields, including cell differentiation and pattern formation, the theory of biochemical networks, protein folding and assembly, and the mechanisms of morphological evolution. He has also written on the social and cultural dimensions of developmental biology. He is coauthor (with Gabor Forgacs) of *Biological Physics of the Developing Embryo* (2005), and coeditor (with Gerd B. Müller) of *Origination of Organismal Form: Beyond the Gene in Developmental and Evolutionary Biology* (2003) and (with Karl J. Niklas) of *Multicellularity: Origins and Evolution* (2016). He is editor-in-chief of *Biological Theory*, the journal of the KLI (Klosterneuburg, Austria).

Arnaud Pocheville is a theoretical biologist and a philosopher of biology, currently a research fellow at the University of Sydney, working on the project 'Causal Foundations of Biological Information' headed by Paul E. Griffiths. He holds a PhD in Life Sciences from the École Normale Supérieure, Paris. He advocates that we need new concepts and mathematical approaches to handle cases of time-scales and levels entanglements in biology. He has worked in particular on the evo/devo entanglement in relation to niche construction and non-genetic inheritance.

Stéphane Schmitt is a research director at the French Centre National de la Recherche Scientifique (Paris). He works on the history of the life sciences, especially in the 18th and 19th centuries. He has published many books and papers on the history of anatomy, embryology, and the sciences of evolution, and is the main editor of Buffon's *Œuvres complètes* (2007–, 9 volumes published to date).

Tobias Uller is an evolutionary biologist. His current work focuses on the relations between development, heredity, and evolution by studying the evolutionary causes and consequences of developmental plasticity and parental effects. He has published over 130 papers on a broad range of topics in ecology and evolution and currently holds a Wallenberg Academy Fellowship and a Senior Lectureship at Lund University, Sweden.

Denis M. Walsh is a professor in the Department of Philosophy, the Institute for the History and Philosophy of Science and Technology, and the Department of Ecology and Evolutionary Biology at the University of Toronto. He is author of *Organisms, Agency and Evolution* (2015).

INTRODUCTION: CHALLENGING THE MODERN SYNTHESIS

Denis M. Walsh and Philippe Huneman

The modern evolutionary synthesis arose out of the conjunction of the Mendelian theory of inheritance and the neo-Darwinian theory of population change early in the 20th century.[1] In the nearly 100 years since its inception, the modern evolutionary synthesis has grown to encompass practically all fields of comparative biology—ecology, ethology, paleontology, systematics, cell biology, physiology, genetics, development. Theodosius Dobzhansky's dictum—"nothing in biology makes sense except in the light of evolution" (Dobzhansky, 1973, p. 125)—aptly encapsulates the centrality of the modern evolutionary synthesis to all biological thought. After 100 years the modern evolutionary synthesis stands as one of the best corroborated theories ever devised. And yet, increasingly, it is facing a battery of challenges. Since the advent of this century the challenges have become both more diverse and more trenchant. It has become commonplace in the last few years to hear claims that the modern synthesis should be wholly rethought (Laland et al., 2014), revised, or extended (Pigliucci & Müller, 2010). The essays collected in this volume all in one way or another discuss these various disparate challenges. Together they constitute a comprehensive survey of some of the more acute challenges to the modern synthesis. Individually, each concentrates on a feature of orthodox modern synthesis and the ways that new thinking in biology, driven by empirical advances, might serve to call into question the heretofore rock-hard certainties of the modern synthesis.

---

1. The first intimations of a grand synthesis appear in Fisher (1918).

1.1. The Modern Synthesis

There is probably no definition of "the modern synthesis." It covers too broad a remit, too wide a range of fields in the life sciences to be encapsulated in an apothegm, or a simple set of precepts.[2] Moreover, any system of scientific thought that survives a century is bound to undergo changes over the course of its development. Modern synthesis thinking has kept abreast of the enormous advances in biology of the 20th century and has adjusted and expanded accordingly.[3] Furthermore, major schools of modern synthesis thought have differed on core theoretical issues, for example, on the primacy of natural selection as the motivating force behind evolution (Depew, 2011), the levels of organization at which selection is manifest (Wilson, 1975). The nature of the modern synthesis is itself part of the narrative of 20th-century biology that is constantly being revised and rewritten. For these reasons we refrain from offering anything resembling a definitive account of the modern synthesis. Nevertheless, there are underlying themes, metaphysical and methodological commitments, preferences and practices, that serve to locate a cohesive if somewhat diffuse body of theory. It is sufficiently unified that the challenges embodied in these essays delineate, address, and call into question an overarching system of thought about evolution.

According to modern synthesis thinking the component processes of evolution—inheritance, development, innovation, and adaptive population change—are discrete and quasi-autonomous. They are discrete in the sense that each has its own proprietary cause. Inheritance is simply the transmission from parent to offspring of replicated materials. Development is the implementation of a program, or at least a set of flexible recipes, that exerts control over the phenotype. Novel evolutionary variants arise ultimately from random changes introduced into the genetic code. And adaptive population change is the change in the relative frequency of replicated entities under the influence of natural selection, mediated by the environment. The component processes of evolution are quasi-autonomous in the sense that each operates more or less independently of the others. The process of inheritance is unaffected by the processes that introduce an adaptive bias to form, and by the process of development. Organisms do not inherit what would be advantageous for them to inherit, instead, for better or worse, they get the traits their parents donate to them at conception (Mameli,

---

2. Alan Love (this volume) argues that any one characterization of the modern synthesis is at best partial.

3. E.g., Motoo Kimura's neutral theory of evolution was inspired by the discovery of the vast amount of genetic diversity harbored by natural populations. That in turn came as a response to the advent of molecular techniques for measuring variation.

2005). Novel evolutionary characters (i.e., mutations) are unbiased by the adaptive demands of the organisms in which they first occur. They are said to occur at random.[4] Neither of the processes of inheritance or development introduces evolutionary changes to biological form. The structure of the inherited material is completely unaffected by the downstream developmental processes that turn programs into organisms. What arises anew in development cannot be genuinely inherited. As neither inheritance, nor development, nor mutation is adaptively biased, there must be another, wholly independent process that introduces adaptive change. Adaptive evolutionary change is the sole province of natural selection.

The component processes of evolution may be discrete and autonomous, but there is a single unit of biological organization that is crucially involved in each. Genes are the units of inheritance inasmuch as genes are the only entities that are replicated and passed from parent to offspring. Likewise they are the units of control over the phenotype in development because genes, and only genes, encode the information for building phenotypes. Besides, the material for evolutionary novelties only arises through genetic mutation and recombination: No novel trait counts as an evolutionary trait until it is underwritten by a new gene or gene combination. So, genes are also the units of evolutionary innovation. Moreover, genes are the units of evolutionary change. No evolutionary change occurs in a population unless it is registered as a change in gene frequencies. Adaptive evolution proceeds through the promotion and accretion of good (fitness-promoting) genes, and the elimination of less advantageous ones. Genes stake their place as the canonical units of biological organization, in modern synthesis theory, by virtue of the fact that in playing a pivotal role in each of the component processes of evolution they also unite those components into a single amalgamated process of evolution.

One particularly significant contribution to the early modern synthesis was the efflorescence of molecular biology. At its inception, molecular biology was not an inherently evolutionary program; it was conducted largely in parallel with the growth of evolutionary theory (Morange, 2009; Muller-Wille & Rheinberger, 2012). Yet, the molecular revolution gave to evolutionary biology a concrete, molecular realizer of the *gene* concept (Waters, 1994) and much more. It also demonstrated how the component processes of evolution could be implemented in the molecular phenomena that underpin life. The discovery of the

---

4. Luria and Delbrück's (1943) experiments provided major empirical support for the randomness of mutation. See Danchin and Pocheville's essay (this volume) for a discussion of what the randomness of mutations entails.

Patrick Bateson's essay addresses the modern synthesis conviction that mutation is random. See also Merlin (2010).

structure and function of DNA and the process of protein synthesis, provided a mechanism by which a single molecular entity could be both the source of phenotypic information that underwrites development and the replicated material that is donated by parents to offspring in inheritance. In doing so, it contributed greatly to the fractionation of evolution into discrete component processes. The central dogma of molecular biology, for example, entrenches the separation of inheritance and development in averring that in protein synthesis information flows from DNA to protein, but not in reverse (Maynard Smith, 1989).

Unlike its Darwinian predecessor, the modern synthesis theory of evolution is based on the conviction that evolution is fundamentally a phenomenon that occurs to assemblages of gene ratios, a phenomenon of "genotype space" in Lewontin's (1974) terminology. All the phenomena of interest to evolutionary biologists—the form, function, behavior of organisms, the inheritance of characters, their changes over evolutionary time, phylogenetic relations among lineages—are ultimately grounded in the dynamics of genes. No one can deny the potency of gene-centered evolutionary thinking in biology, nor can anyone cavil at its record of success. Gene-centered evolutionary biology, propelled by the rediscovery of Mendel's laws in 1901 (Müller-Wille & Rheinberger, 2012), suggested that inheritance, one of the principal problems for Darwinism, was underwritten by the transmission of factors. The Danish botanist Wilhelm Johansen introduced the instrumental concept of the gene for whatever it was that played this role (Roll-Hansen, 2009). More critically, Johansen also forged the distinction between the genotype and the phenotype. The genotype-phenotype distinction served to separate two realms of biological existence. There is a domain of genes and a domain of biological form. The distinction allows evolutionary biology to see through the confusion of form and to concentrate on the dynamics of genes. It also permits biologists to think of evolution as being played out in the realm of genotypes. This primacy of genotype space finds its ultimate expression in Dawkins's (1976) gene's eye view of evolution.[5]

The genetic leitmotif in evolutionary biology has other profound consequences too. The gene became a powerful conceptual tool for streamlining and simplifying the study of evolution. In the process it led to a wholesale redefinition of the basic processes of evolution. Prior to the advent of the synthesis, the component processes of evolution had generally been thought to be complexly intertwined. On pre–modern synthesis views, new heritable characters might originate within development (e.g., Haeckel). Development and inheritance might

---

5. Many, of course, deny that the modern synthesis is committed to the gene's eye view.

introduce adaptive bias into the evolution of form.[6] But the modern synthesis depicts these processes and the relation between them very differently. Under the auspices of the genetic turn, inheritance ceases to be a pattern of resemblance between ancestors and descendants; instead it becomes the process by which DNA is replicated and transmitted. This constitutes a significant change of perspective. There is much more to the pattern traditionally taken to be inheritance than can be accounted for by replication and transmission. Development, traditionally construed as the growth, differentiation, and specialization of tissues, cells, and organs, becomes the expression of information expressed in genes.[7] The origin of novelties becomes synonymous with random genetic mutation. The concept of an adaptation is no longer thought of as the fit between an organism and its conditions of existence. To be an adaptation is to be a trait that has been caused by natural selection in the past (Sober, 1984; Brandon, 1990).[8] Evolution itself takes on a gene-centered redefinition. Since Dobzhansky's (1937) classic *Genetics and the Origin of Species*, it has become commonplace to suppose that evolutionary change consists in change in the genetic structure of a population.[9]

I.2. The Synthesis and Its Discontents

So, while there is no pithy definition of the modern synthesis, there are common themes running through modern synthesis thought. The recent challenges to the synthesis all in one way or another call attention to these themes. They all originate from recent empirical advances, especially those in the study of development and inheritance. Each in its way challenges the reconceptualizations ushered in under the modern synthesis. In particular, each challenge questions either the redefinition of the component processes of evolution *or* the presumptive relation between them. They concentrate on the role of inheritance, the role of development, or the connection between selection and adaptation. The essays in this collection take up these challenges.

---

6. David Depew's essay in this volume discusses Spencer's views on the role of the adaptiveness of organisms in evolution.

7. For instance, in the 1980s the morphogenetic fields, a major developmental concept, was interpreted in terms of genetic regulatory networks (Davidson, 1986).

8. Many philosophers of biology will stand to this view (e.g., Sober, 1984; Brandon, 1990). See David Depew's enlightening discussion of the concept of adaptation in this volume.

9. There is a considerable amount of disagreement among proponents of the synthesis concerning what exactly follows from this characterization of evolution (for a discussion, see Depew, 2011, this volume).

I.2.1. Role of Inheritance

There are two broad categories of challenge to the modern synthesis account of inheritance. The first contends that the transmission of replicators is wholly inadequate to capture the significance of inheritance to evolution (Uller & Helanterä, this volume). Inheritance pluralists (Danchin et al., 2011; Jablonka & Lamb, 2004; Jablonka & Raz, 2009; Lamm, 2014; Mesoudi et al., 2013) generally argue that while gene replication/transmission constitutes a legitimate mechanism of inheritance, there are also many others. Nongenetic modes of inheritance include epigenetic, environmental, and cultural inheritance—these must all be factored into our theory of inheritance. Pluralists advocate a broadening of the concept of inheritance (Pocheville & Danchin, 2015).[10]

Among those putative alternative modes, epigenetic inheritance is undoubtedly the most crucial recent challenge to the modern synthesis. In epigenetic inheritance, an acquired pattern of genome expression can be transmitted to an offspring through various mechanisms such as methylation or chromatin states. In principle, and in practice, environmentally induced changes of gene expression can be transmitted to the offspring and grand-offspring of organisms of the focal generation (Danchin et al., 2011; Sultan, 2015). Epigenetic transmission of this sort occurs frequently in plants, but also figures prominently in metazoans through maternal (and sometimes paternal) imprinting.

Epigenetic inheritance is thought by many researchers to pose a powerful challenge to the modern synthesis. It appears to violate the stricture against the inheritance of novel characters that arise during development. That in turn calls into question the modern synthesis conviction that inheritance is a purely genetic phenomenon. Hence, a major set of debates about the fate of the modern synthesis centers on the status and scope of epigenetic inheritance.

Alternative inheritance theorists face strong opposition from supporters of the orthodox modern synthesis approach. Nongenetic transmission exists, traditionalists say, but it is not *evolutionary* inheritance as such (Merlin, this volume). The mechanisms of nongenetic inheritance are proximate mechanisms (à la Mayr, 1961), for securing the resemblance of parents to offspring, especially in stochastic or unpredictable environments, and for establishing cell differentiation. But they are evolved mechanisms, and they are genetically transmitted from parent to offspring, so just like any other genetic trait, they are subject to an ultimate explanation in the terms of the modern synthesis (Dickns & Barton, 2013). The only contribution that non-Mendelian inheritance could make to evolution, on this view, is to provide a mechanism for organisms to respond to their local

---

10. See, for example, the "inclusive inheritance" of Danchin et al. (2011).

conditions without the need for the responses to be genetically coded. But, on this traditional view, nongenetic inheritance and genetic inheritance are fundamentally different processes; only the latter is really *evolutionary* inheritance.

An alternative challenge argues that nongenetic inheritance and genetic inheritance are in fact contiguous; they are fundamentally the same process. This seems to be the line of thought embodied in developmental systems theory (DST), and related views. Developmental systems theory can be seen as a challenge to the *ontology* of modern synthesis inheritance. According to DST (Griffiths & Gray, 1994), what is inherited are not individual genes, or even traits as such, but "developmental systems." A developmental system is the entire suite of causes that hold in place the resemblance of parent to offspring. These include genes, of course, but also properties of cells, tissues, organisms, physical environments, and cultures. The influence of genes on inheritance, construed in this way, is minimal, and not separable from the entire suite of complexly intermingled causes. DST argues that the modern synthesis redefinition of inheritance overlooks the causal complexity and the interrelation of all these factors.

Should the transmission of epigenetic marks, the causal spread of inheritance, be seen as a refutation of the modern synthesis commitment to inheritance as the transmission of replicators? Conversely, should they only be seen as minor contributions to genetic inheritance? Alternatively, should epigenetic inheritance be treated as an adaptation, allowing genetic information to be adapted to a range of (intracellular and environmental) contexts (Scott-Philips, Dickins, & West, 2011)? Many of the essays in this volume raise the question of the status of the traditional modern synthesis conception of inheritance.

I.2.2. Role of Development

The increased attention paid to organismal development (at all levels from genes to whole organisms) has issued in another form of challenge to the modern synthesis. Under the synthesis, development was marginalized, taken to be merely the expression of a program coded in genes for building a reproductive adult. Development played no role in inheritance, in the adaptive bias of evolution, or in the introduction of evolutionary novelties. Prior to the turn of the current century, it was not uncommon to see development dismissed as a mere irrelevance to the study of evolution. Bruce Wallace captures the prevailing sentiment nicely: "Problems concerned with the orderly development of the individual are unrelated to those of the evolution of organisms through time" (Wallace, 1986, p. 149). The genetic turn, and the impetus it gained from the success of molecular biology, appeared to license the exclusion of development from evolution.

After the publication of Darwin's *Origin of Species*, but before the general acceptance of Weismann's views, problems of evolution and development were inexplicably bound up with one another. One consequence of Weismann's separation of the germline and the soma was to make it possible to understand genetics, and hence evolution, without understanding development. (Maynard Smith, 1982, p. 6)

With the advent of evolutionary developmental biology—"evo-devo" (Hall, 1999)—ontogeny has come to take a central place in the study of evolutionary dynamics.[11] According to the modern synthesis orthodoxy the principal contribution of development to evolution was to provide a check on the adaptation promoting power of natural selection. It has since become obvious that development does much more than merely constrain evolution.[12] Most advocates of evo-devo maintain that the insights generated by the study of development threaten the synthesis.[13] At the very least, it seems to call for a substantial extension of the modern synthesis (Pigliucci, 2010), if not a complete replacement (Newman, this volume).

Many evolutionary developmental biologists emphasize the importance of "developmental constraints" for evolution—a theme made famous by Gould and Lewontin's (1979) celebrated paper "The Spandrels of San Marco." Traditional modern synthesis approaches to modeling population change typically assumed that variation is isometric, unconstrained by organismal development. Where there are no such constraints, selection introduces a systematic adaptive bias to evolutionary change. Only untrammeled selection, the reasoning goes, is capable of producing the exquisite adaptations of each organism to its conditions, and each to its other organisms in its environment, and each part to the whole, that Darwin saw so vividly exemplified in the "entangled bank." In contrast, evo-devo has highlighted the massive constraints exerted on variation by development (e.g., Wake, 1992). They are a pivotal and wholly separate cause of the distribution of form. Only a theory of development can investigate them (see, e.g., Minelli & Pradeu, 2015), which in turn calls for a synthesis between evolution and development). By the close of the 20th century, the traditional marginalization of development (Mayr & Provine, 1980; Wallace, 1986) came under severe scrutiny. An increasing awareness of the importance of developmental constraints (e.g., Maynard-Smith et al., 1985) has been instrumental

---

11. See Alan Love's essay in this volume.

12. See the essays by Depew, Walsh, Bateson (this volume).

13. Here too, the perception of the severity of the challenge varies (see Raff, 1996; Müller, 2007).

in the calls for a synthesis between the study of development and the study of evolution (Hall, 1999).

One of the most conspicuous features of organismal development is its plasticity (Moczek et al., 2011; Sultan, 2015; West-Eberhard, 2003). The plasticity of development buffers organisms against the vagaries of the environment and of their genes. Plasticity is the basis of evolvability (Minelli, this volume). It confers on organisms the capacity to innovate, to initiate novel phenotypes that are adaptive immediately on their first occurrence (Kirschner & Gerhard, 2005; West-Eberhard, 2003). Plasticity can assist in the reliable occurrence of a novel phenotype across generations.

The plasticity of organisms in general includes the capacity to make changes in the face of perturbations that ensure survival and stability. Given the various modes of inheritance that pluralists advocate, these changes can be intergenerationally stable, and can become entrenched over generational time.

The adaptive plasticity of development poses two problems for the modern synthesis approach to selection and adaptation. First, it raises the prospect that a phenotypic novelty can be "an adaptation" at its first appearance. If so, that necessitates a revision of the modern synthesis concept of adaptation. According to modern synthesis orthodoxy, adaptation is an etiological concept. An adaptation is a trait that has been promoted by selection *in the past*. If plasticity can initiate traits that are adaptations on their first occurrence, then the concept of an adaptation is properly not considered an historical one. A further implication of adaptive plasticity is that the production of evolutionary novelties appears to be nonrandom (Wagner, 2014; see also Pocheville & Danchin, in this volume). It is biased by the capacities of organisms to produce adaptive responses to their conditions. However, it is a cornerstone of modern synthesis thinking that evolutionary novelties are produced at random (Merlin, 2010). Mutation may be random, but it does not follow that the production of phenotypic novelties is. There is a long-standing tradition of drawing a distinction between *evolutionary* adaptation (the change of form over time), and *physiological/behavioral* adaptation (the capacity of organisms to respond to local conditions). These are usually held to be fundamentally different processes. Physiological behavioral adaptation is generally treated in the same way as any other adaptive trait, as the effect of natural selection, and not as a cause of adaptive evolution (Pocheville & Danchin, 2015).[14] Development is adaptive because evolution is adaptive. These new considerations make a tantalizing suggestion. If anything, the direction of dependence goes the

---

14. Essays in Depew and Weber (2005) and Bateson and Gluckman (2011) offer attempts to finesse the evident contribution of plasticity, behavioral adaptation, and learning to adaptive evolution.

other way; evolution is adaptive because development is adaptive (Walsh, 2015; West Eberhard, 2005).

Plasticity seems to suggest that development participates in all the component processes of evolution—inheritance, adaptive change, and innovation. Its marginalization from modern synthesis evolutionary thinking is starting to look like a major defect.

However, while it is certainly true that the significance of organismal development to evolution has received heightened attention in recent years, there is no consensus on how development should be incorporated into evolutionary theory (Love, this volume). At one extreme, the study of the way genes contribute to development has invited a renewed, sophisticated form of gene centrism (e.g., Carroll, Grenier, & Weatherbee, 2005). Genes produce phenotypes as parts of complex gene networks (Davidson, 1986). Thanks to developmental genetics, genes are no longer thought of as "units" of phenotypic control, or as a mere recipe or program for producing an organism. This has turned up surprising new results about the process of evolution, for instance that regulatory genes, rather than coding genes, are drivers of evolution. Evolution relies on the dynamics of developmental genes: modifications of gene regulatory networks rather than gene substitution (Wagner, 2011), protein reuse (Wagner, 2014), cis-regulatory evolution (Carroll et al., 2000), and genetic toolkit preservation (Gerhard & Kirschner, 2007). From this perspective, while genes do not contain instructions for discrete traits, they are nevertheless the appropriate unit of study to reveal the evolutionary significance of development.

At the other extreme, some authors advocate a shift away from the significance of genes to an emphasis on the role of whole organisms in regulating developmental processes such as morphogenesis. On this view, genes work in the context of complex, adaptive self-organizing systems. On this approach, the understanding of gene function requires that we take into account the fact that the function of genes is heavily dependent upon their place in the activities of adaptive, plastic organisms (Moczek, 2012). Some researchers propose that the organization and evolution of form is sometimes better understood using the tools of physics rather than genetics (Newman & Bhat, 2009).

Advocates of a new theoretical framework for evolution stress the empirical considerations that motivate their call for change. But the empirical findings alone appear not to be decisive (Pocheville & Danchin, this volume). In a recent publication in *Nature*, defenders of more orthodox styles of evolutionary thinking (Wray et al., 2014) responded robustly to those calling for a root and branch "rethink" of evolutionary theory (Laland et al., 2014). One striking feature of the debate is that both sides agree on the empirical matters of fact. They disagree on the capacity of current evolutionary theory to accommodate them.

"Rethinkers" argue that phenomena such as maternal transmission of proteins or "maternal effects" (Badyaev & Oh, 2008), niche construction (Odling-Smee, Laland, & Feldman, 2003), and ecological development (Gilbert & Epel, 2009; Sultan, 2015), to list but a few examples, fall beyond the framework laid down by the founders of the modern synthesis. Traditionalists hold that these phenomena have long been recognized by evolutionary biologists and, where they are relevant, can be assimilated into evolutionary thinking without any they major adjustment. If the empirical matters of fact are not under dispute, the controversies must be conceptual. Should the documented cases of intergenerational transmission of acquired epigenetic marks (e.g., Weaver et al., 2004), for example, be seen as a reason to give up on the discreteness of development and inheritance? Conversely, should they only be seen as minor contributions to genetic inheritance? Alternatively, should epigenetic inheritance be treated as an adaptation, allowing genetic information to be adapted to a range of (intracellular and environmental) contexts (Scott-Philips et al., 2011)? Many of the essays in this volume take up the conceptual issues at the heart of the debate about the role of development in evolution.[15]

I.2.3. Adaptation and Natural Selection

One of the great dividends of 20th-century modern synthesis evolution is the elegant treatment it gives to the phenomenon of adaptation (and adaptive evolutionary change). The "exquisite contrivances" of organisms, the way they are equipped to survive in their environments, and the way in which all organs in complex organisms are attuned to the functioning of the whole, all result from the cumulative action of natural selection (Darwin, 1862). Modern synthesis versions of Darwinian evolution suggest that adaptive change in populations is gradual, driven by continuous heritable variation, mediated by the external environment. The process is elegantly captured by simple equations like the Price equation (Price, 1970; Uller & Helanterä, this volume) and the breeders' equation. Yet the insinuation of organisms—their development, ecology, and behavior (Gilbert & Epel, 2009; Bateson & Glickman, 2011)—into the account of adaptive evolution complicates the picture. A number of the implications can be seen in contributions to this volume.[16] If adaptation is not the gradual accumulation of minutely favorable random mutations, what is it?

---

15. See essays by Pocheville and Danchin, Merlin, Uller and Helanterä, and Schmitt, in this volume.

16. See essays by Depew, Huneman, Bateson, Newman, and Walsh, in this volume.

This is the principal challenge to the primacy of natural selection as the cause of adaptive evolutionary change (Burian, 1983). The challenge arises from two considerations. The first is that organisms are themselves fundamentally adaptive, able to initiate adaptive innovations that promote their own survival. David Depew (this volume) identifies the problem that the adaptiveness of organisms poses for the conceptual connection between selection and adaptation. If adaptive evolution is biased by the activities of organisms, then it is not exclusively driven by random variation and environmental selection. Patrick Bateson (this volume) stresses the significance of organismal behavior as a buffer and driver of evolution. Behavior, of course, is adaptive in the sense that the behavior of organisms largely consists in the responses by organisms that exploit or ameliorate the affordances of their environment. Pocheville and Danchin (this volume) present a novel account of genetic assimilation that is a direct challenge to the classical view of adaptation. Walsh (this volume) argues that the plasticity of organisms introduces an adaptive bias to evolutionary change. In addition, mechanisms of non-Mendelian inheritance provide other sources of adaptive evolutionary change (Uller & Helanterä, this volume).

Organisms are able to modify their environments. Beavers build damns, termites construct massive mounds, birds manufacture intricate nests. In this way, organisms are capable of altering the very conditions to which biological form evolves. Richard Lewontin (1978, 1983) has steadfastly advocated for including the influence that organisms exert over their environments as one of the principal factors of evolution. Lewontin's challenge has recently found an elegant articulation in niche construction theory (Odling-Smee et al., 2003). Niche construction theory acknowledges a reciprocity between organisms and their environmental conditions. Just as organismal form is adapted by natural selection to meet the exigencies of external environmental conditions, those very conditions are subject to alteration by organisms. This theory consistently draws on seminal ecological studies of organism-environment relations, such as those of Clive Jones, whose "ecological engineering" studies sought to measure the effects of organismal activities on their environments in precise physicochemical terms (Jones, Lawton, & Shachak, 1994.

Niche construction is an old idea. Darwin carefully documented the ways in which the activities of earthworms, especially their excretions, altered the structure, moisture content, and pH of the soil. It has long been understood that the production of O_2 by cyanobacteria the early atmosphere of the Earth. Yet, the significance of niche construction for evolution had been comprehensively overlooked until the advent of the 21st century. Proponents of niche construction argue that it complexifies the idea of natural selection as the "principal agent" of evolutionary change (Huxley, 1940). Niche constructionists argue (as Lewontin does) that concentrating on the way that the external environment affects biological form yields only part of the story of adaptive evolution. Organisms change their environment just as much as environment changes form.

Another general challenge, voiced by Stuart Newman (this volume), is that what we call adaptive evolutionary change is largely the consequence of passive changes in form imposed on organisms by the physics of development. The responses of biological form to its conditions are not guided by genes, but by physical principles governing the behavior of medium scale viscoelastic, excitable materials. From this perspective, it is to be expected that adaptive changes can be rapid and discontinuous, in contrast to the gradualism that is so central to the orthodox modern synthesis.

This is a mere cursory survey of the various ways in which modern synthesis evolutionary theory is being challenged by the empirical advances of 21st-century biology. The variety of causes of intergenerational stability, the adaptive plasticity of organisms, the role of epigenetics in inheritance and development, the alteration of environments by organisms, these factors, and many more, seem to call for a reassessment of the gene-centered evolutionary biology of the 20th century. If the modern synthesis is to survive, it must show itself to be capable of assimilating these empirical advances without forfeiting its guiding principles. It is an important theoretical, historical, and philosophical exercise to hold early 21st-century biology up against 20th-century evolutionary theory and to ask how much the former challenges the latter, and how well, if at all, the latter can accommodate the former.

1.3. The Challenges

The chapters in this volume all address the challenges posed to the modern synthesis by 21st-century biology. The challenges, in one way or another, concern inheritance, development, and adaptation—and the relations between them. The authors do not speak in one voice concerning the severity of these challenges, or the need to reframe or reject the modern synthesis in response. Some of the authors in this volume take a harder revisionary line (Bateson; Newman; Uller & Helanterä; Walsh). Others see the new biology as raising another sort of challenge, a pragmatic one (Depew; Huneman; Love; Merlin; Pocheville & Danchin). Some argue that gene-centered and organism-centered views are complementary perspectives on the same evolutionary phenomena (see especially Love). Other contributors develop a strongly historical perspective on these current debates (Depew; Schmitt). The chapters that follow are organized along the lines of the three general kinds challenges to the synthesis: challenges to adaptation and selection, challenges arising from the consideration of development, and challenges to the modern synthesis conception of inheritance.

I.3.1. Part I: Adaptation and Selection

Darwin's theory of descent with modification is quite obviously a theory of adaptation. As advantageous traits are passed on from generation to generation, and disadvantageous traits winnowed out, a population becomes progressively suited to its conditions of existence. Darwin's theory also establishes a close conceptual connection between natural selection and adaptation. Natural selection, we are told is "daily and hourly scrutinizing, . . . every variation, even the slightest; rejecting that which is bad preserving and adding up all that is; silently and insensibly working, . . . at the improvement of each organic being in relation to its organic and inorganic conditions of life" (Darwin, 1859/1968, p. 133). Yet the relation between selection, adaptation and population change has come under heightened scrutiny in recent years.[17] The authors in this section outline challenges to the traditional modern synthesis view of selection as the principle cause of population change.

I.3.1.1. Chapter 1: David J. Depew—Natural Selection, Adaptation, and the Recovery of Development

David Depew's essay outlines in detail the delicate relation between selection and adaptation. He argues that any account of the relation must observe the "intricate balance" between three aspects of natural selection: the role of chance, pressure from the environment, and telic phenomena (e.g., goodness of fit, function). However we are to understand adaptation, it must not introduce any conceptual incoherence to natural selection.

The issue, it turns out, is as old as the concept of natural selection itself. Depew contrasts Spencer's understanding of the relation between selection and adaptation with Darwin's. While Spencer adopted Darwin's concept of natural selection, it was not, for him, the principal cause of the fit of organisms with their environments. For Spencer, organisms are fundamentally adaptive entities. They respond to the exigencies of their environments by altering themselves. The adaptedness of organisms is the result of "the mutual adjustment of all their parts into a coherent whole rather than as aggregated independent traits" (Depew, this volume, p. 40), through the direct action of the environment. As a result of a

---

17. See Orr (2007) and Orr and Coyne (1992) for an illuminating discussion of the difficult association between selection, as construed population genetics, and adaptation.

general conflation of Darwin's view and Spencer's, the import of Darwin's creative conception of selection was largely lost until its reemergence in the guise of the Weismann-inspired neo-Darwinism around the turn of the century. On the neo-Darwinian reading, the adaptedness of organisms plays no role in the capacity of selection to drive adaptive evolution.

The modern synthesis is often thought to have "undermined the relevance of ontogeny to evolutionary processes" (p. 48). But, Depew stresses, it was never the intention of those who forged the synthesis (e.g. Simpson, Mayr, Dobzhansky) that it should do so. Many early synthesists saw the reliance on population genetics not as a usurpation of organisms but merely as a means of representing, or projecting, the inordinately complex intertwined, developmental, and ecological causes of evolution. It may well be convenient to see organisms as the adventitious construct of selfish genes, and genes as the units of evolutionary adaptation in this way. But as Depew argues, "there is, however, no way to translate this exercise in 'seeing as' into realistic biology without denying, and not merely ignoring, the ontogenetic nature of organisms and without disturbing the delicate balance between chance, determinism, and purpose in the theory of natural selection" (p. 52).

The fanciful nature of this exercise in "seeing as" is brought into focus by the enhanced understanding of the regulatory relations among genes. Depew cites the case of stickleback evolution, in which the same fresh-water adaptation occurs repeatedly and independently through alterations to the *Pitx1* homeobox gene. "This suggests a role for the environment in inducing rapid changes in gene regulation that pushes against or even goes beyond the conceptual limits of the modern synthesis" (p. 56). The upshot is that response to the environment, in the Spencerian sense, rather than the gradual accretion of good gene combinations through iterated selection, is the font of all adaptation. In this way, development is reinsinuated into the explanation of adaptive evolutionary change in a way that appears to challenge the precepts of the modern synthesis.[18] Depew's discussion serves to remind us that the intensified interest in development that has characterized the evolutionary biology of our own century in itself raises no particular challenge to the core of the modern synthesis.[19] Only those developmentalist strains that call into question the modern synthesis commitment "that trans-generational natural selection is the cause of the adaptations

---

18. A remarkably convergent thesis can be found in Bateson claim of the role of the "adaptability driver" in evolution (this volume).

19. See also, the discussion in Huneman (this volume).

that organisms considered as members of adapted populations possess" (p. 29) pose a genuine threat.

I.3.1.2. Chapter 2: Philippe Huneman—Why Would We Call for a New Evolutionary Synthesis? The Variation Issue and the Explanatory Alternatives

Like Depew, Philippe Huneman seeks to articulate the core of the modern synthesis, and to inquire what sort of empirical observations should motivate its rejection. Huneman locates the definitive commitments of the modern synthesis in no less authoritative a source than Julian Huxley. In a letter to Ernst Mayr in 1951, Huxley wrote, "Natural selection, acting on the heritable variation provided by the mutations and recombination of a Mendelian genetic constitution, is the main agency of biological evolution" (Huneman, this volume, p. 77). There are two features to note in this passage. The first is that the raw material of evolution is genetic. The second is the primacy of selection. Huneman takes this latter claim to mean not just that selection is a "principal power" of evolution, but also that it gives us a definition of adaptation. To be an adaptation is to be the effect of selection in the past. Gene-centrism and adaptationism are the heart and soul of the modern synthesis. Any challenge to these claims is a challenge to the theory.

Adaptationism traditionally holds, at least as a working hypothesis, that any trait under investigation is optimally adapted to its circumstances. But there are two well-known kinds of conditions that might prevent a trait from realizing optimality: trade-off and constraint. Trade-offs to adaptedness are readily conceded by advocates of the modern synthesis. But constraint challenges the synthesis in two ways (Gould, 1977, 2002). It challenges both the gene-centric explanation of form and the conviction that adaptedness arises only from selection. The capacities that organisms manifest in development cause the adaptability of form. Adaptation thus comes for free as a consequence of developmental processes.[20] Much of evolutionary developmental biology seems to challenge the second of Huxley's two pillars of the modern synthesis, the efficacy of selection. Natural selection may not be the best way to explain the fit of form to its conditions of existence.

Gene centrism and adaptationism make certain demands on variation, not just on the kind of variation but also on its distribution. If, as Huxley avers, the grist to the evolutionary mill is random mutation, then we should expect that the variation the populations present to selection is isotropic. This is usually

---

20. Newman (this volume) provides an extended discussion.

thought to be achieved by the randomness of mutation. Huneman argues that nonrandom genetic mutation would not pose a particular challenge to the orthodox modern synthesis. However, biased phenotypic variation would. It is here that the recent emphasis on phenotypic plasticity becomes highly relevant to the question of the adequacy of the modern synthesis. Where development is plastic, organisms themselves are the cause of the distribution of variants, and that distribution is biased toward adaptedness. If so, adaptive phenotypes are not fully caused by selection, in violation of Huxley's dictum. Still, Huneman argues, phenotypic plasticity may not precipitate a wholesale rejection of the modern synthesis. The status of plasticity is controversial. It may well itself be an adaptation, to be explained in the orthodox way. "Hence no empirical finding about "plasticity" is likely to settle the issue about a putative explanatory shift, because it already supposes conceptual assumptions about what is plasticity and before that, about what adaptation is" (p. 110). Some authors, notably Stephen Jay Gould (1977), have argued strenuously that heterochrony is a major cause of evolutionary change. Changes in developmental timing alter the distribution of variants in a population. Yet, as Huneman points out, heterochrony is most likely the consequence of changes in the actions of regulatory genes, as emphasized since seminal work by Gehring (1998) and Duboule (1990). This phenomenon can certainly be accounted for by the orthodox modern synthesis. All in all, Huneman concludes, there are very few conceivable empirical discoveries that would force a wholesale revision of the modern synthesis theory of evolution, at least with respect to the issue of directed or undirected variation.

I.3.1.3. Chapter 3: Arnaud Pocheville and Étienne Danchin—Genetic Assimilation and the Paradox of Blind Variation

It has long been recognized that some evolutionary patterns are hard to square with the modern synthesis model of evolution by natural selection on random genetic variations. Genetic assimilation is one of the most commonly proposed ways of reconciling these phenomena with the modern synthesis. Genetic assimilation is "a situation where some phenotypic variations, that are initially caused by environmental events, eventually get caused by genetic variations" (Pocheville & Danchin, this volume, p. 129). Pocheville and Danchin discuss two new models of genetic assimilation, contrasting them with more classical accounts. The principal difference between traditional approaches and those advanced by Pocheville and Danchin is that the latter do not rely ultimately on blind genetic variation. By integrating nonblind epigenetic variation, they develop a series of models in which genetic assimilation accelerates evolution. The first of these models involves "epigenetic mutagenicity," where epigenetic mutations move a population to an adaptive peak, which is later fixed by blind genetic mutation. This model clearly is consonant with modern synthesis thinking.

The second of these models involves "heritable physiological exploration," in which development adapts organisms to their conditions, stabilizes the new adaptive forms, and then transmits them epigenetically. This second model invokes intragenerational variation. Yet, as Pocheville and Danchin point out, this factor is often blackboxed in the standard models of genetic assimilation. Simulations indicate that such systems would accelerate adaptive evolution "by several orders of magnitude" (p. 134). Pocheville and Danchin claim that this is a plausible model for many cases of intertwining between epigenetic variation and adaptive evolution. This model clearly falls beyond the boundaries of the modern synthesis. Yet, it identifies a plausible evolutionary mechanism. There are even putative examples to be found in *Drosophila* and yeast.

It is to be noted that while these models involve "embedded exploration" by the processes of development, they still posit blind variation at the genetic (and even epigenetic) level. "'Selection' . . . of blind variation is still the only explanatory resource accounting for adaptation." Therefore, against some interpretations of the lessons of epigenetic inheritance (especially Jablonka & Lamb, 2002) they do not consider that novel dimensions of inheritance pave the way for a Lamarckian reading of inheritance.

These models raise the question what role the modern synthesis commitment to blind variation is supposed to play. Blindness is notoriously hard to define. Pocheville and Danchin propose a formal characterization of the tenet of blind variation. They argue that the commitment to blind variation should not be understood solely as an empirical claim. It is an indispensable explanatory principle.[21] Blindness allows us to "explain adaptations that literally come from nowhere, in that blindness is precisely that which does not seem to require an explanation. This is probably where the core of the neo-Darwinian paradigm lies" (p. 149).

I.3.1.4. Chapter 4: Patrick Bateson—Evolutionary Theory Evolving

Patrick Bateson surveys the various ways in which the central commitments of the modern synthesis, as it was forged in the 1930s, have been progressively undermined by the evolutionary biology of the late 20th and early 21st century. According to Bateson, the "Neo-Darwinian orthodoxy asserts that speciation results from a slow process of natural selection, that mutations in genes drive evolution with the organism passively playing no role, and that

---

21. See also the essays by Huneman and Walsh (this volume)

developmental processes are irrelevant to an understanding of evolution" (p. 156). Bateson documents how each of these presuppositions has been decisively overturned.

If species diverge through the accumulation of genetic differences of small effect, then we should expect that speciation should be gradual. Evolutionary thinkers of the late 19th and early 20th century proffered arguments in support of discontinuous speciation. Galton and (William) Bateson (who coined the very term "genetics") argued for discontinuous evolution. Bateson's reasoning is particularly prescient. He hypothesized homoeotic changes in development, that is to say developmentally induced changes to serially homologous structures within an organism, as a driving force behind evolution.[22] Homoeotic mutations have since been one of the central areas of research in developmental genetics, thanks to the discovery of homeobox (e.g., *Hox*) genes in the 1980s. Similarly, Goldschmidt (1940) postulated discontinuous mutations as one of the possible sources of the origin of major taxa. But as the synthesis grew (or "hardened" in Gould's 1979 terminology) evolutionary biologists came to concentrate on selection on continuously varying populations as the principal source of divergence. Yet, even here, as William Bateson also demonstrated, where epistatic effects occur between genes, speciation can be rapid.[23]

One of the most prominent trends of late 20th-, early 21st-century biology has been the emphasis on the active role of organisms in constructing, stabilizing, and altering the conditions under which selection happens. This may occur through active mate choice, or behavioral choice, through control of the environment (Odling-Smee et al., 2003), or as Bateson documents, through the "adaptability driver." Under this label, Bateson rethinks what is sometimes called the Baldwin effect. While the adaptability driver was roundly dismissed as either a conceptual impossibility, given the precepts of modern synthesis evolutionary biology, or at best a peripheral influence on evolution throughout much of the 20th century, it is now accepted as a significant contributor to evolution. The learning of complex adaptive behaviors can make conditions propitious for the evolution of developmentally robust behaviors. Bateson argues that one of the principal dogmas of the modern synthesis—namely, that environmentally induced changes cannot propel evolution—is routinely violated with the methylation of genes, a major aspect of what is known as "epigenetic inheritance." Methylation patterns can be environmentally induced, possibly in some rare cases transmitted from generation to generation, and can positively contribute to fitness.

---

22. See Stéphane Schmitt's entry in this volume for an historical discussion.

23. See also Stuart Newman's essay in this volume.

In all, Bateson offers a battery of alternative processes and phenomena that suggest that the role of the environment, of development, of behavior in evolution have been vastly undersold in the orthodox modern synthesis throughout the later part of the 20th century. Whether or not these evolutionary factors are incompatible with modern synthesis thinking is an issue of some considerable debate these days (see Wray et al., 2014, cited earlier). Bateson suggests that 21st-century evolutionary biology is entering a phase of pluralism, in which a multitude of evolutionary processes, tempos, and modes is countenanced, much in the manner of late 19th- and early 20th-century evolutionary biology. Whether the modern synthesis can survive this expansion of the conception of evolution is a question that remains to be answered.

I.3.2. Part II: Development

Many of the challenges currently faced by the modern synthesis arise from the increased emphasis on the study of development that began to coalesce in the later decades of the 20th century. The four essays in this section discuss the various ways in which the emphasis of development has altered the conception of evolutionary theory.

I.3.2.1. Chapter 5: Alan C. Love—Evo-Devo and the Structure(s) of Evolutionary Theory: A Different Kind of Challenge

In the milestone volume *The Evolutionary Synthesis: Essays on the Unification of Biology* (Mayr & Provine, 1980), Michael Ghiselin lamented the neglect of *morphology* in the modern synthesis (Ghiselin, 1980). Ghiselin followed E. S. Russell, who claimed that, since Aristotle, biology had been partitioned into in two traditions, "biology of function," and "biology of form" (Russell, 1911). Darwin's evolutionary biology, and its modern synthesis descendant, it is said, concentrated on the former to the detriment of the latter. Amundson's (2005) historical inquiry into the functionalist/structuralist divide suggests that the breach between the functionalist modern synthesis and the structuralist evo-devo is unbridgeable. Alan Love surveys a loose family of such "developmental form challenges" to the modern synthesis. They are "form" challenges in that each of them charges that the modern synthesis accords too little importance to the explanation of biological form and those processes that generate form.

Love notes the pervasive use of spatial metaphors in articulating these challenges. They are often referred to as potential "expansions" or "extensions" of the modern synthesis. Often the proposed relation between the orthodox synthesis and its amendments is accompanied with a Venn-like diagram, with the putative core commitments of the modern synthesis in the middle, and various of the form-centered evolutionary factors in the periphery. As Love points out, this talk of what evolutionary theory "contains," or "should contain" underestimates the complexity of evolutionary theory, and mischaracterizes the nature of developmental form challenges. These, properly construed, are challenges to the *structure* of evolutionary theory. They pose questions about what the objectives of evolutionary theories should be (e.g., the explanation of form), what the explanatory strategies should be (e.g., causal-mechanistic), and how the various problem agendas of biology (e.g., the origin of novelties, ecology, systematics) are related. But it is not altogether evident that the various form challenges recommend the same, or even compatible, structures for evolutionary theory. Arguably, as Love points out, the structure of evolutionary theory presupposed by Sean Carroll and colleagues, is different from that envisaged by Gerd Müller and his colleagues, although they both fall roughly in the family of "evo-devo" motivated challenges.

Love thinks of these different structures not as theories of evolution per se, but as theory *presentations* (à la Griesemer, 1984). Theory presentations are guides to scientific practice. They show a topology of relations between particular concepts, procedures, inference patterns, and the like. Completeness is not among the objectives of a theory presentation. They are always at best partial. Theory presentations are generated by idealization, by leaving certain details out, by backgrounding others. "Applied to a theory presentation, an idealization intentionally departs from features known to be present *in the theory*. Researchers in the community of evolutionary biology knowingly ignore aspects of the theory, such as excluding variables that are clearly relevant to understanding evolutionary processes" (p. 194). An inordinately complex body of theory, like that of evolution, can sustain any number of workable, complementary, yet seemingly incompatible partial representations. Seeing these developmental form challenges as *structural* challenges reveals something important about the project of scientific inquiry; it is pluralistic in nature. Love quotes Kellert et al. (2006, pp. ix–x): "The explanatory and investigative aims of science [may] be best achieved by sciences that are pluralistic, even in the long run" (Love, this volume, p. 197). This pluralism brings into relief an underappreciated feature of the current conflict between the modern synthesis and its contenders.

> [It] is a kind of pseudo-conflict because it relies on forgetting the idealization choices made in articulating a particular theory presentation. These divergent structures serve different sets of methodological and

epistemological goals in evolutionary theory, and a pluralist stance can recognize the fruitfulness of maintaining each of these reasoning strategies. (p. 196)

I.3.2.2. Chapter 6: Stuart A. Newman—Toward a Nonidealist Evolutionary Synthesis

According to Stuart Newman, the challenge from biological form cuts to the very heart of modern synthesis evolutionary thinking. Newman sees the origin of modern synthesis gene-centered theory growing out of certain foundational convictions. These include the tight correlation between genotype and phenotype, the role of the physics of development as mere background conditions, the insistence on evolutionary change being gradual and incremental, and the conviction that there "is no medium other than genes by which specific organismal features could be transmitted from one generation to the next" (p. 211). These articles of faith, Newman believes, were enacted without sufficient empirical support, and without due heed being paid to way the physics of medium-scale processes influence the structure of form. For these reasons Newman thinks of the modern synthesis as an "idealistic theory." A "philosophically materialist" alternative would give appropriate consideration to the ways that the physics of biological materials influence the evolution of form.

Gene-centered gradualism has difficulty accounting for three puzzles of the large-scale evolution of the metazoans (at least). These are: the striking conservation of toolkit genes over hundreds of millions of years, the almost instantaneous appearance of all the basic forms of metazoans during the Avalon explosion, and the "embryonic hourglass" motif of animal evolution. Newman's "materialist" explanations make sense of them.

Newman's explanation begins with the role of dynamic patterning modules (DPMs) in the construction of metazoan body plans. DPMs are organizing principles that operate at the level of multicellular assemblages. They are the consequence of the coupling of gene products with the generic physical features of the cells and the tissues in which they occur. The DPMs regulate the shape and structure of agglomerations of cells. They cause the formation of lumina, tubes, discrete tissues, tissue layers, segments. The general point is that the basic structures, the motifs that combine to make up the metazoan body types, are not under the exclusive control of genes, but are the consequence of the interaction of gene function and generic physical principles large aggregates of cells.

The Avalon explosion suggests that the entire suite of major metazoan body plans arose abruptly. The sheer rapidity of these occurrences raises a prima facie problem for the gradualism at the heart of the modern synthesis. The puzzle diminishes when we consider the role of DPMs in generating the body plans. The various different phyla appear to have arisen through the use of the same DPMs in different combinations in various ways. It is this differential use of the same conserved toolkit that seems to distinguish one basic metazoan body plan from another.

The DPM hypothesis also helps to explain the "developmental hourglass" seen in metazoans. Egg morphology varies wildly within and between phyla, as do the first products of cell division. Nevertheless, members of each phylum pass through a highly uniform stage, the "phylotypic" stage, from which further divergence issues. The phylotypic stage realizes the arrangement of cells from which DPMs produce the phylum-typical body plan: "The stage of highest similarity is the morphogenetic stage, which represents a recapitulation of the originating events of metazoan evolution—the formation of clusters of similarly sized cells" (p. 219). But why should development start from an egg, and why is egg morphology so variable? Newman argues that the sort of consensus and coordination required for optimal functioning of DPMs is more efficient, reliable, and robust among aggregates of genetically identical cells than in mixed aggregates. Those aggregates that form by cleavage, beginning from a single egg, would have a distinct advantage over those formed by the agglomeration of heterogeneous cells. The egg is thus an evolutionary innovation, a means for securing the reliability of DPM function in the cells of the phylotypic stage.

Newman concludes that the preservation of the toolkit, the convergence of the phylotypic stage, and the sudden emergence of metazoan phyla fail to conform to the modern synthesis "idealistic" picture of the gradual accumulation and alteration of a genetic signal. Rather it points strongly to the role of DPMs in the origination and maintenance of general body plans.

I.3.2.3. Chapter 7: Alessandro Minelli—Evolvability and Its Evolvability

Alessandro Minelli offers a comprehensive overview of the rise to prominence of the concept of evolvability. As Minelli points out *evolvability* encompasses an unwieldy range of meanings and uses. Some evolutionary biologists take evolvability to be a property of populations, or of lineages. There are others who hold that evolvability is a property of the genotype-phenotype map. Some take it to be the capacity of standing genetic variation to influence evolution. Some tie evolvability to the capacity of developmental systems to generate changes independently of natural selection. Others think of evolvability as the ability to generate novelties. Minelli argues for a development-based account of evolvability: "The real heuristic power of evolvability is only found in the evo-devo dimension, that is, when we take into full account the complexities of the genotype→phenotype map" (p. 248).

Evolvability influences the tempo and mode of evolution in a variety of ways. It structures the way that large-scale evolution can proceed from a small number of founders. It facilitates both the convergence of evolution, and large-scale saltations. The last of these is particularly significant:

> Before the advent of evo-devo, the very idea of saltational evolution as an explanation for macroevolutionary transitions (as suggested, for example, by Goldschmidt, 1940), was strictly banned as heretical. However, an appreciation of the nonlinear character of the genotype→phenotype map is enough to realize how major phenotypic changes can be accomplished in a leap. (p. 246)

Facilitated variation permits the adaptive exploration of alternative phenotypes. Minelli documents the way that the genetic assimilation of environmentally induced plastic responses accelerates the rate of evolution. Given the significance of development for evolvability, Minelli argues, we must make sure to "adopt an adequate view of development." "In particular, we need to abandon adultocentrism, the common, somehow finalistic view according to which development is the sequence of processes through which a zygote, a spore or a bud is eventually turned into an adult" (p. 246). In its place Minelli promotes the "principle of developmental inertia" (p. 246). Development should be seen as the iteration and combination of elementary processes such as cell proliferations.

Evolvability is now an area of vital interest for evolutionary biology. It firmly establishes a link between the dynamics of development and the tempo and mode of evolution. Moreover, it would appear that evolvability itself evolves within the history of a lineage. It is therefore of crucial importance to understand how evolvability impacts on the evolution of a lineage and how in turn, the evolution of a lineage influences its evolvability. The centrality of the concept of evolvability underscores the significance of evolutionary developmental biology for the understanding of evolutionary dynamics. Minelli cites approvingly the Hendrikse et al. (2007) conviction that "evolvability is the proper focus of evolutionary developmental biology" (p. 248).

I.3.2.4. Chapter 8: Denis M. Walsh—"Chance Caught on the Wing": Methodological Commitment or Methodological Artifact?

According to the modern synthesis, evolution is ineluctably chancy, "chance caught on the wing," in Jacques Monod's resonant phrase. This is a foundational commitment of the synthesis (Merlin, 2010).[24] Denis Walsh's essay challenges it. Monod's influential book *Chance and Necessity*, traces the commitment to ineluctable chance back to its source. According to Monod, it is imposed on evolutionary theorizing by the demands of the scientific method.

In arguing for the ineliminability of chance, Monod channels the pre-Socratic philosopher Democritus: "Everything in the world is the fruit of chance and necessity." Monod quite astutely notices that explanations in Democritean science are very much like our own. Just like modern mechanists, the pre-Socratic Atomists explain the properties of complex entities by adverting to the properties of their parts and the interactions between them. And they do so for pretty much the same reasons, principal among them being the avoidance of any appeal to purpose. Democritus's great opponent in Antiquity is Aristotle. Aristotle's rich and thoroughly unmodern approach to explanation holds that some things happen in the natural world because they fulfill purposes. These cannot be adequately explained without recourse to "final cause" explanations. In fact, Aristotle motivates the need for "final cause" explanations by showing that without them perfectly regular, predictable, and explainable natural occurrences looks like chance events. This is an error that the sciences should not fall prey to.

Walsh exploits the Aristotelian approach to explanation to argue that the modern synthesis commitment to ineluctable chance is one such error. In arguing against the role of chance, Walsh is seeking to cast doubt on the fractionation of evolution that is such a foundational feature of the modern synthesis. Development is not fundamentally conservative. It is a source of non-random adaptive bias. Nor is it wholly independent of inheritance. Nor is selection the only source of adaptive bias. For its part, mutation may well be random, but it is not the sole, perhaps not even the principal, source of evolutionary novelty. According to this "neo-Aristotelian" take on biology, purposive development is the source of change and stasis in evolution. Organisms as purposive entities participate in, indeed drive, adaptive evolution.

I.3.3. Part III: Inheritance

The rediscovery of Mendel's ratios of inheritance was one of the principal theoretical contributions to modern synthesis evolutionary biology. Mendel's ratios suggested that the resemblance of parent to offspring was underwritten by the inheritance of discrete and independent factors. Mendelism set the scene for

---

24. See Pocheville and Danchin (this volume) for a discussion of what the putative randomness of mutation might entail.

perhaps the greatest shift in evolutionary thinking since Darwin. It offered a wholly new proprietary definition of inheritance. Inheritance became defined as the transmission of factors (later genes). Although it took some time for biologists to realize it (Provine, 1971), this reconceptualization of inheritance was tailor made for the synthesis of inheritance with neo-Darwinian natural selection (Gayon, 2000). The essays in this section revisit the modern synthesis conception of inheritance in the light of recent advances in biology. The central question is whether heredity should be redefined, once again to suit the demands of the new evolutionary thinking.

I.3.3.1. Chapter 9: Francesca Merlin—Limited Extended Inheritance

The concept of inheritance has evolved over time to suit the requirements of the theory of biological form in which it is embedded (Müller-Wille & Rheinberger, 2012). Whereas for Darwin inheritance constituted a gross pattern of similarity and difference—offspring tend to resemble their parents more than they resemble members of other lineages—the theory of inheritance that arose with first Mendelism and then the modern synthesis completely transmogrifies the concept. Under the modern synthesis theory of evolution, inheritance has become the transmission of replicated genes from parent to offspring. Francesca Merlin notes that this conception of inheritance is one of the most common sources of challenge to the modern synthesis.

Merlin canvasses the proposed revisions to the notion of inheritance. These fall into two rough classes: those that seek to expand the number of modes or channels of inheritance and those that seek to redefine the concept of inheritance. The first retains the idea that inheritance is fundamentally the transmission of materials, but countenances the transmission from parent to offspring of heritable materials other than DNA. The proposed revisions posit a multiplicity of discrete channels of inheritance: epigenetic, environmental, cultural. The second class of challenges contends that inheritance is not simply particulate but also holistic, in the sense that the control over inheritance is distributed throughout the gene/organism/environment system. Merlin agrees that the traditional modern synthesis account of inheritance is in need of revision, but she does not endorse any of the proposals currently on the table. They share a flaw; they conflate transmission with inheritance. While each correctly identities an evolutionarily important form of transmission of features, there is little reason to regard them as "heredity" transmission. "The fact that some form of transmission is evolutionarily important does not imply that it is a form of inheritance" (p. 286).

The failure of these attempts at revising inheritance teaches us four lessons. They are: (1) Not all causes of intergenerational similarity are part of inheritance,

(2) transmission is inheritance only when it is "vertical," (3) inheritance always involves material overlap, and (4) inheritance is a cause, not a mere effect. With these lessons in mind, Merlin is able to articulate her account of inheritance, "limited extended inheritance." Inheritance, Merlin says, is "the intergenerational transmission, via reproduction, of the set of (genetic and nongenetic) factors and mechanisms conferring on the new entity the capacity to acquire, to have, the capacity to reproduce" (p. 298).

I.3.3.2. Chapter 10: Tobias Uller and Heikki Helanterä—Heredity and Evolutionary Theory

The modern synthesis arose out of a détente between Mendelian theory and neo-Darwinian theory. These are very different kinds of theories. One is a theory of hereditary transmission, and the other a theory of population change. Uller and Helanterä ask whether inheritance construed as gene transmission is appropriate for a theory of evolutionary population change. They contend that inheritance as gene transmission is descriptively adequate only if "nongenetic interactions between parents and offspring . . . do not contribute to the rate or direction of phenotypic evolution" (p. 306).

Uller and Helanterä rehearse some well-known arguments to the effect that gene transmission is deficient as a conception of inheritance for models of phenotypic evolution. These revolve around the issue that gene transmission as inheritance leaves out a range of nongenetic causes of parent-offspring resemblance, and interlineage difference. This deficiency motivates a search for an alternative characterization of inheritance.

Uller and Helanterä survey three promising ones: heredity as parent-offspring covariance, heredity as information transfer, and heredity as development. They argue for the last of these. Heredity as parent-offspring covariance is an important feature of quantitative genetic models, and figures in various versions of the breeders' equation and the Price equation. The latter represents phenotypic evolution as the sum of two discrete effects, heredity (covariance of offspring phenotype and parental phenotype) and selection (covariance of fitness and phenotype value). But as Uller and Helanterä demonstrate, a nonlinear relation between parent and offspring phenotypes can produce a "spurious response to selection." Nongenetic inheritance factors (e.g., epigenetic and ecological inheritance) appear more likely to produce such transmission biases, which in turn affect the rate and direction of phenotypic evolution. Because the processes of development (including niche construction and extragenetic inheritance) contribute to the resemblance of parents to offspring, heredity must be fundamentally a developmental process. These processes in turn, do not only contribute to inheritance but also contribute to the origin and spread of phenotypes. Uller and Helanterä conclude that heredity "cannot be reduced to genetic inheritance, and the causal-mechanical perspective offered by heredity-as-developmental-process is the only one of the four concepts of heredity that we have discussed that is also causally and explanatorily sufficient in evolutionary biology" (p. 305). On this view, there is no distinction between the processes that secure inheritance and those of development. The discreteness and the autonomy of inheritance and development, however, is one of the central tenets of the modern synthesis theory of evolution. Uller and Helanterä conclude with a call for abandoning the orthodox modern synthesis conception of inheritance.

I.3.3.3. Chapter 11: Stéphane Schmitt—Serial Homology as a Challenge to Evolutionary Theory: The Repeated Parts of Organisms from Idealistic Morphology to Evo-Devo

Stéphane Schmitt surveys the concept of serial homology as it appears in evolutionary theories and their "fixist" predecessors. Serial homologues are repeated parts of an organism that have the "same fundamental plan—such as the arms and the legs of a human being, the vertebrae of a vertebrate, or the segments of an arthropod" (p. 343). Schmitt uncovers a quite surprising theoretical continuity in the concept of serial homology between fixist and evolutionary theories. Explaining the existence of "striking morphological similarities between parts of one and the same individual" (p. 362) has always posed a problem for comparative biology.

Serial homologues have been invoked in the defense of any number of accounts of the relatedness among organisms. Schmitt documents the pivotal role that the concept played in the idealist morphologies of Félix Vicq d'Azyr and Johann Wolfgang von Goethe. Despite their differences, each sought to demonstrate that organisms are built out of fundamental building blocks, each of which could be traced back to some common ideal type or structure. Goethe, for instance, insisted that all floral parts are variations on a basic organ type. Similarly, the vertebrate skull, Goethe argued, is constructed out of modified vertebrae. The vast variety of organismal form is thus built on the repetition and transformation of the same morphological theme. The repetition and of parts was to bolster the unificationist theories of Étienne Geoffroy Saint Hilaire. The idea that organisms are constructed of series of highly modified parts of similar basic structure helped to promote the understanding of ontogeny as a source of information concerning the similarity among organisms of different taxa. The repetition of parts within organisms served to motivate an alternative conception of the nature of complex organisms as coalescences of simple, small organisms.

The concept of homology in general, and serial homology in particular, had an important place in post-Darwinian biology. Haeckel for instance used homology to argue for the primacy of the morphological sciences in the study of evolutionary relations. The evolutionary morphology of the late 1860s onward expanded and enriched the concept of homology. The importance of homology in general, and serial homology in particular, for the understanding of evolution has been severely challenged by the genetic turn in evolution. Serial homologues were thought by Boyden to be "the results of the actions of identical genes in different zones of a polarized cytoplasm" (Schmitt, p. 361, this volume). But as Gavin De Beer pointed out, "characters controlled by identical genes are not necessarily homologues" so that "homology of phenotypes does not imply the similarity of genotypes" (p. 362). The definition of homology in general and serial homology in particular has experienced heightened interest since the advent of evolutionary developmental biology and the discovery of the genetic tool kit. And yet despite the challenges, the significance of homology in general, and serial homology in particular, for comparative biology continues to be recognized.

> Work on repeated parts thus represents an important element of continuity in the life sciences starting from the end of the 18th century; and serial homology has succeeded in redefining itself many times over within different theoretical contexts and different disciplines leading up to the preset day—despite the considerable transformations biology underwent in the course of that long period. (p. 363)

Acknowledgments

This book has been supported by the ANR grant 13-BSH3-0007. It benefited from the Laboratoire International Associé CNRS France-Canada ECIEB (Evolutionary and Conceptual Issues in Evolutionary Biology).

References

Amundson, R. (1994). Two concepts of constraint: Adaptationism and the challenge from developmental biology. *Philosophy of Science, 61*, 556–578.

Amundson, R. (2005). *The changing role of the embryo in evolution.* Cambridge, UK: Cambridge University Press.

Bateson, P., & Gluckman, P. (2011). *Plasticity, robustness, development and evolution.* Cambridge, UK: Cambridge University Press.

Brandon, R. (1990). *Adaptation and environment.* Princeton, NJ: Princeton University Press.

Cain, J. (2009). Rethinking the synthesis period in evolutionary studies. *Journal of the History of Biology, 42*, 621–648.

Carroll, S. B., Grenier, J. K., & Weatherbee, S. D. (2000). *From DNA to diversity: Molecular genetics and the evolution of animal design*. London, UK: Wiley-Blackwell.

Danchin, E., Charmontier, A., Champagne, F. A., Mesoudi, A., Pujol, B., & Blanchet, S. (2011, July). Beyond DNA: Integrating inclusive inheritance into an extended theory of evolution. *Nature Reviews Genetics, 12*, 475–486.

Darwin, C. (1968). *The origin of species*. London, UK: Penguin. (Original work published in 1859.)

Darwin, C. (1862). On the various contrivances by which British and foreign orchids are fertilised by insects, and on the good effects of intercrossing. London, UK: John Murray. Retrieved April 8, 2016, from http://darwin online.org.uk/content/

Davidson, E. H. (1986). *Gene activity in early development*. Orlando, FL: Academic Press.

Dawkins, R. (1976). *The selfish gene*. Oxford, UK: Oxford University Press.

Depew, D. (2011). Adaptation as a process: The future of Darwinism and the legacy of Theodosius Dobzhansky. *Studies in the History of Biology and the Biomedical Sciences, 42*, 89–98.

Depew, D., & Weber, B. (1995). *Darwinism evolving: Systems dynamics and the geneaology of natural selection*. Cambridge, MA: MIT Press.

Dickins, T. E., & Barton, R. A. (2013.) Reciprocal causation and the proximate-ultimate distinction. *Biology and Philosophy, 28*, 747–756.

Dickins, T. E., & Rahman, Q. (2012). The extended evolutionary synthesis and the role of soft inheritance in evolution. *Proceedings of the Royal Society. Series B, 279*, 2913–2921. doi:10.1098/Rspb.2012.0273

Dobzhansky, T. (1937). *Genetics and the origin of species*. New York, NY: Columbia University Press.

Dobzhansky, T. (1973). Nothing in biology makes sense except in the light of evolution. *American Biology Teacher, 35*, 125–129.

Duboule, D. (Ed). (1994). *A guidebook to homeobox genes*. Oxford: Oxford University Press.

Fisher, R. A. (1918). The correlation between relatives on the supposition of Mendelian inheritance. *Transactions of the Royal Society of Edinburgh, 52*, 399–433.

Fisher, R. A. (1930). *The genetical theory of natural selection*. Oxford, UK: Clarendon Press.

Gayon, J. (2000). From measurement to organization: A philosophical scheme for the history of the concept of inheritance. In P. J. Beurteon, R. Falk, and H.-J. Rheinberge (Eds.), *The concept of the gene in development and evolution: Historical and epistemological perspectives* (pp. 60–90). Cambridge, UK: Cambridge University Press.

Gehring, W. (1998) *Master control genes in development and evolution: The homeobox story*. New Haven, CT: Yale University Press.

Gerhard, J., & Kirschner, M. (2005). *The plausibility of life: Resolving Darwin's dilemma*. New York, NY: Norton.

Gerhard, J. M., & Kirschner, M. (2007). The theory of facilitated variation. *Proceedings of the National Academy of Sciences, 104*, 8582–8589.

Ghiselin, M. (1980). The failure of morphology to assimilate Darwinism. In E. Mayr & W. B. Provine (Eds.), *The evolutionary synthesis: Perspectives on the unification of biology* (pp. 180–193). Cambridge, UK: Cambridge University Press.

Gilbert, S., & Epel, D. (2009). *Ecological developmental biology*. Sunderland, MA: Sinaeur.

Goldschmidt. R. (1940). *The material basis of evolution*. New Haven CT: Yale University Press.

Gould, S. J. (1977). *Ontogeny and phylogeny*. Cambridge MA: Harvard University Press.

Gould, S. J. (1983). The hardening of the modern synthesis. In M. Grene (Ed.), *Dimensions of Darwinism* (pp. 71–93). Cambridge, UK: Cambridge University Press.

Gould, S. J. (2002). *The structure of evolutionary theory*. Cambridge, MA: Belknap Press.

Gould, S. J., & Lewontin, R. C. (1979). The Spandrels of San Marco and the Panglossian paradigm: A critique of the adaptationist programme. *Proceedings of the Royal Society London*, Series B, 205, 581–598.

Griesemer, J. (1984). Presentations and the status of theories. *PSA: Proceedings of the Biennial Meeting of the Philosophy of Science Association* (Vol. 1, pp. 102–114).

Griffiths, P. E., & Gray, R. D. (1994). Developmental systems and Evolutionary Explanation. *Journal of Philosophy*, 91, 277–304.

Hall, B. K (1999). *Evolutionary developmental biology*. Amsterdam, The Netherlands: Kluwer.

Hendrikse, J. L., Parsons, T. E., & Hallgrímsson, B. (2007). Evolvability as the proper focus of evolutionary developmental biology. *Evolution and Development*, 9, 393–401.

Huxley, J. (1942). *Evolution: The modern synthesis*. London: Allen Unwin.

Jablonka, E., & Lamb, M. (2002). The changing concept of epigenetics. *Annals of the New York Academy of Sciences*, 981, 82–96.

Jablonka, E., & Lamb, M. (2004). *Evolution in four dimensions: Genetic, epigenetic, behavioral, and symbolic variation in the history of life*. Cambridge, MA: Bradford Books.

Jablonka, E., & Raz, G. (2009). Transgenerational epigenetic inheritance: Prevalence, mechanisms, and implications for the study of heredity and evolution. *Quarterly Review of Biology*, 84, 131–176.

Jones, C., Lawton, J., Shachak, M. (1994). Organisms as ecosystem engineers. *Oikos*, 69, 373–386.

Kellert, S. H., Longino, H. E., & W aters, C. K. (2006). Introduction: The pluralist stance. In S. H. Kellert, H. E. Longino, & C. K. Waters (Eds.), *Scientific pluralism* (pp. vii–xxix). Minnesota Studies in the Philosophy of Science (Vol. 19). Minneapolis: University of Minnesota Press.

Kimura, M. (1973). *The neutral theory of molecular evolution*. Cambridge, UK: Cambridge University Press

Kirschner, M., & Gerhard, J. (2005). *The plausibility of life: Resolving Darwin's dilemma*. New Haven, CT: Yale University Press.

Laland, K., Uller, T., Feldman, M., Sterelny, K., Müller, G. B., Moczek, A., Jablonka E., & Odling-Smee, J. (2014). Does evolutionary theory need a rethink? Yes: Urgently. *Nature*, 514, 161–164.

Lamm, E. (2014). Inheritance systems. In: E. N. Zalta (Ed.), *Stanford Encyclopedia of Philosophy* (Spring 2012 Edition). Plato.Stanford.Edu/Archives/Spr2012/Entries/inheritance-Systems

Lewontin, R. C. (1974). *The genetic basis of evolutionary change.* New York: Columbia University Press.

Lewontin, R. C. (1978). Adaptation. *Scientific American, 239,* 212–230.

Lewontin, R. C. (1983). Gene, organism, and environment. In Bendall, D. S. (Ed.), *Evolution: From Molecules to Men* (pp. 273–285). Cambridge, UK: Cambridge University Press.

Luria, S. E., & Delbrück, M. (1943). Mutations of bacteria from virus sensitivity to virus resistance. *Genetics, 28,* 491–511.

Mameli, M. (2005). The inheritance of features. *Biology and Philosophy, 20,* 365–399.

Maynard Smith, J. (1982). *Evolution and the theory of games.* Cambridge, UK: Cambridge University Press.

Maynard Smith, J. (1989). *Evolutionary genetics.* Oxford, UK: Oxford University Press.

Maynard Smith, J., Burian, R., Kauffman, S., Alberch, P., Campbell, J., Goodwin, B., Lande, R., Raup, D. & Wolpert, L. (1985). Developmental Constraints and Evolution. *Quarterly Review of Biology, 60,* 265–287.

Mayr, E. (1961). Cause and effect in biology. *Science, 134,* 1501–1506.

Mayr, E., & Provine, W. B. (Eds.). (1980). *The evolutionary synthesis: Perspectives on the unification of biology.* Cambridge, MA: Harvard University Press.

Merlin, F. (2010). Evolutionary chance mutation: A defense of the modern synthesis consensus view. *Philosophy and Theory in Biology, 2.* dx.doi.org/10.3998/ptb.6959004.0002.003

Mesoudi, A., Blanchet, S., Charmentier, A., Danchin, E., Fogarty, L., Jablonka, E., ... Pujol, B. (2013). Is non-genetic inheritance just a proximate mechanism? A corroboration of the extended evolutionary synthesis. *Biological Theory, 7,* 189–195.

Moczek, A. P. (2012). The nature of nurture and the future of evo-devo: Toward a theory of developmental evolution. *Integrative and Comparative Biology, 52,* 108–119.

Moczek, A. P., Sultan, S., Foster, S., Ledón-Rettig, C., Dworkin, I., Nijhout, H. F., ... Pfennig, D. W. (2011). The role of developmental plasticity in evolutionary innovation. *Proceedings of the Royal Society B, 278,* 2705–2713. doi:10.1098/Rspb.2011.0971

Monod, J. (1971). *Chance and necessity: An essay on the metaphysics of life.* (Trans. A. Wainhouse). New York: Schopf and Sons.

Morange, M. (2009). *Life explained.* New Haven, CT: Yale University Press.

Müller, G. B. (2007). Evo-devo: Extending the evolutionary synthesis. *Nature Reviews Genetics, 8,* 943–949. doi:10.1038/Nrg2219

Müller-Wille, S., & Rheinberger, H.-J. (2012). *A cultural history of heredity.* Chicago, IL: Chicago University Press.

Newman, S., & Bhat, R. (2009). Dynamical patterning modules: A "pattern language" for development and evolution of multicellular form. *International Journal for Developmental Biology, 53,* 693–705.

Odling-Smee, F. J., Laland, K., & Feldman, M. (2003). *Niche construction: The neglected process in evolution*. Princeton, NJ: Princeton University Press.

Orr, H. A. (2007). Theories of adaptation: What they do and don't say. *Genetica*, *123*, 3–13.

Orr, H. A., & Coyne, J. A. (1992). The genetics of adaptation revisited. *The American Naturalist*, *140*, 725–742.

Pigliucci, M. (2010). Genotype-phenotype mapping and the end of the "genes as blueprint" metaphor. *Philosophical Transactions Royal Society B*, *365*, 557–566.

Pigliucci, M., & Müller, G. (Eds.). (2010). *Evolution: The extended synthesis*. Cambridge, MA: MIT Press.

Price, G. R. (1970) Selection and covariance. *Nature*, *227*, 520–521

Pocheville, A., & Danchin, É. (2015). Physiology and evolution at the crossroads of genetics and epigenetics. *Journal of Physiology*, *9*, 2243–2243.

Provine, W. B. (1971). *The origins of theoretical population genetics*. Chicago, UK: University of Chicago Press.

Raff, R. (1996). *The shape of life: Genes, development and the evolution of animal form*. Chicago, IL: Chicago University Press.

Roll-Hansen, N. (2009). Sources of Wilhelm Johansen's genotype theory. *Journal of the History of Biology*, *42*, 457–493.

Scott-Phillips T., Dickins, T., & West, S. (2011, January) Evolutionary theory and the ultimate–proximate distinction in the human behavioral sciences. *Perspectives on Psychological Science*, *6*, 38–47.

Sober, E. (1984). *The nature of selection*. Cambridge, MA: MIT Press.

Sultan, S. (2015). *Organism and environment: Ecological development, niche construction, and adaptation*. Oxford, UK: Oxford University Press.

Wagner, A. (2011). *The origin of evolutionary innovations: A theory of transformative change in living systems*. Oxford, UK: Oxford University Press.

Wagner, A. (2014). *The arrival of the fittest: Solving evolution's greatest puzzle*. New York, NY: Current Books.

Wallace, B. (1986). Can embryologists contribute to an understanding of evolutionary mechanisms? In W. Bechtel (Ed.), *Integrating scientific disciplines* (pp. 149–163). Dordrecht, The Netherlands: M. Nijhoff.

Waters, C. K. (1994). Genes made molecular. *Philosophy of Science*, *61*, 163–185.

Weaver, I. C. T., Cervoni, N., Champagne, F. F., D'Alesssio, A. C. V., Sharma, S., Seckl, J. R., Dymov, S., Szyf, M. & Meamey, M. J. (2004). Epigenetic Programming by Maternal Behavior. *Nature Neuroscience*, *7*, 847–854.

West-Eberhard, M. J. (2003). *Developmental plasticity and evolution*. Oxford, UK: Oxford University Press.

Wilson, D. S. (1975). A theory of group selection. *Proceedings of the National Academy of Sciences*, *72*, 143–146.

Wray, G. A., Hoekster, H. E., Futuyma, D. J., Lenski, R. E., Mackay, T. F. C., Schluter, D., & Strassman, J. E. (2014). Does evolutionary theory need a rethink? No, All is well. *Nature*, *514*, 161–164.

ADAPTATION AND SELECTION

1

NATURAL SELECTION, ADAPTATION, AND THE RECOVERY OF DEVELOPMENT

David J. Depew

In this chapter, I review the history of some recurrent questions about the idea of adaptation. What is its relationship to the process of natural selection? Are organisms or traits adapted when they first appear, natural selection retaining those that are already "fit" and eliminating the others? Or does natural selection evolve adaptations only as selectively favored heritable variation is propagated through and maintained in populations? Answers to this question turn on another. What sorts of entities do (or should) we say are adapted? Traits? Organisms? Populations of organisms that have traits whose effects on environments structure these populations? Gene pools whose replicative prowess defines populations of organisms equipped with good-making traits?

I use these questions to frame something of a criterion, or at least touchstone, for resolving them. If natural selection is a leading factor in adaptation, a bad criterion for calling any of the entities mentioned above adapted (or *an* adaptation if the entity in question is a trait) will be one that introduces a degree of incoherence into the very idea of natural selection by disturbing the intricate balance among its three conceptual aspects: chance in the origin of variation, pressure from the environment, and a variety of subtly different properties—goodness, purpose, design, end directedness, goal orientation, functionality, and perhaps a few others—that collectively I call "telic." A good account of adaptive natural selection will equally honor all of these aspects.

A good account will also honor the fact that organisms are developmental entities. They come into being by a process of cellular differentiation. They have life cycles and life histories. They have evolved so many layers of feedback that they can be said to be self-regulating and in some sense self-organizing. The recent (re)turn to development's role in evolution takes these facts about organisms to be important. Some versions of the new developmentalism are indifferent or hostile to natural selection as a source of significant evolutionary change. Their advocates tend to speak ill of Darwinism. Others find the proximate source of the variation on which natural selection works in developmental processes. Their aim is to pry genetic Darwinism loose from the stress on mutation and recombination of structural genes that has dogged it since the mid-20th-century rise of molecular genetics. In this case the turn toward development is largely a turn toward regulatory gene sectors as a proximate source of selectable variation. Stronger versions of this option implicate developmental processes not just in generating the variation on which selection works but also in inducing adaptedness and spreading it through populations. My aim is not to decide among these alternatives—this may be premature—but to insist that, if it is to be persuasive, a favored alternative must refrain, whether in attacking its opponents or defending itself, from introducing conceptual incoherence into the idea of natural selection by slighting one or more of its jointly defining aspects. In making this point, I first review some of Darwinism's history. To draw attention to the present state of the question about whether the modern synthesis is to be affirmed, expanded, extended, or replaced, I consider some facts about a case that is tossed around in the literature of evolutionary biology today: the natural history of stickleback fish.

1.1. Natural Selection: From Spencer to Darwin and Back

Herbert Spencer's core ideas about biological evolution preceded the *Origin of Species*. For Spencer, organisms are adaptive as individuals apart from natural selection. He means that in virtue of the plasticity of living matter they tend to adjust or accommodate to their environmental circumstances. Because these circumstances are defined by Malthusian scarcity, only fit or adapted individuals—individuals that rise to the occasion, as it were—survive. Others perish without issue. When Spencer's *Principles of Biology* appeared in 1864 (as part of his *Principles of Synthetic Philosophy*) it showed the influence of Darwin's *Origin of Species*, but only insofar as Spencer saw in the *Origin* an opportunity to press opinions he already had.

Running hard with the Malthusianism he shared with Darwin, Spencer identified natural selection as a force impelling individual organisms to adapt to their circumstances or die. It was Spencer who came up with the phrase "survival of the fittest." At some point he became aware that Darwin's model of natural selection differed from his own. He had his doubts about how many adapted traits are evolved by gradual, transgenerational, directional natural selection, the leitmotif of Darwin's account. He came to think that this might be true of organs of defense, such as the fishing rods that grow out of the foreheads of anglerfish (Spencer, 1887, p. 7). But he declared that even if natural selection on this understanding of it is a way of evolving adapted parts it has no way of evolving adapted organisms, since what makes them adapted for Spencer, as for Cuvier and others but more dynamically, is the mutual adjustment of all their parts into a coherent whole rather than a sum of aggregated traits (Spencer, 1887, p. 40). It is this harmony that makes for the survival of the fittest—and for adaptedness.

Darwin was aware that Spencer's approach differed from his own. In an interesting passage in the manuscript of *Natural Selection*, the sprawling notebook that he mined for material we find in the *Origin*, he portrays a scenario of the sort Spencer, among others, had in mind:

> Seeing how absolutely necessary whiteness is in the snow-covered arctic regions [. . .] we might attribute the absence of color [in a white Arctic bird] to a long course of selection. But it may be that whiteness is the direct effect of cold and that the struggle for life has only so far come into play that colored animals in arctic regions live under a great disadvantage. (Stauffer, 1975, p. 377)

It is noteworthy that Darwin does not call the trait that results from this process an adaptation, nor does he call the process itself natural selection, even if in one sense it is. To be sure, Darwin's characteristic combination of truthfulness and timidity increasingly led him to acknowledge that "the direct effect of the environment" portrayed here is a real evolutionary process, and a common one too: so much so that at the end of his life he told a correspondent,

> There can be no doubt that species may become greatly modified through the direct action of environment. I have some excuse for not having formerly insisted more strongly on this head in my *Origin of Species*, as most of the best facts have been observed since its publication. (Darwin to M. Neumayer, March 9, 1877, in F. Darwin, 1887, vol. 2, p. 232)

But, he insisted, this is not adaptive natural selection. Why?

Darwin distinguishes in various places between the fact, the power, and the theory of natural selection (Darwin, 1868, vol. 2, p. 430; Gayon, 1997). The first refers to nonrandom differential survival of individuals of the same species in a single generation. The elimination of nonwhite birds in the arctic exhibits natural selection in this sense. The power of natural selection, by contrast, refers to nonrandom differential reproductive success over transgenerational time. It refers, that is, to what in the passage about Arctic birds he calls "a long course of natural selection." Over many generations natural selection has the power to evolve adaptations and new species out of small but heritable variations that initially appear in individuals independently of the utility they subsequently come to acquire. It is this sense of natural selection that figures in Darwin's *theory* of natural selection. Its "paramount power," he argues in the second half of the *Origin*, perspicuously and parsimoniously explains a large range of biological facts. This is why Darwin says that his theory would be falsified if its gradualist axiom were ever proven to be unsound (Darwin, 1859, pp. 189, 194, 471).

The point is subtle. Darwin recognized that the effect of environments in physiologically, even biochemically, inducing traits that protect animals from the intense struggle for existence in regions like the arctic is different from, even if it is concomitant with, the power of natural selection to evolve traits whose function is to protect these animals in and from their environments. It might be temperature that makes arctic birds white. It might do so with such rigor that it quickly reduces, suppresses, or eliminates the trait variation on which natural selection feeds in evolving adaptations. If so, however, the whiteness of arctic birds will not be the accumulated consequence of a process in which organisms born with heritable colorations differentially survive and reproduce. To be sure, whiteness considered as a direct effect of the environment on a developing organism will have the *effect* of protecting arctic birds. Still, we will be unable to claim that they are white *because* whiteness has this protective effect. Citing the physiological causes of how organisms respond to the properties of their environments fails to forge a connection—an evolved connection—between cause and effect, as Darwin's theory of adaptive natural selection requires. This is why adapted traits are for Darwin irreducibly *for* some good, goal, purpose, or function. In much that passes as Darwinism the absence of this extended process weakens, undermines, or eliminates the telic aspects of natural selection in favor of a fruitless dialectic between chance and determinism (Depew 2013).

In Spencer's case the element of environmental force prevails over the elements of chance and purposiveness. For Spencer, transgenerational natural selection is conceptually assimilated to single-generation environmental selection. For Darwin, it is the other way around. The upshot is that "selection for" takes center stage in Darwin's thought, while Spencer's stress falls on "selection against." To be sure, differential elimination is for Darwin a necessary feature of the process of natural selection. But rather than being only an executioner of the antecedently maladapted, he sees selection as gradually working to produce exquisitely adapted traits, organisms, races, species, and ecologically coadapted communities. The "paramount power" of natural selection is from this transgenerational perspective a "creative" power.

How could theories so different have become so conflated? I suspect the answer is that Darwin failed too often to signal when he was using "natural selection" to refer to what Robert Brandon calls single-generation "environmental selection" and when he was referring to the "long course of selection" that figures in his notion of the "paramount power of natural selection" and his theory of evolution by natural selection (Brandon, 1990). The Malthusianism to which both authors appeal as the source of deterministic environmental pressures that constantly bear down on organisms, as well as their equal devotion to the novel proposition that all life on earth evolved from a common ancestor, meant in practice that the more or less simultaneous reception of Darwin and Spencer's writings stressed their overlaps, not their differences. Darwin's hesitant use of Spencer's phrase "survival of the fittest" in later editions of the *Origin* confused the issue even more.

The moral of the story is this. The entanglement of Darwinism and Spencerism in the last quarter of the 19th century was a disaster for the accurate transmission of a great scientific doctrine both to the public and within the sphere of expert opinion itself. By the time of his death, Darwin's "creative" view of adaptive natural selection, as the makers of the modern synthesis called it, had to all intents and purposes been coopted by Spencerian survival of the fittest. The upshot was that people who called themselves Darwinians in, say, 1880, ascribed adaptedness or fitness to the responsiveness of organisms to environmental challenges independently of the workings of natural selection, which they took to be a pruning process. Beginning in the 1890s, however, and culminating in the modern evolutionary synthesis of the 1940s, a revival of Darwin's conception of adaptive natural selection as the creative factor in evolution became the foundation of 20th-century evolutionary science. In the following section I laud the virtues of this family of research programs, but also lament their increasing distance from the developmental locus of evolutionary dynamics that Spencer, for all his deficiencies, rightly stressed.

1.2. Setting Development Aside: From Weismann to Dawkins

In the mid-1880s, Spencer was challenged and defeated by August Weismann, giving birth to "neo-Darwinism," as George Romanes called it. Neo-Darwinism forbids attempts like Spencer's to combine natural selection with the inheritance of acquired characteristics. Acquired characteristics, said Weismann, are not heritable in the sense required of any coherent theory of Darwinian evolution.

Adapted traits, he insisted, must arise instead from natural selection working exclusively on germ-line variation.

The rise of Weismannian hard inheritance forms a watershed in the history of Darwinism. Because it eventually gave rise to statistical, population-level accounts of natural selection that are remote from and increasingly indifferent to the developmental process, it is worth pointing out that Weismann himself assumed no less fully than Darwin that the defining conditions of the transgenerational process of natural selection—variation, heritability, and differential retention in the next generation—occur only in beings that come into existence by way of an ontogenetic process of successive differentiations and integrations. As an embryologist, Weismann simply located the processes of competition, selective elimination, and heritable retention of fit variants deeper and earlier in the developmental process than Spencer or for that matter Darwin himself. Nor, accordingly, can Weismann be called a trait atomist: one who regards organisms as assemblies of independently evolved adaptive traits. Locating the process of natural selection and adaptation in embryological differentiation implies a certain degree of dynamic, diachronic holism in organisms.

In fact, no one working on these questions in the early decades of the 20th century denied the ontogenetic-environmental locus of evolutionary processes or that natural selection occurs only in entities that are developmental systems. True, some of them did set aside this fact in order to pursue new lines of inquiry. Thomas Hunt Morgan, for example, the father of transmission genetics, set aside the embryology in which he had been trained in order to determine how Weismann's germ-line determinants, once he had identified them with Wilhelm Johannsen's genes and physically located them on chromosomes, are inherited. He said the bracketing was temporary. In fact, it endures to this day, since it is a generative, even constitutive, condition for the "population thinking" on which the modern synthesis is based.

Like most genetic evolutionists in the early 20th century, Morgan thought of mutation as the direction-giving, innovative factor in evolution. Natural selection served for him merely as a sieve to get rid of what are bound to be preponderantly harmful mutants (Beatty, 2016; Morgan, 1916, 1932). In this stress on "selection against," Morgan was a bit of a Spencerian even if he categorically rejected Spencer's ideas about inheritance. We can also find turn-of-the-20th-century Darwinians, however, who championed natural selection as the innovative, directional, and in that sense creative factor in evolution. The so-called biometricians brought sophisticated forms of statistical measurement developed by Francis Galton and Karl Pearson to the study of natural populations. By using these measures, they found empirical cases of what Darwin had postulated: a continuum of slight phenotypic variation and, under conditions of environmental stress, gradual directional natural selection for traits better adapted to novel conditions. More persuasively than anyone in their ranks, Walter F. R. Weldon confirmed the existence of natural selection in the wild by correlating morphological change in the width of crabs' carapaces with the environmental degradation of Plymouth Bay over transgenerational time (Weldon, 1893).[1] Biogenetic Darwinians like Weldon were hostile both to the single-generation macromutationism of Hugo de Vries and William Bateson and to the eliminative view of selection entertained by neo-Mendelians like Morgan. In fact, in the first third of the 20th century something of a thirty years' war broke out between Mendelian mutationists and neo-Darwinian adaptationists (Provine, 1971).

The conflict was eventually resolved by the ingenious Ronald A. Fisher (Fisher, 1918, 1930). By abandoning the crude assumption that there is one gene for one trait he was able to use statistical analysis not simply to measure transgenerational phenotypic change in populations, like the biometricians from whose ranks he had risen, but to picture, describe, and define with mathematical precision the dynamics of Mendelian populations. Arrays of different genotypes, Fisher argued, are very much like arrays of atoms in the statistical-mechanical theory of gases. Any system of freely interbreeding alleles, like randomly colliding atoms in contained gases, will (counterfactually) remain macroscopically in equilibrium generation after generation until a force exerted somewhere inside the system changes its state (Fisher, 1930, p. 36). Natural selection at the trait–environment interface is such a force. Its effect is to change the relative frequencies of genotypes. The rate of natural selection in populations of freely recombining genotypes, Fisher proclaimed, is directly proportioned to the degree of additive genetic variance in the population as a whole (Fisher, 1930, pp. 35, 46).[2] This picture of the dynamics of populations of allelic differences ended the

---

1. Transgenerational, to be sure, but nonetheless rapid, given the intensity of the degradation of the environment in Plymouth Bay, where these crabs lived. The degradation was caused by silting up of the bay under the influence of flows of industrial effluent carried by rivers and streams emptying into it. It has long been a matter of dispute whether Darwinism would ever have been formulated without the ideological conflicts of the industrial revolution to lend it a vocabulary. It is less disputable that natural selection might not have been confirmed in the wild in the absence of an ecocide rapid enough to make it perceptible. Another case is melanism, or the increasing prevalence of dark forms, in pepper moths (*Biston betularia*). Under the influence of industrial soot, dark forms are less visible to birds that prey on pepper moths. So these spread through populations. The phenomenon was confirmed in the 1950s by Bernard Kettlewell, an heir to the biometrical tradition in its population-genetic reformulation (Kettlewell, 1955, 1956; see Rudge, 1999).

2. Additive genetic variance is the mean value of the portion of total genetic variance in a population that excludes pleiotropy and epistatic interaction among genes.

conflict between Darwinian naturalists and Mendelian geneticists by inscribing Mendelian genetics into what was fundamentally a restatement of the biometrical hypothesis (Provine, 1971).

The modern evolutionary synthesis of the 1940s, announced under that name in Julian Huxley's *Evolution: The Modern Synthesis* (1942), was founded on Fisher's *Genetic Theory of Natural Selection*, albeit as amended by Sewall Wright's stress on genetic drift—the fixation of genotypes in small interbreeding populations by pure chance—and by the phenomenon of gene flow between populations due to in and out migration. Its founders, Theodosius Dobzhansky, Ernst Mayr, and George Gaylord Simpson, insisted, however, that the formalisms of mathematical population genetics have value only insofar as they are useful tools for framing and solving real biological problems. Real biological problems are problems faced by biologists because more fundamentally they are faced by organisms as they endeavor to thrive as individuals and as populations to adapt over transgenerational time to environments that are constantly changing, not least by the agency of organisms themselves as they degrade resources to make a living. At the core of the modern synthesis is a recommendation that by looking at these processes from the perspective of measureable changes in the genetic composition of populations it is possible to restore Darwin's intuition about the transgenerationally creative power of natural selection, retrospectively liberating him from his own muddles and from the benighted company he was forced to keep (Mayr, 1991).[3]

It is often said that by looking at organism-level processes from the perspective of gene frequency dynamics the makers of the modern synthesis undermined the relevance of ontogeny to evolutionary processes. The only relevant moment in the life cycle is reproducing. The fact is, however, that when Dobzhansky, his colleague L. C. Dunn, and their student Richard Lewontin construed population genetic shifts as readable shadows projected by entwined ontogenetic and ecological processes they were far from indifferent to development (Lewontin, 1974, pp. 12–13). Dobzhansky took the fact that organisms are developmental systems so seriously that he modeled natural selection's various modes on aspects of ontogeny (Dobzhansky, 1951, 1970). He followed his friend and fellow Russian émigré I. Michael Lerner and the Russian geneticist Ivan I. Schmalhausen's efforts to import population genetics into developmental biology by recasting

---

3. The founders of the modern synthesis respected the geneticist Hermann Muller but declined to recognize him as adhering to their principles. They took his notion that natural selection favors one and only one allele as fit in a given environment, slowly discarding the rest, as too resonant with a non-creative view of selection and with Morgan's idea that mutation is the source of evolutionary novelty or creativity. (Morgan was Muller's mentor.) They also regarded his ideas as entangling the modern synthesis with eugenics.

eliminative natural selection as normalizing selection, a process that evolved in order to discard developmental failures in ontogeny, and by thinking of stabilizing selection as a population-level phenomenon that mirrors the ontogenetic process that keeps development on track (Depew, 2011; Gilbert, 1994; Lerner, 1954; Schmalhausen, 1949). When something begins to go wrong evolved mechanisms push the restart button. Similarly, balancing selection maintains in populations the homeostatic equilibrium we see in healthy organisms by selectively favoring heterozygotes, which enhances the probability of adaptation to future environments by retaining in recessives a balanced multiplicity of slightly different alleles. The evolution of diploid chromosomal structure is traceable to the advantages of balancing selection. Directional selection, for its part, uses recombination of the considerable genetic variation already stored in natural populations more than new mutations to meet changed environmental pressures.[4]

The ontogenetic locus of the evolutionary process underlies Dobzhansky, Mayr, and Simpson's insistence that interactions between genes and environments are so dynamic that except under highly constrained circumstances it is a fool's errand to attribute an observed difference between individuals to this much nature and that much nurture (Keller, 2010). The fact that organisms are developmental systems means that, "What is inherited is not this or that trait, but the manner in which the organism responds to the environment" at each phase of its life cycle (Dunn & Dobzhansky 1946, p. 134). This diachronic and holistic view of organisms both reflects and grounds the perception that in an ecological theater as mutable as the earth natural selection's finest products will be adaptations for enhancing developmental flexibility and future evolvability (Dobzhansky, 1937, 1951, 1970). An evolutionary perspective is required even to identify phenomena like these. They can be seen only "in the light of evolution" (Dobzhansky, 1973).

In explaining phenomena of this sort the analogy between artifacts and organisms, whose part–whole relations are not as decomposable as those found in machines, is not as heuristically productive as it is in searching for ecological rationales for particular traits, or as probative. The function of the fishing hooks of anglerfish might as readily have been worked out by the natural theologian William Paley, who assumed a fixed, creationist world, as by Darwin or Dobzhansky, who did not. It follows that thinking of natural selection as "design without an intelligent designer," as practitioners of Darwinian research

---

4. In making these claims, Dobzhansky was taking his cue from Fisher, who in 1932 asked his fellow geneticists to inquire into how and why natural selection had evolved evolution's entrenched mechanisms, such as dominance, hybrid vigor, and the diploid or polyploid chromosome itself (Fisher, 1932).

programs seeking to identify evolved functions for particular traits do, encourages inferences that tend to underestimate the dynamic interaction among organism, gene, and environment in and beyond ontogeny. The very fact that the design-without-a-designer trope comes trailing the theological clouds of glory that gave rise to it forces those who use it as an "intuition pump" constantly to exorcise its creationist ghost by invoking mechanistic materialism to give themselves metaphysical aid and comfort. In doing so, they put the developmental nature of evolved and evolving beings even further into the background and at risk.

Just such risks were incurred by the "ecological genetics" that E. B. "Henry" Ford established at Oxford in the 1950s. Ecological genetics is certainly a research program within the modern synthesis. It assumes the same tool kit of mathematical population genetics as the largely American counterpart I have described above, albeit with a greater stress on additive genetic variation and directional natural selection. The original aim of ecological genetics was to demonstrate that many of the species-identifying traits that systematists long presumed to be adaptively neutral (because ex hypothesi they stay put long enough to be classified) are actually adaptations. A paradigm case is David Lack's work on the beaks of the finches Darwin collected in the Galapagos Islands, which turn out to be finely adapted to food sources specific to different islands (Lack, 1947; Weiner, 1994). This point having been made by the 1950s, the research program morphed into a search for adapted traits that are antecedently presumed to pervade, even to constitute, organisms. It morphed, that is, into what Lewontin and Stephen Jay Gould call the "adaptationist" research program (Gould & Lewontin, 1979). Still, neither Lack nor his Oxford-schooled colleagues denied that Darwin's finches develop and interact with their environments as whole organisms. They merely implied that this fact is irrelevant in posing and answering the particular questions they were asking about how organisms fit into and make use of their environments.

It was only with the introduction of game theory into evolutionary biology in the post-Watson-and-Crick climate of the late 1960s and the 1970s that the strategic indifference of adaptationists to development turned into something more substantively at odds with the fact that organisms interact with environments as ontogenetic wholes. To be sure, the application of cold war game theory to evolutionary problems gave Darwinism an explanation of cooperative behavior in social insects impressive enough to erase a reputation for glorifying selfishness that had long compromised its authority (Axelrod & Hamilton, 1981; Dawkins, 1976; Hamilton, 1964; Maynard Smith, & Price, 1973; Wilson, 1975; but see Nowak, Tarnita, & Wilson, 2010). The guiding principle of this proposal is that cooperation at the level of phenotypes is a good way for genes to optimize their individual fitness or replicative power. This posed a problem, however, for the research program of Dobzhansky, Mayr, and Simpson, which focused on speciation and the origin of higher taxa rather than on searching for adaptive scenarios for particular traits. Whole organisms, not genes "for" traits, are the targets of natural selection. Accordingly, "A gene is never visible to natural selection and in the genotype it is always in the context with other genes, and interaction with those other genes, that a particular gene is either more favorable or less favorable" (Mayr, 2001; see also Mayr, 1963, p. 263, on "beanbag genetics").

In order to press the (Fisherian) claim that fitness can indeed be ascribed to individual genes the Oxford biologist Richard Dawkins reversed the received interpretation of population genetic Darwinism. From the claim that nothing is more natural for DNA than to make more of itself he inferred that it is not organisms that are the beneficiaries of gradual natural selection, but genes considered as chunks of DNA that stay intact through repeated rounds of meiotic division (Dawkins, 1976). Regulatory genes, which express themselves with exquisite sensitivity to environmental circumstances, are as "selfish" in this sense as protein-coding structural gene sequences. Dawkins may have been speaking metaphorically, hyperbolically, and merely persuasively when he said that in "genic selectionism," as his hypothesis is called, organisms are treated as aggregations of optimally adapted traits and so have no more internal complexity or agency than "clouds in the sky or dust storms in the desert" (Dawkins, 1976, p. 34). However, I can think of no way to translate this exercise in "seeing as" into realistic biology without denying and not merely ignoring the ontogenetic holism of organisms and, by taking (co) adapted genes rather than adapted organisms as natural selection's beneficiaries, unsettling Darwinism's delicate balance between chance, determinism, and teleology.

In the 1970s, hypotheses emerged that extended the game-theoretical view of natural selection as optimizing the adaptedness of genes "for" particular traits to the human sciences, first in the form of sociobiology and later as evolutionary psychology (Barkow, Cosmides, & Tooby, 1992; Wilson, 1975). These programs for unifying the natural and social sciences announce that gene-selectionist adaptationism is the culmination of the modern evolutionary synthesis and the final justification and vindication of the materialistic Darwinian worldview. In recent decades views of this stripe have passed in public as the sum and sustance of Darwinism, not just a version of it. Just as economists take it as axiomatic that the choices of economic actors will optimize their self-interest, sociobiologists and evolutionary psychologists take it as axiomatic that gene frequencies will in the long run follow the contours of the optimal payoff matrices of game theory. This assumption places lower burdens of proof on adaptive hypotheses, a bias that Lewontin and Gould noted by famously calling these scenarios "just-so stories" (Gould & Lewontin, 1979). Many of today's most publicly visible Darwinians call on genic selectionism to naturalize traditional gender roles with the same passion that led earlier generations of self-proclaimed Darwinians to proclaim the naturalness of unequal classes and races (McKinnon, 2005). That human males are by nature promiscuous and females on the lookout for a hard-working monogamous mate is traced to the fact that males have more sperm than they know what to do with, while females must protect their "parental investment" in a few eggs (Gangestad & Scheyd, 2005; Trivers, 1972).

This contention is dubious not just because it attempts to make biology testify on the conservative side in contemporary disputes about gender and sexuality, but also because it assumes rather than proves that behavioral traits are biological adaptations rather than cultural practices (Sahlins, 1976). It assumes, too, that the genotype of population genetics can be identified with the gene of molecular biology and that allelic variation in the molecular gene can be identified with point mutations in genes. These claims conflate genotypes as populational differences with genes as ontogenetic causes (Moss, 2003). By contrast, the maturation of the study of regulatory sectors of the genome and with it the return of dynamically developing and adapting organisms to center stage in evolutionary science suggests that the supposedly new Darwinian social sciences may have saddled themselves with outdated genetics in the same way Social Darwinians once saddled themselves with Spencer's commitment to Lamarckian inheritance and a sieve-like view of natural selection.

1.3. At the Boundaries of the Modern Synthesis: Sticklebacks

To explore the impact of the revolution in regulatory genetics on the Darwinism of the modern synthesis, I propose briefly to review some salient facts about the natural history of sticklebacks, a well-studied family of about sixteen species of smallish (3–4 inches), scaleless fish that nearly (but not quite) circle the globe in subarctic regions.[5] Some stickleback species live in saltwater, others in fresh. Hybrids are found in tidal estuaries. Saltwater sticklebacks are ancestral to freshwater, freshwater species having adapted to lakes and streams with the receding glaciers. Some freshwater species live in the depths of lakes, others at the surface, and others still in streams. A salient fact is that stickleback

5. More facts about sticklebacks are explored in Depew and Weber, in press.

speciation is rapid—sometimes in as few as 10 generations (Gelmond, Hippel, & Christy, 2009; McKinnon & Rundle, 2002)—and is associated with more or less simultaneous changes early in the life cycle. Genotypically, specific base pairs in the homeobox regulatory gene *Pitx1* are deleted, alleles coding for the protein ectodysplasin, which are found in ocean-going sticklebacks in much lower frequencies, are amplified, and a variety of recombination-blocking inversions facilitating reproductive isolation are put in place (Chan et al. 2010; Colosimo et al., 2005; Colosimo et al., 2004; Jones et al., 2012; McKinnon & Rundle, 2002). Phenotypically, in freshwater species the dorsal spines that give these fish their name are quickly reduced or not expressed at all. An even more salient fact is that these shifts occur in freshwater species without benefit of gene flow or adaptive radiation between geographically isolated lakes (Colosimo et al., 2004). Independently evolved freshwater stickleback populations can establish reproductive isolation from other populations in the same body of water, but can successfully mate with their counterparts in other lakes (Schluter & Nagel, 1995). It is relevant that sexual selection, and with it the possibility of sympatric speciation, plays a role in the reproductive isolation of many stickleback species (Bakker, 1993; Blouw, 1996). Males make nests, attract mates, and raise offspring.

These are not things partisans of the orthodox modern synthesis expect to see. True, these phenomena do not violate its core principles. Gene frequency changes still underlie phenotypic changes, and although the shifts occur in a few generations they do require more than one. The ways of sticklebacks can be accommodated to orthodoxy, moreover, by appealing to several accepted notions. One is the phenotypic plasticity of genotypes with wide norms of reaction, that is, genotypes that have become fixed in populations to meet a range of shifting environmental circumstances. It is possible that these sudden changes express genotypes with very wide norms of reaction. Another notion is the concept of parallel evolution. In contrast to convergent evolution, in which populations with different ancestors light on the same solutions to similar environmental problems, parallel evolution occurs when populations sharing the same ancestors evolve the same adaptations in similar environments or, in parallel speciation, so many of the same genes that they can interbreed. Still, experimental data showing that dorsal spine expression is almost immediately affected in ocean-going sticklebacks when their diet is changed suggests that the case of stickleback evolution approaches, even if it does not cross, the conceptual boundaries of the modern synthesis "by emphasizing the importance of environmentally initiated change" rather than what is happening genotypically (Pfennig et al., 2010, pp. 459, 466). This is particularly challenging to currently ascendant gene-centered versions of the modern synthesis.

The developmental biologist Scott Gilbert, writing with David Epel, reflects on the shift of the causal-explanatory focus in sticklebacks to the interface between the environment and development:

> Parallel evolution can result from the independent recruitment of similar developmental pathways by different organisms. Instead of the view that extrinsic selection pressures play the dominant role in such phenomena, the current view is that intrinsic developmental factors are critical in producing these parallel variations. Such parallel evolution was once the justification for the "creativity" of natural selection. Now we can see that development is what is creative. (Gilbert & Epel, 2009, p. 341, citing Gilbert, 2006)

As an accomplished historian of evolutionary biology, Gilbert knows how charged "creative" or "creative factor" have been in evolutionary theory ever since Spencer's 1887 *The Factors of Organic Evolution*. Whenever it is said that natural selection rather than Spencer's environmental pressures or Morgan's mutations or anything else is the novelty-producing, direction-giving, and so "creative" factor in evolution, the implication is that natural selection working over multiple generations is the source, usually in conjunction with other factors, of the adaptations that allow a subpopulation to gain a reproductive edge. This is as true of Mayr's organocentric interpretation of the modern evolutionary synthesis as it is of Dawkins's genocentric version. In denying that natural selection is the creative factor in stickleback evolution, Gilbert and Epel are sending a message that a case like this cannot even be properly described, let alone explained, without detailed ecologically context-dependent knowledge of gene regulation and gene expression and that this knowledge that might run afoul of the basic principles of the modern synthesis.

Gilbert and Epel do not deny that natural and sexual selection play a positive role in spreading the suppression of dorsal spines in freshwater sticklebacks. On the contrary, they characterize ontogenetic changes as the source of the variation on which selection can work in fairly orthodox ways. They explicitly assert, "If the loss of *Pitx1* expression in the pelvis occurs this trait can be readily selected" (Gilbert & Epel, 2009, p. 341, citing Colosimo et al., 2004). Nor do they deny that this trait is adaptive, although they would probably greet with skepticism most selectionist just-so stories about the origin or loss of dorsal spines in stickleback. (A favorite has it that in saltwater environments predatory fish are deterred from lacerating themselves on these spiny protrusions whereas in freshwater, where no such predators exist, they are not needed and so are not maintained.) What they are denying is that dorsal spine reduction in freshwater sticklebacks is adaptive *because* a force flowing from the environment to the organism—what they call an "extrinsic selection pressure"—slowly spreads an initially rare genetic variant through an interbreeding population until over many generations it becomes an adaptation. This cannot be the case because, far from being rare, the simultaneous shifts that occur in stickleback populations between salt and fresh water environments, including deletion of gene segments, recur regularly and rapidly in noncontiguous freshwater environments. It is more likely, they conclude, that the proper cause is change in the timing, placement, or rate of enzymatic action of gene regulation under changed environmental circumstances. (I prescind from the possible involvement of gene duplication and splicing, short segments of RNA, and epigenetic markings of DNA in mediating the environment–development interface; my present interest is in drawing attention to the general shift to development as a source of variation and cause of adaptation.)

If so, at least three conceptual implications follow. First, due to the responsiveness of the regulatory genetic system to environments, the production, and not just the uncovering, of variation should not be characterized as random or fortuitous. This renders the notion of "chance mutation" ambiguous. Second, the selective retention of this variation should not be portrayed as an uphill battle. This makes the notion of "selection pressure" misleading. Mechanisms for adapting and evolving ("evolvability") obviate characterizations like Dawkins's "Climbing Mt. Improbable" (Dawkins, 1996; see Reiss, 2009, for a refutation of this idea). Third, and most importantly for our main concern, predication of the process of adaptation and its product, adaptedness, is *shifted forward* from the posterior position it occupies in orthodox population genetic Darwinism, in which adaptations evolve gradually over many generations, to the responsiveness of each organism to its environment in virtue of "intrinsic developmental factors." When they talk somewhat loosely about "the arrival of the fittest" in contrast to "the survival of the fittest," Gilbert and Epel are not harking back to Spencer, but gesturing toward the idea that evolved mechanisms for evolving tend to make the production of variation coincide with the onset of adaptedness (Gilbert & Epel, 2009, p. 324n). If natural and sexual selection help spread new adapted traits around it is not because they are becoming adaptations in and through this process, but because they are already adaptive. Gilbert and Epel say that adaptive novelties can be selectively reinforced, not that selection is the cause of adaptation.

It follows from this analysis that populations are adapted only if their constituent organisms are. This proposition puts us in territory explored by Spencer, although Gilbert and Epel put the causal accent on internal shifts in regulatory genes rather than on the impinging environment and avoid Spencer's stress on "selection against." Nonetheless, their way of affirming "selection for" arguably does not quite remain within the bounds of the modern evolutionary synthesis. So dead set is the synthesis against the idea that adaptedness is transferred from organisms to populations that it can be said to be committed at a definitional level to the opposite view. According to the synthesis, adaptedness can be predicated of organisms and their traits only insofar as they are members of populations that are in the process of becoming or have become adapted by the reproductive success of heritable variants which when they first occur ("by chance") are not adaptations. Gilbert and Epel do not see it this way. Of course they might stipulate that an organism cannot be said to be adapting or adapted or that one of its traits is an adaptation until a genetic underpinning or mechanism for this claim has been identified. But rather than being a necessary and sufficient condition for predicating adaptation, as the makers of the modern synthesis required—in part to make sure Lamarckism didn't sneak in—this requirement can be demoted to a necessary condition. The proposal also has the advantage of corresponding to the common meaning of "adapted." Attributing adaptedness to organisms before predicating it of populations seems to view things right side up.

1.4. Adaptation, Development, and the Boundaries of the Modern Synthesis

Gilbert and Epel's is not the only way the new developmentalism can or already has affected evolutionary theory. I have focused on it because it is so diagnostic of the conceptual issues currently in play in evolutionary theory. Accordingly, in the final section I situate their "ecological developmental biology" in a wider range of theorizing in order to assess how deep the problem of conceptual continuity or discontinuity between the new developmentalism and the modern synthesis runs.

The recovery of development's relevance to evolution through the maturation of regulatory genetics has stimulated a wide range of competing proposals aimed at defending (Coyne, 2009; 2014; Lynch, 2007; Wray et al., 2014), expanding (Gould, 1980), extending (Pigliucci & Müller, 2010; Laland et al., 2014), or replacing (Gilbert & Epel, 2009) the modern evolutionary synthesis. Evaluating these proposals depends on two issues: what the current return to development is telling us about matters of biological fact and what advocates of defending, expanding, extending, or replacing the modern synthesis take the synthesis to be, mean, and imply. It is possible that definitions vary so widely that what look like contradictions between these four proposals are compatible and what look like agreements might be contradictory. In assessing this possibility, it is important to bear in mind that the synthesis is flexible enough to have accommodated in its increasingly long life span a range of new facts, mechanisms, interpretations, and research programs (Depew & Weber, 1995, 2011, 2013). It has been capacious enough to embrace a Dobzhansky, a Mayr, a Ford, a Dawkins, and even a Gould.[6] Nonetheless, the modern synthesis does have conceptual boundaries. Chief among them, I submit, is its insistence that transgenerational natural selection is the proper cause of the adaptations that organisms considered as members of adapted populations evolve.

To confirm this claim we must turn to the testimony of the makers of the modern synthesis: Dobzhansky, Mayr, and Simpson. They articulated the synthesis in the 1940s by showing how mathematical population genetics can help solve the problems faced by organisms and naturalists alike. They took as central the problems of trait fixation, biogeographical dispersion, speciation, cladogenesis, evolutionary grades, and classificatory practice (Cain, 2009). Their approach to solving these problems was built on a new conception of fitness that had already emerged from mathematical population genetics. Simpson puts the point clearly:

> Darwin emphasized the survival of favored (or the early death of unfavored) individuals.
>
> The survivors were, for the most part, the "fittest," in the sense of qualification for success in competition, the "struggle for existence." That is still the usual nonscientific understanding of the process, but to specialists in the study of evolution "natural selection" now means the average production of more offspring by such organisms in a population as are distinguished by any particular heredity factors. "Fitness" is now defined solely as relative success in reproduction. (Simpson, 1959, xi)

---

6. Gould and Niles Eldredge's proposal in the early 1980s to expand the synthesis envisioned natural selection and other recognized evolutionary "forces" operating in different proportions at multiple levels of the biological hierarchy: genes, organisms, groups, and species (Gould, 1980). Development played a large role in this proposal, but it is treated more as a constraint on selection than as a source of the variation on which selection plays. Gould monkeyed with the level and rate of natural selection, but not with its core principles. My focus is on developmental processes taken as the origin of the variation on which selection works in a manner that is the very opposite of constrained and on adaptedness considered as a context-dependent property of such variation. So I do not discuss Gould and Eldredge's multi-level "expansion of the modern synthesis." Nor do I discuss overtly anti-Darwinian forms of evolutionary theory that (mis)characterize natural selection as necessarily uncreative zero-sum competitions, sometimes calling on objections to Spencerism to serve as (misplaced) objections to Darwinism (Goodwin, 2001, for example).

Depending on one's epistemology, fitness so construed is either synonymous with the relative adaptedness of a population that possesses heritable factors favoring higher reproduction rates in comparison to closely related populations or an effect of (perhaps supervening on) these factors, which are in most cases myriad, context-dependent, and interactive. With their stress on predictive sufficiency, neopositivists tend toward the first view. Realists, who are willing to acknowledge that most adaptations and other evolutionary effects are identifiable only in retrospect, tend to prefer the second.

In response to his patron and sparring partner Morgan's insistence that mutation is the creative factor in evolution (Morgan, 1916, 1932), Dobzhansky argued that the new conceptions of fitness and adaptedness, however analyzed, require the honorific "creative" to be assigned to natural selection working transgenerationally on mutation or more often on other genetically based sources of variation such as recombination, and doing so in conjunction with auxiliary factors such as genetic drift and gene flow. In a late, somewhat excessive, but telling formulation, Dobzhansky wrote:

> Evolution is a creative process in precisely the same sense in which composing a poem or symphony . . . are creative acts. . . . How can selection, a process devoid of foresight, be the composer of biological symphonies? The idea is preposterous if selection is represented only as a preserver of a lucky minority of random mutations. Selection is, however, much more than a sieve retaining lucky and losing the unlucky mutations. (Dobzhansky, 1970, pp. 430–431; see Dobzhansky, 1964, pp. 150–151)

Other founding fathers said much the same thing. In the book that gave the modern synthesis its name, Julian Huxley wrote, "The statement that selection is a destructive agency is not true if it is meant as merely destructive. . . . It has a share in evolutionary creation. Neither mutation nor selection alone is creative of anything important in evolution; but the two in conjunction are creative" (Huxley, 1942, p. 28). Using the dichotomy between typological essentialism and population thinking that he so often wielded as a weapon, Mayr wrote, "Causal essentialists saw selection as a purely negative process that could eliminate unfit deviations from the type but could not play a constructive role in evolutionary advance" (Mayr, 1980, pp. 18, 2).

Aided by the skills of philosophers of biology in analyzing concepts, champions of the modern synthesis hold that its "creative" way of looking at natural selection is uniquely capable of maintaining the proper balance between its chancy, deterministic, and telic aspects (Brandon, 1990). They assign the element of chance to mutation and recombination, the element of determinism to impinging environmental forces that selectively amplify variation, and the telic element to the functional parts and goal-oriented behaviors that result from transgenerational natural selection. In doing so they make a legitimate claim to have separated the wheat from the chaff in Darwin's own thinking.

The discovery of the role played by regulatory genes in generating variation tied to changes in developmental timing, placement, and rate has put pressure on the modern synthesis as I have demarcated it. In particular, it has called into question whether Dobzhansky's claim that natural selection's way of responding to "challenges that arise from environments" deserves to be called creative after all (Dobzhansky, 1964, pp. 150–151). On Dobzhansky's view, problems are posed to organisms by environments. They do not solve these problems directly, as in Lamarckism or any other way of contravening Weismann's hard hereditary. Instead, solutions to the problems organisms face arise indirectly by way of transgenerational responses to environmental contingencies by the populations to which organisms belong. Over time, the environment selectively amplifies fitter variants until a population's state of adaptedness has been optimally, if transiently, adjusted. It can be argued that this orthodoxy still assigns the lead role to external forces and in so doing portrays organisms as too passive to capture the agency that they exert in responding to and making their own environments (Lewontin 1982). If so, the long shadow cast by Spencer's assignment of the leading role in evolutionary change to a demanding environment continues to fall over the modern synthesis. This objection is especially persuasive when we see regulatory genes setting off spasms of genome-wide variation that almost seem calculated to respond in short order to environment changes, as in the case of stickleback speciation. This being so, we may well ask whether a picture like Gilbert and Epel's, in which the generation of heritable variation by "intrinsic developmental factors" coincides with adaptedness, does a better job integrating the ontogenetic nature of organisms with natural selection than the modern synthesis under any interpretation of it.

Gilbert and Epel seem to think so. It is not surprising to read, "The fifty year . . . Mayr-Dobzhansky Synthesis is due for a makeover" (Gilbert & Epel, 2009, p. 398). Even if the extent of their anticipated conceptual shift is a bit unclear, they seem to be calling for replacement rather than expansion. In spite of their reassuring view that successive Darwinian theories are continuous enough to retain scientific knowledge that originated in earlier but generally inadequate conceptual frameworks while tossing out misfires, and in spite, too, of their willingness to assign a degree of cocausality to natural selection in spreading adaptations through populations, I believe that the single-generation link they postulate between adaptation and variation transgresses the conceptual bounds of the modern synthesis (Gilbert & Epel, 2009, p. 397). I interpret their call for a "third evolutionary synthesis" (to succeed the "Fisher-Wright" and "Mayr-Dobzhansky" syntheses) as acknowledging that they think so too (Gilbert & Epel, 2009, pp. 290, 398). Their way of putting this call, however, in which they urge refocusing on the "arrival" rather than the "survival of the fittest," is a bit infelicitous, since it calls unwarranted attention to Spencer's way of framing the issues. Gilbert and Epel themselves reject the Spencerian environmental determinism that we criticized early in this chapter as both false and not authentically Darwinian (Gilbert & Epel, 2009, p. 324).

Calls for a new synthesis generally share Gilbert and Epel's desire to bring organisms back into evolutionary biology by treating them as complex, self-formative, self-organizing developmental systems. In contrast to the burden Spencer placed on them to overcome Malthusian environmental pressures, organisms so construed can generally be counted on to adjust their relationship to the environments in which they are embedded and to which they are entrained. Organisms, that is to say, are inherently adaptive. Advocates of this view realize that this is a return to the main line of biology that runs from Aristotle to Cuvier minus the antievolutionary animus that gave Darwinism its chance (Reiss, 2009; Walsh, 2006). Recent discoveries about the sensitivity of regulatory genes to environmental contingencies that impel organisms to "recruit" helpful variation, including something like directed mutation in microbes (Cairns, Overbaugh, & Miller, 1988) and epigenetic inheritance (Jablonka & Lamb, 2005), have reinforced the perception that if evolutionary history had not bequeathed to organisms an inherent capacity for adapting to constant change rather than merely accommodating to it the species-rich and increasingly powerful lineages we see around us would long since have disappeared or never have evolved at all.

In any version of this approach the telic dimension is returned to the individual organism in the form of end-oriented development, where Aristotle first placed it. To deprive the modern synthesis of its claim to find the causes of adaptation, adaptedness, and adaptations at the population-genetic level some philosophers of biology who find these causes in the "intrinsic developmental factors" of individual organisms have argued for purely "formalist" versions of population genetics (Ariew & Matthen, 2002; Huneman, 2014; Walsh, 2006, 2015). If population genetics explains, they say, it is only in the weak sense of parsimoniously redescribing a confusing array of particular facts by using statistical summaries of events whose proper causes are ontogenetic. In this spirit, Denis Walsh writes that when population geneticists project phenotype space onto genotype space in the hope of learning something about evolution's causes they are "chasing shadows," not reading them (Walsh, 2000).[7]

Arguments of this sort are effective against the trait-atomistic adaptationism that Gould saw as the culmination of "the hardening of the Modern Synthesis" in the 1950s and 1960s (Gould, 1983). Gould was right to distinguish the hardened synthesis from the synthesis as such, even if he did not identify the difference perspicuously enough by failing to note that the hardened synthesis pays no attention to organisms as developmental systems in ways I have described its main line does (Depew, 2011). Are there proposals that appeal to the new knowledge of regulatory genetics, epigenetics, and other aspects of development that keep the ontogenetic origin of variation and its differential retention within the boundaries of the modern synthesis?

There are. Working with Carl Schlicting, Jonathan Kaplan, and Gerd Müller, the quantitative geneticist Massimo Pigliucci has called for "extending" rather than replacing the modern synthesis by taking into account the discovery that ontogeny is the site at which variation originates, becomes adaptive, and gives rise to the "endless forms most beautiful" that systematists classify (Carroll, 2005; Pigliucci & Kaplan, 2006; Pigliucci & Müller, 2010). Point mutations in structural genes can disrupt proper functioning, as medical geneticists know all too well, but the probability that they will ever lead to adaptation or speciation is vanishingly small. Pigliucci and Müller are entitled to say that they are extending, not replacing, the modern synthesis because they retain the transgenerational, population-level view of natural selection as the creative factor in evolution and proper cause of adaptation. They do so, while at the same time repudiating trait-by-trait adaptationism, especially in its gene-selectionist version, by reframing phenotypic plasticity, which became recessive in the period of the hardening, as developmental plasticity (Pigliucci, 2001).

Phenotypic plasticity, as we noted when we were discussing sticklebacks, refers to the width of responses to different environments of the same genotypes, their "norms of reaction." First mentioned by Johannsen and articulated by his

---

7. Some of the ensuing argument about this proposal has turned on whether it is appropriate to call genetic drift an evolutionary "force." If genetic drift seems to be no more than a statistical artifact of what might be called the law of small numbers, the same thing might be true of other evolutionary "forces," including natural selection. In response, Roberta Millstein and her colleagues have made a strenuous effort to prove that even genetic drift, which seems at first sight to be a causally empty notion, cannot be predicated unless it has causes as physical as those of natural selection (Millstein, Skipper, & Dietrich, 2009). The resolution of this dispute is important in determining whether evolutionary theory will remain conceptually continuous with the modern synthesis. What at first sight looks like scholastic nitpicking carries large implications for and addresses pressing problems in evolutionary biology.

Russian mentors, Dobzhansky tied phenotypic plasticity to his claim that heterozygotes are adaptively superior and so will be found at most chromosomal loci (Beatty, 1987). The waning fortunes of that hypothesis negatively affected those of phenotypic plasticity. It came to be regarded not as a property of particular genes, but as an epiphenomenal "by-product of evolution at numerous loci to produce appropriate trait means in different environments" (Schlichting & Pigliucci, 1993, p. 366, paraphrasing Via & Lande, 1985). Pigliucci and his collaborators call this interpretation into question. They argue that there are indeed phenotypically plastic genes, but that they are not causally anchored in random mutations in genes that code for structural gene products. They are the environmentally sensitive regulatory sectors of the genome that we earlier supposed to be at work in, among other things, stickleback development and stickleback evolution (Pigliucci, 2001; Schlichting & Pigliucci, 1993).

I like to imagine that Dobzhansky would have been pleased by the recovery of phenotypic plasticity as developmental plasticity. In retrospect, he might follow Mary Jane West-Eberhard in repudiating the stipulation that genetic variation must temporally precede natural selection in the evolution of adaptedness, in retrospect judging this demand to have been an overreaction to the hardening of the synthesis and the reductionism of early molecular geneticists (West-Eberhard, 2003). Contrary to the molecular dogmatism of Crick and Watson, which does not have a developmentalist bone in its body, and even of Jacques Monod and François Jacob, who discovered the first regulatory genes, West-Eberhard argues that genes are following, not leading, indicators of evolutionary change. In defending this view she reworks the old idea of the "Baldwin effect," which the makers of the modern synthesis initially accepted in principle but subsequently repudiated when their fear of Lysenko's Lamarckism, among other things, led them to trim their theorizing to conform to the central dogma of molecular biology (Simpson, 1953). First presented at the turn of the 20th century as a via media between Spencer and Weismann, the Baldwin effect, anachronistically expressed, asserts that phenotypes permitted by existing genotypic plasticity can be reliably reconstructed over a succession of generations by the agency of organisms in making their environments (including social environments) until more supportive genotypes shift norms of reaction in ways that better defend populations against environmental buffeting (Baldwin, 1896). Dobzhansky might appreciate mathematical models showing that lineages with a history of phenotypic plasticity adapt more readily to changing environments (Fierst, 2011). He might even be persuaded that this discovery updates the hypothesis that variation already hidden in gene pools makes parallel evolution in sticklebacks and other lineages not just possible but probable.

In contrast to Gilbert and Epel's, there is nothing in Pigliucci or West-Eberhard's proposals that contradicts the claim that adaptation is a multigenerational affair, even if the number of generations required is, at least in principle, as few as two. If niche construction leads the way there is nothing odd in rapid adaptation and speciation (Laland et al., 2014). Perhaps the most striking, if infrequently acknowledged, implication of calls for extending the synthesis along these lines, however, is that genocentric adaptationism of Dawkins's sort can no longer count as a coherent research program within the modern synthesis, let alone as its acme or even less the culmination of the entire Darwinian research tradition (Pigliucci & Kaplan, 2006, pp. 74–77). To extend the synthesis, it seems, requires throwing something out and freeing what remains from the influence of the dogma that evolution reduces to natural selection considered as a mechanistic process of optimization ranging over random mutations in self-replicating molecules. This conception of natural selection does scant justice to variation's many sources, natural selection's many modes, the roles of nonselectionist processes in evolution, and the telic nature of agency-enhancing adaptation. It does even less justice to human behavioral evolution.

Let me draw attention, finally, to a third tendency among evolutionary biologists, whose partisans are sometimes critical of genocentric adaptationism but who nonetheless resist calls for an extended synthesis, sensing in these calls that more dubious ones lurk around the corner. This group is composed of, and to some extent speaks for, working evolutionary biologists who have a strong desire to affirm the scientific status of a discipline they are already practicing. This discipline, they point out, has shown its scientific character by its steady progress. Starting from a toolkit that originally contained little more than one-locus-two-allele models of population genetics, professional evolutionary biologists have accumulated an ever-growing array of techniques, scenarios, models, and mechanisms that have proven capable of explaining an ever increasing range of biological phenomena. Chief among these is still natural selection ranging gradually over populations with different gene frequencies (Coyne, 2009; Wray et al., 2014).

To those who promote this view, evolutionary science rests on empirical, largely experimental norms of adequacy of a sort that will (and if it really is a mature science must) be recognized and acknowledged by all good scientists. In biology at least, if not in physics, empirical sufficiency of this sort is consistent with the great diversity of biological phenomena because the generalizations that proximally guide its predictions and explanations of cases are context-dependent models, not context-independent laws. Scientific knowledge of evolution grows with the expansion of evolutionary biology's methods and models and with the swelling library of cases to which professional students of evolutionary biology are introduced in their apprenticeships. Jerry Coyne's slow but sure articulation and validation of Dobzhansky's midcentury intuitions about speciation is a case in point (Coyne & Orr, 2004). This knowledge was achieved by hard work at the bench and in the field (Coyne, 2009). Coyne's worry is that stressing discontinuity with or wholesale revision of the modern synthesis will result in discounting what is already known and decrease the likelihood of getting more of it.

Coyne's former mentor Lewontin generally shares his high standards of empirical proof and distrust of speculation (Lewontin, 1974, pp. 5–8). Decades ago, Lewontin attempted to make even clearer than his own mentor, Dobzhansky, that natural selection, operating as it does in many modes on populations of developmental beings, sometimes decreases adaptedness (Gould & Lewontin, 1979; Lewontin, 1970). This fact alone, he argued, falsifies the "adaptationist program." In making this case, Lewontin was critiquing not just adaptationism, but the very idea of separate traits as biological adaptations. It is, he argued, a notion imported by Darwin into evolutionary biology from natural theology and social ideology. The heuristic value of this sally in suggesting Leibnizean sufficient reasons for the evolution of this or that trait, he suspects, is greatly outweighed by its distorting effect on biological inquiry. The tropology of design-without-a-designer is inseparable from the theological idea of design-with-a-designer and from an analogy between artifacts and organisms so strong that it runs roughshod over the dynamic interaction among environments, genes, and organisms in ontogeny. Its advocates make things worse by appealing for support to mechanistic metaphysics, whose rise, persistence, and dominance in our culture is best explained by its ideological role in projecting onto nature an economically unequal and undemocratic way of distributing power (Levins & Lewontin, 1985; Lewontin, 1978, 1991). The reductionism and determinism of molecular genetics, and its recent monetization in genetic biotechnology, are from Lewontin's perspective the latest chapter in an old story that has given materialism a bad name.[8]

Lewontin finds the defects of adaptationism not only in the focus of Darwinian research programs on traits rather than organisms, but also in Dobzhansky's description of populations of organisms as responding to problems set by their environments (Levins & Lewontin, 1985, pp. 67–73, 98, 154–155). This characterization, he argues, promotes a crypto-Spencerian view of organisms as passive beings rather than agents that make their own environments (Lewontin 1982). It

---

8. Adaptationists, including Lewontin's bête noir E. O. Wilson, misunderstand Lewontin's oft professed Marxian materialism as an ideological distortion more egregious than anything they could possibly be accused of (Lumsden & Wilson, 1983, p. 4). Actually, historical materialism affords Lewontin a way of placing a strong burden of proof on speculative ideas. Unless empirically proven, ideas that circulate widely in a society should presumptively be seen as ways of making unequal distributions of power seem natural and hence legitimate. Showing them up as ideology clears the way for science rigorously to prove what is actually the case. Adaptation is high on Lewontin's list of contaminated ideas.

is a background picture that led Dobzhansky in his last decade to entanglements with behavioral genetics that flirted with genetic determinism and rode rough-shod over his earlier insights into evolution and ontogeny:

> An irony of the intellectual history of [Dobzhansksy's 1937] *Genetics and the Origin of Species* is that Dobzhansky came into evolutionary genetics from the study of morphological diversity in nature and so was able to relate the abstractions of genetic theory to the biology of organisms, yet in the end he and the field he founded became captives of abstractions that . . . speak the language of gene frequencies . . . [and await the day when population genetics will] finally fold developmental biology into its considerations. (Lewontin, 1997, pp. 354–355)

This may or may not be a fair characterization of Dobzhansky. Still, remarks like this show that Lewontin has already penetrated far enough into the circle of ideas associated with hopes for an extended synthesis to suggest, first, that evolutionary biologists who accept trait adaptationism and genic selectionism as legitimate research programs within the modern synthesis cannot possibly be reliable critics of proposals for extending it; and, second, that Lewontin's own hesitancy to jump on the bandwagon sounds a cautionary note that hypotheses advanced by advocates of an extended synthesis must be rigorously validated before they can be accepted. It almost follows from Lewontin's view of organisms as agents in and makers of their own worlds that they are adaptive in an active sense because they are developmental systems. But just saying this does not free one from the obligation to prove it in particular cases by revealing complex natural histories in which genes, organisms, and environments interact in and beyond ontogeny.

For my part, I am confident that adaptation is a real process in nature—real enough to require explanation rather than being either trivially true of organisms considered as inherently adaptive or trivially false in virtue of its entanglement with the design paradigm. The process of adaptation can and has been analyzed in ways that are not dependent on the design paradigm even as a heuristic "intuiton pump" (Brandon, 1990; Depew, 2015).

I am also confident that, although it has been derailed on several occasions, a developmentalist strain has existed in the modern synthesis from the beginning and that the current explosion of knowledge about the role of regulatory genetics and developmental dynamics in the generation and retention of variation will show that calls for extending it are more continuous with its core conceptual commitment to the creativity of natural selection than opponents and advocates of extending the synthesis sometimes suppose. The pursuit of further research under the presumption of an extended synthesis may eventually result in a shift of the sort anticipated by Gilbert and Epel. But I suggest that the most reliable way of determining whether the creative factor in evolution should be resituated without buying a pig in a poke is at present to follow the lead of the extended synthesis.

References

Ariew, A., & Matthen, M. (2002). Two ways of thinking about fitness and natural selection. *Journal of Philosophy*, *49*, 55–83.

Axelrod, R., & Hamilton, W. (1981). The evolution of cooperation. *Science*, *211*, 1390–1396. doi:10.1126/science.7466396

Bakker, T. (1993). Positive genetic correlation between female preference and preferred male ornament in sticklebacks. *Nature*, *363*, 255–257.

Baldwin, J. M. (1896). A new factor in evolution. *American Naturalist*, *10*, 354, 422, 441–451, 536–553.

Barkow, J., Cosmides, L., & Tooby, J., eds. (1992). *The adapted mind: Evolutionary psychology and the generation of culture*. New York, NY: Oxford University Press.

Beatty, J. (1987). Weighing the risks: Stalemate in the classical/balance controversy. *Journal of the History of Biology*, *20*, 289–319.

Beatty, J. (2016). The creativity of natural selection? Part 1: Darwin, darwinism, and the mutationists. *Journal of the History of Biology*, *59*, 659–684.

Blouw, D. (1996). Evolution of offspring desertion in a stickleback fish. *Ecoscience*, *3*, 18–24.

Brandon, R. (1990). *Adaptation and environment*. Princeton, NJ: Princeton University Press.

Cain, J. (2009). Rethinking the synthesis period in evolutionary studies. *Journal of the History of Biology*, *42*, 621–648.

Cairns, J., Overbaugh, J., & Miller, S. (1988). The origin of mutants. *Nature*, *335*, 142–145.

Carroll, S. B. (2005). *Endless forms most beautiful: The new science of evo devo and the making of the animal kingdom*. New York, NY: Norton.

Chan, Y., Marks, M., Jones, F., Villarreal, G., Jr., Shapiro, M., Brady, S., . . . Kingsley, D. (2010). Adaptive evolution of pelvic reduction in sticklebacks by recurrent deletion of a *Pitx1* enhancer. *Science*, *327*, 302–305.

Colosimo, P., Hosemann, K., Balabhardra, S., Villareal, G., Jr., Dickson, M., Grimwood, J., . . . Kingsley, D. (2005). Widespread parallel evolution in sticklebacks by repeated fixation of Ectodysplasin alleles. *Science*, *307*, 1928–1933.

Colosimo, P., Peichel, C., Nereng, K., Blackman, B., Shapiro, M., Schluter, D., & Kingsley, D. (2004). The genetic architecture of parallel armor plate reduction in threespine sticklebacks. *PLoS Biology*, *2*, E109.

Coyne, J. (2009). Are we ready for an "extended evolutionary synthesis"? *Evolution Is True* (Blog). Retrieved February 10, 2016, from http://whyevolutionistrue.wordpress.com/2009/02/16/are-we-ready-for-an-extended-evolutionary-synthesis

Coyne, J., & Orr, A. (2004). *Speciation*. Sunderland, MA: Sinauer.

Darwin, C. (1859). *On the origin of species*. London, UK: John Murray.

Darwin, C. (1968). *Variation of animals and plants under domestication*. 2 vols. London, UK: John Murray.

Darwin, F. (1887). *Life and letters of Charles Darwin*. 3 vols. London, UK: John Murray.

Dawkins, R. (1976). *The selfish gene*. Oxford, UK: Oxford University Press. Rev. ed. 1989.

Dawkins. R. (1996). *Climbing mount improbable*. New York, NY: Norton.

Depew, D. (2011). Adaptation as process: The future of Darwinism and the legacy of Theodosius Dobzhansky. *Studies in the History and Philosophy of Biology and the Biomedical Sciences, 42*, 89–98.

Depew, D. (2013). Conceptual change and the rhetoric of evolutionary theory: Force talk as a case study and challenge for science pedagogy. In K. Kampourakis (Ed.), *The philosophy of biology: A companion for educators* (pp. 121–141). Dordrect: Springer.

Depew, D. (2015). Accident, adaptation, and teleology in Empedocles, Aristotle, and Darwin. In P. Sloan, K. Eggleson, & G. McKenny, *Darwin in the twenty-first century: Nature, man, and god* (pp. 116–143). Notre Dame, IN: University of Notre Dame Press.

Depew, D., & Weber, B. (1995). *Darwinism evolving*. Cambridge MA: MIT Press.

Depew, D., & Weber, B. (2011). The fate of Darwinism: Evolution after the modern synthesis. *Biological Theory, 6*, 89–102.

Depew, D., & Weber, B. (2013). Challenging Darwinism: Expanding, extending, or replacing the modern evolutionary synthesis. In M. Ruse (Ed.), *The Cambridge encyclopedia of Darwin* (pp. 405–411). Cambridge, UK: Cambridge University Press.

Depew, D., & Weber, B. (in press). Developmental biology, natural selection and the conceptual boundaries of the modern evolutionary synthesis. *Zygon*.

Dobzhansky, T. (1937). *Genetics and the origin of species*. New York, NY: Columbia University Press.

Dobzhansky, T. (1951). *Genetics and the origin of species* (2nd ed.). New York, NY: Columbia University Press.

Dobzhansky, T. (1964). *Heredity and the nature of man*. New York, NY: Harcourt Brace.

Dobzhansky, T. (1970). *Genetics of the evolutionary process*. New York, NY: Columbia University Press.

Dobzhansky, T. (1973). Nothing in biology makes sense except in the light of evolution. *American Biology Teacher, 35*, 125–129.

Dunn, L. C., & Dobzhansky, T. (1946). *Heredity, race, and society*. New York: New American Library.

Fierst, J. (2011). A history of phenotypic plasticity accelerates adaptation to a new environment. *Journal of Evolutionary Biology*, 24, 1992–2001.

Fisher, R. (1918). The correlation between relatives on the supposition of Mendelian inheritance. *Transactions of the Royal Society of Edinburgh*, 52, 399–433.

Fisher, R. (1930). *The genetic theory of natural selection*. Oxford, UK: Oxford University Press.

Fisher, R. (1932). The evolutionary modification of genetic phenomena. *Proceedings of the 6th International Congress of Genetics*, 1, 165–172.

Gangestad, S., & Scheyd, G. (2005). The evolution of human physical attractiveness. *Annual Review of Anthropology*, 34, 523–548.

Gayon, J. (1997). The paramount power of selection: From Darwin to Kauffman. In Dalla Chiara, M.L., Doets, K., Mundici, D., van Benthem, J. (Eds.), *Structure and norms in science. Dordrecht: Springer*, (pp. 265–282). *Synthese*.

Gelmond, O., Hippel, F. V., & Christy, M. (2009). Rapid ecological speciation in threespined stickleback *Gasterosteus aculeatus* from Middleton Island, Alaska: The roles of selection and geographical isolation. *Journal of Fish Biology*, 75, 2037–2051.

Gilbert, S. (1994). Dobzhansky, Waddington, and Schmalhausen: Embryology and the modern synthesis. In M. Adams (Ed.), *The evolution of Theodosius Dobzhansky* (pp. 143–154). Princeton, NJ: Princeton University Press.

Gilbert, S. (2006). The generation of novelty: The province of developmental biology. *Biological Theory*, 1, 209–212.

Gilbert, S., & Epel, D. (2009). *Ecological developmental biology*. Sunderland, MA: Sinauer.

Gould, S. (1980). Is a new and general theory of evolution emerging? *Paleobiology*, 6, 119–130.

Gould, S. (1983). The hardening of the modern synthesis. In M. Grene (Ed.), *Dimensions of Darwinism* (pp. 71–93). Cambridge, UK: Cambridge University Press.

Gould, S. J., & Lewontin, R. (1979). The Spandrels of San Marco and the Panglossian paradigm: A critique of the adaptationist programme. *Proceedings of the Royal Society of London B*, 205, 581–598.

Goodwin, B. (2001). *How the leopard changed its spots: The evolution of complexity* (Rev. ed.). New York, NY: Scribners.

Hamilton, W. D. (1964). Genetic evolution of social behavior. *Journal of Theoretical Biology*, 7, 16–52.

Huneman, P. (2014). Formal Darwinism as a tool for understanding the status of organisms in evolutionary biology. *Biology and Philosophy*, 29, 271–279.

Huxley, J. (1942). *Evolution: The modern synthesis*. London, UK: Allen and Unwin.

Jablonka, E., & Lamb, M. (2005). *Evolution in four dimensions: Genetic, epigenetic, behavioral, and symbolic variation in the history of life*. Cambridge, MA: MIT Press.

Jones, F., Grabherr, M., Chan, Y., Russell, P., Mauceli, E., Johnson, J., . . . Brady, S. (2012). The genomic basis of adaptive evolution in threespine sticklebacks. *Nature*, *484*, 55–61.

Keller, E. (2010). *The mirage of a space between nature and nurture.* Durham, NC: Duke University Press.

Kettlewell, H. (1955). Selection experiments on industrial melanism in the *Lepidoptera. Heredity*, *9*, 323–342.

Kettlewell, H. (1956). Further selection experiments on industrial melanism in the *Lepidoptera. Heredity*, *10*, 287–301.

Lack, D. (1947). *Darwin's finches.* Cambridge, UK: Cambridge University Press.

Laland, K., Uller, T., Feldman, M., Sterelny, K., Müller, G., Moczek, A., . . . Odling Smee, J. (2014). Does evolutionary theory need a rethink? Yes. Urgently. *Nature*, *514*, 161–162.

Lerner, I. M. (1954). *Genetic homeostasis.* New York, NY: Wiley.

Levins, R., & Lewontin, R. (1985). *The dialectical biologist.* Cambridge, MA: Harvard University Press.

Lewontin, R. (1970). The units of selection. *Annual Review of Ecology and Systematics*, *1*, 1–18.

Lewontin, R. (1974). *The genetics of evolutionary change.* New York, NY: Columbia University Press.

Lewontin, R. (1978). Adaptation. *Scientific American*, *239*, 212–228.

Lewontin, R. (1982). Organism and environment. In H. Plotkin (Ed.), *Learning, development and culture: Essays in evolutionary epistemology* (pp. 151–170). New York, NY: Wiley.

Lewontin, R. (1991). *Biology as ideology.* New York, NY: HarperCollins.

Lewontin R. (1997). Dobzhansky's *Genetics and the Origin of Species*: Is it still relevant? *Genetics*, *147*, 351–355.

Lumsden, C., & Wilson, E. (1983). *Genes, mind, and culture: The coevolutionary process.* Cambridge, MA: Harvard University Press.

Lynch, M. (2007). *The origin of genomic architecture.* Sunderland, MA: Sinauer.

Maynard Smith, J., & Price, G. (1973). The logic of animal conflict. *Nature*, *246*, 15–18.

Mayr, E. (1963). *Animal species and evolution.* Cambridge, MA: Harvard University Press.

Mayr, E. (1980). Prologue. In E. Mayr & W. Provine (Eds.), *The evolutionary synthesis* (pp. 1–48). Cambridge, MA: Harvard University Press.

Mayr, E. (1991). *One long argument.* Cambridge, MA: Harvard University Press.

Mayr, E. (2001). What evolution is. In Brockman, J. (Ed.), *Edge* (p. 92).

McKinnon, S. (2005). *Neo-liberal genetics: The myths and metaphors of evolutionary psychology.* Chicago, IL: Prickly Paradigm Press.

McKinnon, J., & Rundle, H. (2002). Speciation in nature: The threespine stickleback model system. *Trends in Ecology and Evolution*, *7*, 480–488.

Millstein, R., Skipper, R., & Dietrich, M. (2009). (Mis)interpreting mathematical models: Drift as a physical process. *Philosophy and Theory in Biology, 1*, e002.

Morgan, T. H. (1916). *A critique of the theory of evolution*. Princeton, NJ: Princeton University Press.

Morgan, T. H. (1932). *The scientific basis of evolution*. New York, NY: Norton.

Moss, L. (2003). *What genes can't do*. Cambridge, MA: MIT Press.

Nowak, M., Tarnita, A., & Wilson, E. (2010). The evolution of eusociality. *Nature, 466*, 1057–1062. doi:10.1038/nature09205

Pfennig, D., Wund, M., Snell-Root, E., Cruickshank, T., Schlichting, C., & Moczek, A. (2010). Phenotypic plasticity's impacts on diversification and speciation. *Trends in Evolution and Ecology, 25*, 459–467.

Pigliucci, M. (2001). *Phenotypic plasticity: Beyond nature and nurture*. Baltimore, MD: Johns Hopkins University Press.

Pigliucci M., & Kaplan, J. (2006). *Making sense of evolution*. Chicago, IL: University of Chicago Press.

Pigliucci, M. & Müller, G. (2010). Elements of an extended evolutionary synthesis. In M. Pigliucci & G. Muller (Eds.), *Evolution: The extended synthesis* (pp. 2–17). Cambridge, MA: MIT Press.

Provine, W. (1971). *The origins of theoretical population genetics*. Chicago, IL: University of Chicago Press.

Reiss, J. (2009). *Not by design: Retiring Darwin's watchmaker*. Berkeley and Los Angeles: University of California Press.

Rudge, D. (1999). Taking the peppered moth with a grain of salt. *Biology and Philosophy, 14*, 9–37.

Sahlins, M. (1976). *The use and abuse of biology*. Ann Arbor: University of Michigan Press

Schmalhausen, I. I. (1949). *Factors of evolution: The theory of stabilizing selection*. Philadelphia, PA: Blakiston.

Schlichting, C., & Pigliucci, M. (1993). Control of phenotypic plasticity via regulatory genes. *American Naturalist, 142*, 366–370.

Schluter, D., & Nagel, L. (1995). Parallel speciation by natural selection. *American Naturalist, 146*, 292–301.

Simpson, G. G. (1953). The Baldwin effect. *Evolution, 7*, 110–117.

Simpson, G. G. (1959). Forward. *The life and letters of Charles Darwin* (pp. v–xvi). New York, NY: Basic Books.

Spencer, H. (1887). *The factors of organic evolution*. New York, NY: Appleton.

Stauffer, R. (Ed.). (1975). *Charles Darwin's natural selection*. Cambridge, UK: Cambridge University Press.

Trivers, R. (1972). *Parental investment and sexual selection*. Chicago, IL: Aldine.

Via, S., & Lande, R. (1985). Genotype-environment interaction and the evolution of phenotypic plasticity. *Evolution, 39*, 505–522.

Walsh, D. (2000). Chasing shadows. *Studies in the History and Philosophy of Biological and Biomedical Sciences, 31*, 135–153.

Walsh, D. (2006). Organisms as natural purposes: The contemporary evolutionary perspective. *Studies in the History and Philosophy of the Biological and Biomedical Sciences, 37a*, 771–791.

Walsh, D. (2015). *Organisms, agency, and evolution.* Cambridge, UK: Cambridge University Press.

Weldon, W. F. R. (1893). On certain correlated variations in *Carcinus moenas. Proceedings of the Royal Society, 54*, 318–329.

Wiener, J. (1994). *The beak of the finch.* New York, NY: Knopf.

Wilson, E. (1975). *Sociobiology: The new synthesis.* Cambridge, MA: Harvard University Press.

West-Eberhard, M. J. (2003). *Developmental plasticity and evolution.* New York, NY: Oxford University Press.

Wray, G., Hoekstra, H., Futuyma, D., Lenski, R., Mackay, T., Schluter, D., & Strassman, J. (2014). Does evolutionary theory need a rethink? No, all is well. *Nature, 514*, 161, 163–164.

2

WHY WOULD WE CALL FOR A NEW EVOLUTIONARY SYNTHESIS?

THE VARIATION ISSUE AND THE EXPLANATORY ALTERNATIVES

Philippe Huneman

It is pervasively claimed that the framework of evolutionary theory, the Modern Synthesis (MS), built between the 1930s and the 1950s through the synthesis of Mendelian genetics and Darwinian theory, has to be deeply revised, changed, "extended" (Danchin et al., 2011; Odling-Smee, Laland, & Feldman, 2003; Pigliucci & Muller, 2011), or "expanded" (Gould, 2002). Empirical and conceptual reasons are supposedly converging to support such a move. Supporters of a new synthesis (e.g., Danchin et al., 2011; Gilbert, Opitz, & Raff, 1996; Jablonka & Lamb, 2005, Pigliucci & Muller, 2011; Odling-Smee et al., 2003, Oyama, Griffiths, & Gray, 2001, etc.) may disagree on many things, but they generally think that the MS, as a scientific theory through which explanations for biological phylogeny, adaptation, and diversity should be sought, gave too much importance to natural selection or that its focus on genes, exemplified by the textbook definition of evolution as a change in genotypic frequencies, was exaggerated.

In this chapter I question what kind of empirical findings would be likely to force a dramatic theoretical change, challenging the core claims of MS regarding natural selection and genetics. I focus on one strain of critiques that target the explanatory weight of natural selection as selection for the fittest traits (provided that variant traits arise by mutation and recombination). I distinguish several critiques according to their content and to the strength of the challenge: Whereas some challenges necessitate a complete rethinking of the core concepts of evolutionary theory, others are solved by amending the main concepts or modeling assumptions. This chapter aims to capture at the most abstract level the structure of such issues in order to facilitate assessment of how present and future empirical findings bear on the controversies. It questions the relationships between empirical research and theoretical models, asking what kinds of empirical facts regarding variation, would force us to strongly reconceptualize evolutionary theory. I eventually argue that, even if the explanatory role of selection can be constrained by the structure of variation, and such constraint compels us to reshape the explanatory factors in evolutionary biology, very few kinds of variations are likely to play this role.

The first section provides a systematic overview of the variety of "extended" syntheses, showing their place in the general space of controversies in evolutionary theory and the empirical findings that motivate them. The second section considers one line of controversy, the status of natural selection, and focuses on the concept of "constraints" as advanced by Gould and Lewontin (1978), which triggered many debates. Since constraints reveal that variation may have other explanatory roles than those defined by MS (i.e., fuel for selection), taking them into consideration entails changing the classical selection-based model of explanation of traits. Thereby the third section describes two distinct explanatory schemes for a theory, one proper to MS and an alternative one; yet many *weak* revisions of the MS scheme are still possible even though they will not turn it into an alternative explanatory scheme. The fourth section considers which types of variation should motivate a departure from the MS scheme, and which changes to the explanatory scheme would in this case occur. It formulates in a very general manner what random, directed, biased, blind, and adaptive variation mean in order to decide which of these could support the alternative explanatory scheme. It then considers some cases of random and biased and directed variation, both allelic and phenotypic, and argues that in many cases an appeal to an alternative explanatory scheme needs additional principled reason to be legitimate.

2.1. The Rationale(s) for Extending the Synthesis

The general idea of an extension of the evolutionary synthesis comes naturally to mind if one considers that many of these projects stem from disciplines that had been arguably left aside from the MS: developmental theory, as it has been constantly argued by evo-devo theorists (e.g., Amundson, 2005; Burian, 2005), morphology (Ghiselin, 1980; Love, 2003), and ecology. Indeed, many of the calls for extension come from the side of developmental theory (e.g., Newman & Muller, 2003; Lewontin, 2000), from morphology, or from ecology (Odling-Smee et al.,

2003). There are many ways of understanding these claims,[1] but a common theme is precisely that they challenge the MS in order to unite all biological fields of research related to evolution, adaptation, and diversity in a single theoretical frame.[2]

Especially, there are different views about what exactly would change in the structure of evolutionary biology, were we to adopt an "extended" synthesis. Some are arguing that an extended synthesis would explain *form*, and more generally, the fact that some overarching forms, such as the tetrapod limb, are pervasive in the biological world, whereas MS focused on adaptation and genes (Pigliucci, 2007; also see Amundson, 2005, who places the controversy in the light of a long-standing contrast between functionalism and structuralism). In this perspective, it is not obvious whether what we are dealing with is an *extension*, or just another kind of science; therefore the very project of an "extended *synthesis*" as compared to the previous synthesis is doubtful (Amundson, 2005), and assessing what "extended synthesis" means includes deciding about this ambiguity.

Others think that the extended synthesis deals with same *explananda* as the MS, but has to change *explanantia* or add new ones. This would happen because natural selection is not the only major *explanans* of evolution, adaptation, or diversity, since other processes, such as self-organization (Kaufmann, 1993; Walsh, 2003), adaptive variation (Jablonka & Lamb, 2005), or niche construction (Odling-Smee et al., 2003) are plausible *explanantia* of adaptation and evolution. Finally, it may be that, even if natural selection is still the main *explanans*, it has to be understood differently: by changing the units of selection (switching from genes to "developmental cycles," as advocates the DST, e.g., Griffiths and Gray, 1994), by integrating the effects of plasticity (Love, 2010; West-Eberhard, 2003), or by focusing on its effects on developmental genes or regulatory genes rather than coding genes generally (Carroll, 2005). Other very different theoretical elaborations of a new evolutionary framework would still concur on an extension of our concept of selection: allowing for sexual selection, signal selection (Zahavi & Zahavi, 1997), social selection (Roughgarden, 2009), multilevel selection (Wade, 1977; Damuth & Heisler, 1988; Goodnight, 2005; Okasha 2006; Sober & Wilson, 1998, etc.), species selection (Gould & Lloyd, 1999; Rice, 2004), and so on. Actually, many of these approaches are reflected in the contents of the book *Extended Synthesis* (Pigliucci & Müller 2011).

---

1. Love (this volume) presents an interesting approach based on the idea of problem agendas.

2. See Gayon (1998), Smocovitis (1996), and Mayr and Provine (1980), on the rise of the MS through population genetics and then its import to other disciplines.

There is no doubt that MS as a historical stage in science was much more complex than the theory targeted by those biologists looking for an alternative framework (e.g., Depew, 2011, this volume); moreover it is plausible that these biologists frame the MS in their own way in order to make their own case more striking. Several core commitments that one could consider as crucial to the MS define what will be at stake afterward, and they could be wrapped up in what Huxley wrote to Mayr in 1951: "*Natural selection*, acting on the heritable variation provided by the mutations and recombination of a Mendelian genetic constitution [a], is the *main agency* [b] of biological evolution." Let us unpack these claims further.

a. "The heritable variation provided by the mutations and recombination of a Mendelian genetic constitution": Notwithstanding their divergences, MS biologists agreed that inheritance is mostly genetic (the variance not due to the environment is genetic, and mostly additive genetic variance). As a corollary, evolution can be defined as a change in gene frequencies in a population; change of traits, species transformation, and other evolutionary features supervene on it. Hence population (and quantitative) genetics appears as the science—*sensu* mathematical modeling—of the process of evolution by natural selection. One must subsequently decouple *inheritance* (pattern transmission of traits from one generation to another in a population) and *development* (the way a genotype gives rise to a genotype), this process underlying the transformation of the zygote into an adult organism, only the former pertaining to evolution.

b. "Natural selection is the main agency": selection defines what an adaptation is, and therefore explains why adaptation is so pervasive in the biological world, even though population genetics per se builds many models, in which non-selective forces like mutation, migration, or drift are in principle as likely as selection to contribute to the diversity patterns of populations.

These core commitments have never been absolutely uncontroversial. On the contrary, since the times of Fisher and Wright, there has always been disagreement about which claims to accept, even though mainstream evolutionary biology since these days seems to be strongly committed to Huxley's dictum. Yet the degree to which one accepts claims a and b determines the space of controversies in evolutionary biology,[3] so that the calls for an extended synthesis, fueled by new findings in evolutionary and developmental biology, still occur in this definite

---

3. For a detailed presentation of the space of controversies, see Huneman (2014).

space. According to whether what is questioned is rather the genetic nature of the variation (claim a) or the role of natural selection (claim b), different types of extended synthesis can be proposed.

Notice in this context that the issues of *adaptationism* (about b), which is the name given by controversies over the weight of natural selection, and issues of *gene centrism* (about a) are not directly or necessarily related. From the beginnings of the MS up to now we can find instances of all combinations: Some like Mayr would accept Huxley's claim about selection as the "main agency" but not the idea that evolution is a change in gene frequencies.[4] Others would precisely accept the centrality of genes, like Michael Lynch, who ironically said, "Nothing in evolutionary biology makes sense except in the light of population genetics" (Lynch 2007), yet he would not ascribe a prominent role to optimization since his own models show that drift is one of the main agents of the architecture of the metazoan genome.

However many of the "extended" theories are indeed often challenging both the crucial role of natural selection and the gene centrism of MS: That is, they extend evolutionary processes beyond population genetics, and they extend the set of putative explanantia or processes beyond natural selection. In each case, some empirical findings seem to legitimate those extensions, for instance cases of nongenetic inheritance (Hellantera and Uller, 2010; Danchin et al., 2011; Jablonka & Lamb, 2005) or apparently directed or nonrandom variation (e.g., Jablonka & Lamb, 2005; Moxon, Raineyt, Nowak, & Lenski, 1994; Walsh, 2013). The question raised by calls for a revised, extended, or new synthesis is therefore to assess in principle to what extent these findings compel us to overcome one or both of the claims summarized by Huxley, and then which transformations of the explanatory structure of evolutionary biology would follow.

2.2. Trade-Off Adaptationism

To formulate this question I'll start by reconstructing a major target of the challengers to the MS synthesis, which is what I call the "trade-off adaptationism", after some remarks in Gould and Lewontin (1978).

---

4. "Evolution is not a change in gene frequencies, as is claimed so often, but the maintenance (or improvement) of adaptedness and the origin of diversity. Changes in gene frequency are a result of such evolution, not its cause" (Mayr, 1997). The focus on "adaptedness," which is a property of organisms, and contrasts with "adaptation" as the character of a trait, introduces the hotly debated question of the relation between traits as adaptations and the overall adaptedness of whole organisms.

2.2.1. Constraints and the Explanatory Switch

The "spandrel paper" (Gould & Lewontin, 1978, hereinafter GL) aimed at stating the limits of the MS and introducing key concepts for alternative approaches ("developmental constraints," *Bauplan*, etc.), relying on the work by Alberch, Gould, Hall, and other evolutionary developmental theorists (see Alberch et al., 1979). Even though it occurred long before the explicit idea of an "extended synthesis," it provides an interesting starting point for a reflection about what one should change in evolutionary theory *if some claims turn out to be empirically true.*

The GL targeted the atomism and adaptationism of evolutionary biologists who decompose an organism into separate traits and then explain each trait as an optimal response to environmental demands. According to GL, the adaptationist, after having dissociated the organism into traits, explains the (value of) the trait under study as an optimal answer to an environmental demand. For instance, fitness or some proxy such as energy intake is a function of the value of a trait x; there is an optimal value (Xo), which is the one selection would reach, at least if it acts alone (Potochnik, 2009). If the actual value $X = Xo$, then the adaptationist would say that selection caused this trait, and would explain it as an adaptation for an environmental demand; the likelihood of the explanation is indeed maximal, since the conditional probability of the data $(X = Xo)$ given the hypothesis (natural selection) is 1. But if X differs from Xo, GL argues that the adaptationist can always update the story by assuming a new selective pressure that would change the predicted outcome of natural selection (Figure 2.1), since the optimum is now the result of a trade-off between these two environmental demands.

More precisely, trade-offs may first hold between selective pressures, like a selective pressure to display few colors in order to avoid predator, and a sexual selection pressure, which stems from the preference of females toward conspicuous color. Second, trade-offs may hold between the traits themselves. For example, if selection optimizes the speed race, regarding predators as selective pressure, it will also raise the costs paid by another trait, for example metabolic efficiency or limb robustness; this means that the genuine optimum will be a trade-off between selection for speed and something else, for example, metabolic cost paid by *other* organs, some limb fragility, and so forth. Adaptationism is always a trade-off adaptationism, namely fitness trade-offs representing the fact that the traits of an organism are integrated, and thereby actually evolving new traits values is only possible for selection under the condition that other traits of the organisms are not too much hindered. Hence, such trade-offs somehow import the integrated character of organisms within the optimality modeling that characterizes the adaptationist method of behavioral ecologists.

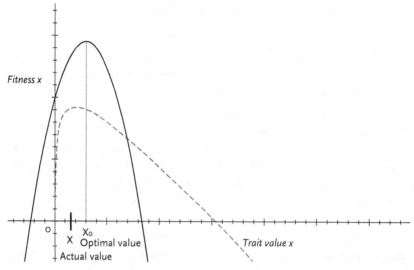

FIGURE 2.1 Trade-off between hypothesized environmental demand (bold curve) and other primarily unknown environmental demands (whose taking into account results into the dashed curve) explains the difference X-Xo between optimal value (regarding hypothesized demand in bold) and actual value of the trait.

Finally, the distance to the optimal trait can also be due to the fact that x is linked, genetically, developmentally, or morphologically, to trait value Y that has been selected for, therefore the value of x is the one that is induced by the value of y that has been selected.[5] Here, the integrated character of organisms is crucial, since what is explanatory of the nonoptimality pertains to the structure of the organism—however ultimately selection is what explains the trait values (as one says, X is a "by-product of selection"). In a word, from selective pressures trade-offs on one trait, to fitness trade-offs on several traits, and then to selection by-products, we have seen an increasing mitigation of the purely adaptationist model, and an increasing acknowledgment of the integrated, structured, or holistic character of organisms (in the sense that traits are in principle linked to one another).

Granted, an explanation of X ≠ Xo may also be the lack of genetic variation (but this would mean that we can wait for the optima to be reached, hence x is just a transient phenotypic state), or genetic drift. Drift, and hidden environmental demands, organismic cost, or by-product selection are competing explanations, but this question might be solved by either doing replicated experiments,

---

5. The example of the chin given by GL perfectly illustrates this. Another classic case is the existence of male nipples.

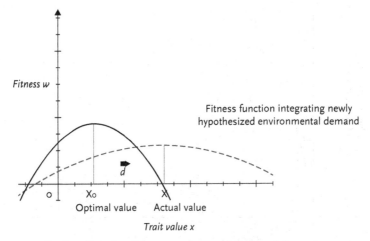

FIGURE 2.2 Drift, and trade-off between rival environmental demands, as competing explanations for the difference between actual and optimal trait values X-Xo. If the correct model of environmental demands is the bold curve, then drift may explain the difference d. Otherwise X is an optimal trait value given a new hypothesis on environmental demands.

or considering the size of the population (small size increasing the plausibility of drift as an explanation) (Figure 2.2). Hence claiming that adaptationism is methodologically flawed may be too strong.

But there is another possible explanation of X ≠ Xo, which considers the integrated character of organisms in another way than the fitness trade-offs I just mentioned, and which GL calls "developmental constraints." As Amundson (1994) emphasized, these constraints are not so much about *selection*, but about *variation*. Suppose, indeed, that (Figure 2.3), because of the developmental processes producing phenotypic variants on the basis of the set of genotypes, the phenotypic trait cannot take any value, but that all values are *clustered around X*. Thus, instead of any trade-off between selective pressures or selected trait values, this mere fact would explain that the actual phenotype is not the highest fitness trait value Xo, simply because no variant can have Xo or values close to Xo.

From a logical viewpoint, constraints can be seen either negatively—they prevent the reaching of Xo—or positively—they somehow canalize or bias the variation around X. As Maynard Smith et al. (1985) define it, "a developmental constraint is a bias on the production of variant phenotypes or a limitation on phenotypic variability caused by the structure, character composition or dynamics of the developmental system."

Indeed, in Figure 2.3, the fact that X is not Xo is explained by the fact that *no variation reaches Xo*; hence the next question is, Why can't X reach Xo? And

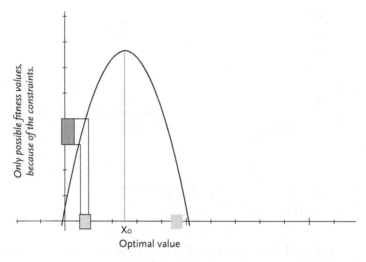

Xo
Optimal value

Only values reached by variation, because of the constraints.

FIGURE 2.3 Constraint on variation, as alternative non-adaptationist explanation for X-Xo.

here—this is the crucial point—the answer concerns *the causes of variation*, and not selection, therefore no trade-off adaptationism is allowed. Such cases, indeed, weaken the explanatory role of selection; thereby in order to capture the reasons for the lack of optimality in the species under focus, evolutionary explanation should integrate the explanations of variation. Given that developmental processes are likely to impinge on heritable variation (Cheverud, 2007), these should be studied in order to understand evolution and adaptation.

In some cases, what underlies constraints on variation precisely resort to the integrated character of organisms. Suppose that the trait x under focus is correlated to many traits, including trait y, and that the variation on y is limited either for genetic or for morphological reasons; then it will not be possible to vary x in any way, since it would reach the limits of y's variation. Therefore asking about the constraints on variation as the cause of the fact that a trait value is X and not Xo entails asking about the connection between x and y and the reasons of y's limited range of variation. In this case, contrary to the "by-product selection" scenario, selection does not play an explanatory role, because the trait linkage bears on variation. This is why, unlike the alternative explanations of a nonoptimal trait surveyed earlier, such a situation would challenge the selectionist explanatory perspective.

In essence, constraints concern the covariance between traits. "Genetic change can alter the covariance among traits of the organism and that covariance may itself be subject to selection. But there *is* covariance and we can use its strength as a rough local measure of the strength of constraint" (Maynard Smith et al., 1985). Constraints can be explored by exploring these covariances; however, an obvious epistemic problem is that such covariances may not be distinguishable from the covariance due to the trade-offs that selection instantiates, as seen above. One hope for disentangling these two covariances consists in experimentally investigating possible variations (Brakefield, 2006).

Yet the constraint issue is more general than cases where the adaptationist prediction *fails*. Considering this makes salient the meaning of "constraints" as *positive* biases. Suppose that X = Xo, but that variation is clustered around Xo. In this case, selection as an optimizing process does not seem to explain the value of the trait, because it leaves unanswered the prior question why variants far from Xo do not occur. From a logical point of view, it is similar to the case of the flying fish falling back into the water, which Williams (1966) cited as a nice example of something that does not need a selectionist explanation (since what else could the fish do?).[6]

Granted, there exists a long-term controversy regarding whether selection or variation—and especially mutation, as a major cause of genetic variation— accounts for (adaptive) evolutionary change. The first Mendelians supported (macro)mutations as the major cause of evolution, while the MS put the causal weight on selection, forging several arguments—including Fisher's geometric

---

6. An additional important distinction holds between relative and absolute constraints: Whereas the latter makes it impossible to reach some variant trait, the former makes it mostly difficult as regards another cluster of traits values. This latter term provides an obvious instance of a mostly positive meaning of constraint, since here some variations are privileged, and therefore some evolutionary pathways become more likely than others. (Of course, the distinction is itself not absolute and is rather scale relative, and universal constraints may be overcome on long-scale evolution.) As Maynard Smith et al. (1985) explained, "One factor influencing the pathway actually taken is the relative ease of achieving the available alternatives. By biasing the likelihood of entering onto one pathway rather than another, a developmental constraint can affect the evolutionary outcome even when it does not strictly preclude an alternative outcome." This phenomenon is the rationale for holding phenomena of parallel evolution at any level as an effect of constraints. In order to display it, a figure should also represent a probability distribution on variation, rather than a continuum of phenotypic values like Figure 2.3 above.

Another limitation of the presentation made here is that variation appears as quantitative, trait values being numerical values plotted on an axis. But many concerns about constraints arise regarding qualitative phenotypic differences, where precisely only several qualitative traits, such as possible folding of a shell, are reachable. This shortcoming affects the graphical presentation but not the logics of the argument.

Moreover, what was depicted in the optimality model in Figure 2.1 & 2.2 is a seemingly isotropic variation—all possible variants are reached. However it assumes some equiprobability, which is a form of randomness; but as developmental biases constraints often take the form of departures from equiprobability.

model—to show that actual beneficial mutations should mostly be micromutations and therefore are unable to cause adaptive evolution by themselves. The abovementioned explanatory switch to the causes of variation therefore instantiates anew the claim that selection should not be considered as the main cause, and variation in turn is to be explained if one wants to explain evolution and adaptation. Yet this is mostly based now on our knowledge of developmental processes, as evo-devo's emphasis on heterochrony could illustrate (Gould, 1977)—and later on, our empirical knowledge of genome regulation (Carroll, 2005; Davidson, 1986). The critique of trade-off adaptationism instantiates this controversy between primacy of variation versus selection that constantly affected the history of evolutionary biology since Darwin (e.g., Beatty, 2016). Recast in these terms, GL simply argues that one should not assume that selection is the most explanatory factor in any context, something that many founders of the MS, and all population geneticists (for which selection is one mathematically possible outcome among others), would have agreed on (but few behavioral ecologists).

Many constraint-emphasizing conceptions of evolution convey the message that organisms, through their development and therefore the variation it favors, precludes, or allows, play an active role in evolution and should be therefore put to the fore of evolutionary thinking (Alberch et al. 1979; Bateson, 2005; Walsh, 2015). They could be subsumed under the call for considering organismic *Bauplan*, which is the last message of the "spandrels" paper.[7] Yet the view that constraints are developmental processes supervening on genes or acting on genes does not exhaust the general theme of the power of constraints: Those can also stand at the genetic or genomic level, for instance when pleiotropy imposes constraints on the possible variation that underlies a given phenotypic trait. We know now how cell differentiation and pattern formation occur through the concerted action of genes embedded in network structures such as gene regulatory networks (Davidson, 1986, Oliveri et al., 2008). In this context, many of the constraints are proper to the genetic level because they stem from features of the gene regulatory network, or more generally from the principled features of genome architecture. For instance, regulatory genes are often pleiotropic, because a same transcript is likely to affect the expression of many coding genes, and hence many traits; to this extent, they play a crucial role in evolution (Carroll, 2005). This especially concerns the evolution of new forms because on the one hand, mutations to regulatory genes are less detrimental than those that occur to coding genes—since this does not hinder

---

7. Claiming that some constraints are physical constraints, due to physicochemical processes that play at the level of molecules which constitute the cells—as exemplified by some of Newman's theories (Newman, this volume)—also rejects the gene centrism, but in a way that is less directly committed to the organismic stance.

any functional protein—but on the other hand they can affect the organism in a much more integrated way (due to their pleiotropic nature) (Love, 2008; Müller & Newman, 2005 for this difference between genes viz. evolution).[8]

In the last decade we may arguably have moved beyond the concept of constraint as an operational concept, and turned toward a more fine-grained appreciation of the various variation-supporting processes that operate in evolution. Speaking of genetic channeling, facilitated phenotypic variation, facilitated variation, and other formulations not only is more precise but allows one to directly consider the positive meaning of constraint.

2.2.2. Explaining Diversity Alternatively: Constraints and Clustering Morphospace

The connection between constraints—understood positively—and the patterns of diversity of life forms clearly appears when we consider a modified case of trade-off adaptationism and its critique.

Suppose, as in the case of Figure 2.3, that we have *two clusters* of possible variations, around X and X', and that the extant trait value is X; then another question to ask is: Why is x = X rather than x = X'?, X' being situated in the other cluster. Given that both traits have the same fitness, selection cannot be the answer. Rather, only the study of the reasons for variation patterns can illuminate this issue, for instance the unraveling of a switch between these two clusters, which could be affected by some other properties of the system, or just sensitive to slight boundary conditions, or a phylogenetic legacy that accounts for the fact that only the X-cluster of variations is instantiated in the species or clade under focus.

This case directly impinges on explaining *diversity*, which comes at any level (species, genomes, cell metabolic networks) and according to all evolutionary biologists such as Mayr or Lewontin, is in general one of the major explananda of the theory. Here, only several small clusters (X, X', in light gray in Figure 2.3) of variation are possible, because of reasons related to developmental processes, namely, the way in which, at all levels, from molecules to cells and organisms, the zygote develops phenotypic traits. Therefore, given that lots of species share a large part of their genetic make-up, epigenetic settings, and developmental patterns, we might see that a set of populations of parent species divides into several small groups defined by the X and X' values. What explains diversity belongs to the study of developmental mechanisms, and especially the possible connection

---

8. It is an open experimental issue about whether genetic channeling accounts for any developmental bias (Brakefield, 2006).

between sets of genotypes and possible phenotypes (e.g., genotype-phenotype maps; Huneman, 2010), as governing the possible clusters of traits—rather than to the study of natural selection dynamics in population genetics. In this manner, Wake (1991) analyzed salamanders, also relying on experimental embryology, and pointed out that there is a developmental switch that can either produce five or four digits. In the relevant lineages, four-toe variants are produced because, due to increase in genome size or decrease in body size, the switch is activated.[9] So variation is discrete and restrained, which is why these many lineages converge to the four-toe option—without a direct causal or explanatory role for gradual evolution by natural selection yielding design (Huneman, 2015).

Yet classical evolutionary theory often considers another explanandum, tied to diversity, which is the *unity across diversity*. While life displays amazing patterns of diversity, patterns of unity across this diversity are also striking: Fins of fish and wings of birds are similar; many diverse lineages have eyes. Following a long-standing tradition that goes back at least to Owen, MS reemploys two categories to understand this: homology (a), that is, the traits are similar because both stems from a common ancestor; and analogy or homoplasy (b), that is, adaptive convergence: Different species exhibit a same trait because they faced common selective pressures.[10] Here, as Darwin already noticed it in chapter 6 of the *Origin*, under this latter kind of sameness one will find fine-grained differences in the mechanisms underlying the focal trait, because here selection started from different points or used different materials to build the common features.

Now, considering the case of Figure 2.3, where the causes of variation directly explain the trait, what could explain this unity across diversity? Suppose that all variation is slightly clustered around X. Then in many parent species, what explains the fact that organisms exhibit the same trait X is precisely the fact that they share these developmental features. If now many species or clades have the same trait, then they do not need to have a common ancestor: It might be that the molecular processes underlying development are such that they necessarily give rise to variation severely clustered across trait X. Convergent evolution has often been seen as a witness of clustered genotypic variation, for whatever reason, and therefore

---

9. The number of cells is the proximate cause of the switch here, and can be reached either by decreasing the size of the animal then the limb size, or increasing the size of the genome, both things occurring in the phylogeny of *Plethodontidae*. "This is a direct example of design limitation, in which alternative states are sharply defended" (Wake, 1991, p. 549). Notice that in the case of the genus *Myriapoda*, variation seems undeniably discontinuous since only even numbers of pairs of legs are reachable.

10. Convergence is used for similar traits in independent lineages, whereas one talks of parallelism about lineages that have a common ancestor and evolve similar traits that are not here because of common ancestry. (See Pearce, 2012, on these concepts.) This is not important here.

as a support for theories that favor variation- and especially mutation-producing processes over selection (as major explanantia for adaptive evolution).[11]

Beyond or below the gene level, often people will point out self-organization processes within the chemicophysical underlying of cells as the source of the clustering of variation (e.g., Hall, 2003; Walsh, 2003), and therefore as the reason why so many species will have a trait in common. The point here is that if a feature seems "advantageous," which is common to many species, biologists will be tempted to explain it by selection, but they should first control whether this feature could not arise universally "for free."[12]

When these self-organizing processes responsible of variation yield several small intervals of possible outcomes (like the light gray intervals in Figure 2.3), the explanation of diversity is connected to explanation of unity across diversity; thus, the explanation of evolution is highly concerned by the explanation of the switches between possible outcomes of these processes. For instance, in recent work Newman and colleagues have established the surprisingly few molecules involved in many developmental processes (Newman & Bhat, 2009; Forgacs & Newman 2005), which combine into a limited set of what they call "dynamical patterning modules" (DPMs). This explains that there will be in general very few developmental modules highly conserved, very common, across many clades. But the fact that several (few) combinations are possible explains crucial differences in clades, it explains, in a word, the diversity between high-level taxa.[13] Thus the emerging picture is that high-level diversity may rely on bifurcations between DPMs, whereas within-clade diversity is explained by more traditional selectionist accounts of diversity and unity across diversity.[14] In this case, physics rather than network theory explains the clustering of variation; and this clustering in

11. See Lenormand, Chevin, and Bataillon (2016) for an assessment of claims that parallelism may either support mutationism or selectionism, see also Depew (this volume), with the case study of sticklebacks.

12. A recent example is given by modularity in metabolic network of cells. Modularity is in general said to be advantageous, especially because it allows a system to remain functional when one module is ineffective, whereas nonmodular systems are much more fragile. It is a fact that cell metabolic networks are remarkably modular; therefore, we tend to think that selection shaped them, in many clades, according to this perspective. However, Solé and Valverde (2009) have shown that if a network is made up through very simple rules of adding new nodes, the chances of coming up with a modular network are in fact extremely high when you get to high-dimensional networks; this means that actual metabolic networks in cells were not always the result of selection, because the variation among possible networks spontaneously favors modular networks. Here, the unity is explained by an almost mathematical fact, which concerns the limited range of ways to build networks beyond a certain size.

13. See also Newman (this volume).

14. This is perfectly in line with Gould and Eldredge's major theory, the punctuated equilibria, since here there is a discrepancy between low level taxa (species), which diversify quickly but

turn explains several features concerning patterns of biodiversity in the theoretical morphospace.

If in many cases diversity and unity of life forms and properties are explained by considering underlying features of genetic mutation and phenotypic variation in general, then one can make sense of the way Pigliucci and Muller (2011) present the currently desired "extended synthesis": They say that Darwin and the MS had to use statistical thinking to understand evolution, since they had a very incomplete empirical knowledge, but because of our deeper understanding of molecular and developmental mechanisms governing heritable variation, we can design mechanistic explanations of what statistical explanations were just a place-holder. Of course, this may leave open the question of whether these mechanistic explanations indeed replace traditional ones. But the point is that, if we have to move to the causes of variation to understand traits in cases like Figure 2.3, we should consider the mechanisms of producing variation—as Newman & Bhat (2009) perfectly exemplifies it—rather than use population thinking,[15] to the extent that the causes of variation are mostly responsible of the pervasiveness of a given set of variants. The kind of explanation at stake is not statistical explanation but mechanical causes.[16]

In order to understand the competition between these mechanistic-variation-oriented vs. population-selection-oriented explanations, and the sense in which an "extended synthesis" approach intends to replace MS explanation, I turn to the general explanatory frameworks that are supported by these two approaches, regarding the explananda here described (adaptation, diversity, unity). The alternatives to the MS will appear to be as diverse as the content of the concept of constraint was heterogeneous.

2.3. The Alternative Explanatory Schemes

Should empirical evidence supporting the constraints model against the trade-off adaptationism bring a *radical change in the explanatory structure* of evolutionary

---

not much via natural selection, and diversification across phyla, which is discontinuous and may require another account.

15. Because population-thinking, in essence statistical, has been forcefully opposed by Mayr as the landmark of evolutionary thinking to typological thinking, it could be expected that an "extended synthesis," advocating a switch in the concept of causality, from a statistical to a mechanical understanding, would go with an attempt to make sense of typology (Lewens, 2009; Love, 2009).

16. And in fact the philosophical distinction between process-based concepts of causation and difference-making-based accounts (Hall, 2003), which encompass all probabilistic accounts, could clearly be used here.

theory? I now examine such change more systematically, and then in the next section turn to assessing of the call for an extended synthesis with regard to the patterns of variation empirically detected.

2.3.1. Explanatory Schemes

Within the space of controversies I highlighted already, it has been a long-standing critique of the MS that it focuses on the adaptive *differences*, whereas other perspectives would investigate the *commonalities* across clades. Amundson (2001), in order to characterize the divide between MS evolutionary theories and developmental theorists, said that the former emphasize "the importance of adaptation over homologies" whereas the latter emphasize "the importance of homology over adaptation." In MS indeed, commonalities are either analogies, hence adaptations, or homologies, namely they arose from a common ancestor that evolved the homologous trait via natural selection. In an alternative framework, homology (namely, homologous traits) would not in essence derive from adaptation and natural selection. This is precisely what appears in the case of a constrained variation demonstrated at the basis of the evolution of a trait. If variation is clustered because of the developmental processes, then it is plausible that in many species the variation will undergo the same constraints because developmental processes are to some extent shared by many species in a clade, and that therefore the trait will be common across species or clades. This is all the more possible that the mechanisms of constraints lie at the molecular level, which is widely common—as it is instantiated in Newman's studies of DPM. It is also the case with some hugely conserved features of cell metabolism (Kirschner & Gerhart, 2005). So here, as soon as in one species the variation is clustered because of these reasons, we can be sure that other species and even clades will display the same or corresponding traits. Switching to the study of causes of variations therefore entails that commonalities across clades may become as such an important explananda, with no tie to adaptation, either in the form of adaptive convergence or in the form of homologies to a trait that first emerged through selection. To this extent, a same developmental process can even be detected at the origin of different phenotypic outcomes, because those would also depend on distinct features of the species and its environment. As Raff said, making this link between the focus on universal patterns and the developmental, often molecular, mechanisms: "To developmental biologists, there is a mechanistic universality in developmental processes despite any diversity of ultimate outcome" (Raff, 1992).

Take the tetrapod limb, which has been a major object of study for developmental biologists (Hall, 2003; Raff, 1996; Wake, 1991). The object itself pertains to the category of commonalities across several clades. Understanding that a same set of physical processes is at stake to produce as a unique outcome a general pattern of connection between parts provides ipso facto an explanation of its universality across clades, even though no mention is made of selection, either convergent, or at the origin of a first character (plesiomorphic). Whereas commonalities are somehow derived explananda in MS (from the explanation of the first, plesiomorphic state brought about by natural selection), they are explananda per se in this perspective, hence it is legitimate to claim that here we have switched explananda, as do Amundson (2005) (diversity vs. commonalities) or Pigliucci (2007) (genes vs. form). What differs seems therefore to be the explanatory architecture of the theory.

Thus, in this section I argue that the challenge to the MS is less about proposing new explanantia, than advancing a new explanatory scheme for phenomena that pertain to adaptation and diversity of organisms. By "explanatory scheme," I mean an organization of explananda and typical explanantia that allows some combinations and some questions, such as "Is x more responsible than y for z?," and specifies which families of alternative hypotheses should be empirically conceived of when a class of phenomena is encountered as well as what should be taken as a default. For instance, Mayr's distinction between proximate and ultimate causes illustrates an explanatory scheme: The explanation for a given phenomenon can be searched at two levels, before and during the lifetime of the individual, and these explanations are articulated in a specific way (population /individual, evolutionary/developmental, etc.). Moreover, explanatory schemes can be defined at various theoretical levels: within a discipline (e.g., behavioral ecology, molecular evolution), or within a theoretical structure that encompasses disciplines, such as the MS and some of its challengers. Noticeably, the explanatory scheme proper to behavioral ecology takes natural selection as acting to maintain traits, until another explanation may be given. Explanatory schemes also prioritize explananda: MS, as Amundson emphasized, put "adaptation" as a privileged explanandum with respect to commonalities and unity of type.

A real revision of an explanatory scheme thus consists in a reordering of explanantia and explananda of the MS, possibly adding novel explananda or explanantia if needed. Therefore, the competing claims suggested by different authors about which explananda/antia to add make sense only when replaced in such perspective. I argue that in order to understand the extended synthesis project, a general change in explanatory schemes of evolution (from MS to an alternative scheme) allows for focusing either on an explanandum switch, or on an explanans switch, hence encompassing various formulations of this project.[17]

---

17. This suggests the kind of pluralism discussed by Love (this volume.)

2.3.2. Modern Synthesis Explanatory Scheme, and Its Alternative

Let us start with the MS explanatory scheme, which is, as we know, centered on selection or adaptation. To sum up, natural selection causes adaptations; adaptation, because of adaptive radiation, accounts for diversity. Natural selection more generally yields diversity also via disruptive selection, or the maintenance of polymorphisms between strategies in evolutionary game theory (e.g., when a mixed strategy is stable). More generally, some diversity is also accounted for by drift, for instance the nucleotide diversity in populations, as Kimura has forcefully argued; yet this does not transform the explanation of diversity, since population genetics is still what accounts for it, thus conforming to a commitment of the MS. And then unity across diversity is accounted for either by parallel adaptations, or by homologies, while in the last instance, homologies are traced back to adaptations: The first character (of which all others are homologous states) was an adaptation. This is exactly the reason why for Darwin the "principle of natural selection" was higher than the "principle of common descent," since the latter presupposes the former[18]: "The law of the Conditions of Existence is the higher law; as it includes, through the inheritance of former adaptations, that of Unity of Type." So the MS explanatory scheme would be the following (Figure 2.4). Notice that as an explanandum adaptation is logically prior to diversity and unity.

Now, if natural selection is explanatorily downplayed because of the constrained structures of variation, as advocated by critiques of trade-off adaptationism, what explanatory scheme would we have? It would surely be less homogeneous: Granted, adaptation is still often explained by natural selection, and adaptive radiation as well as drift or balancing selection explain some diversity. But now, homologies, which underlie unity of type, namely, unity across lineages, are not in principle relying on natural selection; they are often based on common developmental mechanisms of variation; and this also explains, as we said, diversity, via switches, bifurcations, and so forth. Finally some adaptations can be understood

---

18. "It is generally acknowledged that all organic beings have been formed on two great laws Unity of Type, and the Conditions of Existence. By unity of type is meant that fundamental agreement in structure, which we see in organic beings of the same class, and which is quite independent of their habits of life. On my theory, unity of type is explained by unity of descent. The expression of conditions of existence, so often insisted on by the illustrious Cuvier, is fully embraced by the principle of natural selection. For natural selection acts by either now adapting the varying parts of each being to its organic and inorganic conditions of life; or by having adapted them during long-past periods of time: the adaptations being aided in some cases by use and disuse, being slightly affected by the direct action of the external conditions of life, and being in all cases subjected to the several laws of growth. Hence, in fact, the law of the Conditions of Existence is the higher law; as it includes, through the inheritance of former adaptations, that of Unity of Type" (*Origin of Species*, 1859, chap. VI, summary)

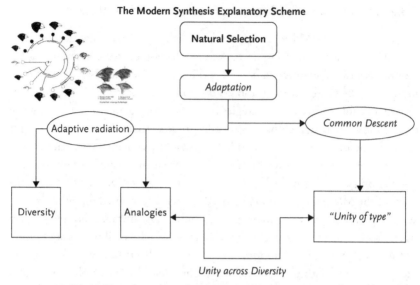

FIGURE 2.4 The MS explanatory scheme. Note that diversity can be explained by natural selection via adaptive radiation (illustrated by Darwin's finches; image from Schluter, 2000).

in relation to the developmental mechanisms directly producing variations in a very restricted range, as we saw—be it through adaptive variations, or through more complex mechanisms like phenotypic plasticity plus genetic accommodation as advocated by West-Eberhard (2003), or finally via niche construction, which is seen by its proponents as an alternative pathway to adaptation (Odling-Smee et al., 2003) (see also below, section 2.4).

It is clear then, that one could see this alternative explanatory scheme either as focusing on another kind of explanandum (since unity and diversity stand as explananda by themselves, irrespective of adaptation)—namely, unity of type, thereby commonalities, pervasiveness of few forms, and so forth—or as the emergence of new explanantia, situated at the level of variation (mechanisms that produce variations, channel them, constrain them, etc.)—hence, processes other than natural selection (Figure 2.5).

Because this new level of explanantia comes into play, a core commitment of the MS (the centrality of population genetics, see [a] above) cannot be supported any more, since the explanatory role of selection is affected by the developmental and molecular mechanisms of variation, which stand in principle outside population genetics. The two constitutive claims of MS (natural selection is the main *explanans*, see [b] above, and development does not matter to evolution, see [a] above) are obviously rejected in this explanatory scheme. And finally the

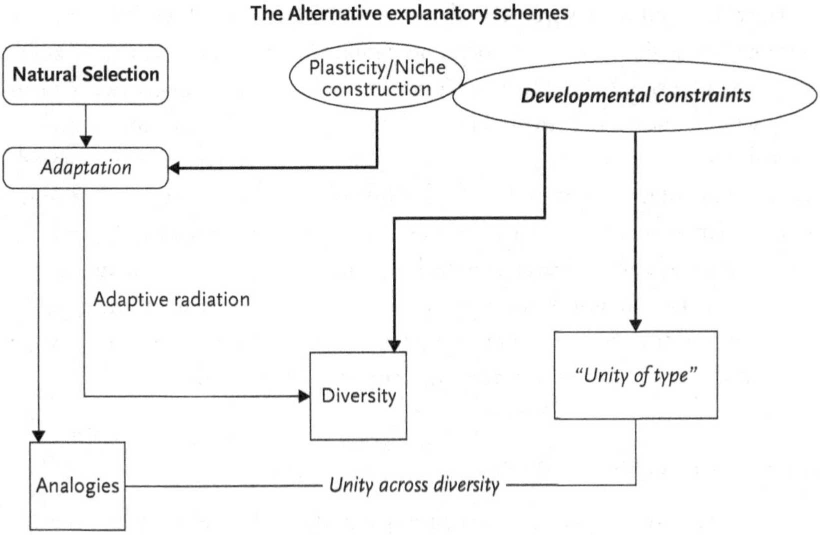

The Alternative explanatory schemes

FIGURE 2.5 Alternative explanatory scheme (AMS) summarizing several critiques to the MS.

extrapolation thesis is challenged here: Microevolution can still be explained by natural selection acting on gene mutations, and the same processes also account for the divergence due to adaptive radiation, but macroevolutionary phenomena such as large-scale universality across clades or clade divergence may be explained by other processes (the right column in the diagram of Figure 2.5).[19]

It may not be accurate to talk of an "extended synthesis" in the case of many authors who reject the core commitments of the synthesis, since an "extension" supposes that, precisely, one adds to a constant core some novelties. However, what appears here is that taking very seriously the switch in explanation induced by an acknowledgment of the primary explanatory role of variation-producing processes in evolution leads to a reshuffling of the core of the synthesis. Amundson (2005) seems to have seized an important truth regarding the current challenges for the MS, namely, that pace Pigliucci (2007) or Carroll (2012) the putative extension would bring about a major rebuilding of the pillars of the theory.

Considering the issue of trade-off adaptationism versus constraints studied in the previous section, it is plausible to think that both selection and developmental biases contribute to a given trait, and that in some cases one or the other

---

19. As I noticed it when considering Newman's views, this fits well with Gould and Eldredge's punctuated equilibria thesis about phylogenetic patterns, even though it is not entailed by this thesis.

explanation should be favored, based on the available empirical evidence concerning population size, lineage history, and genomic and developmental make-up. However, adopting either the MS explanatory scheme or the alternative explanatory scheme means that the default states, the weight of each explanantia and the ordering of the favored explanantia will change in each of them. Conversely, various states of the empirical knowledge will also grant preferentially one or the other explanatory scheme—and the MS challengers concur in claiming that the current state of our empirical knowledge grants the alternative scheme; hence, the whole epistemic justification process is somehow dual. Part of the question therefore is to specify which empirical facts about variation should grant which explanatory scheme, and this is the endeavor of the last section.

2.3.3. Intermediate Positions

But the alternative explanatory scheme is not the only way to doubt the MS explanatory scheme; one can think of intermediaries between the two schemes— where the accounts of unity and diversity will not directly shift explanantia, but rather integrate combinations of selection, developmental constraints, plasticity, within the explanans, together with natural selection. Many mixed explanatory schemes can indeed be conceived of, and those will define grades of strength in explanatory divergence.[20] At the extreme distance from the MS explanatory scheme, variation-producing processes as an explanans are explanatory per se, instead of being combined with natural selection, as it appears in the alternative explanatory scheme: This defines the strong revision of MS that some call for. But as long as the variation processes are still integrated with natural selection into an explanation of adaptation, unity or diversity, this does not dismantle the spine of the MS explanatory scheme, namely, the connection between selection and adaptation, and then the explanatory role of adaptation regarding other explananda: We still use a merely weak refinement of the MS.

So theoretical views genuinely alternative to the MS are not necessarily suggesting new explanantia compared with those of the MS; rather, they radically challenge the overall structure of explananda and explanantia. For instance, adaptive radiation will still be a possible explanans for some facts of divergence, but it would be included in a set of distinct possible explanantia—some selection-based, some

---

20. Such a view resonates with the way Walsh (2008) conceives of the grades in involving development within evolutionary explanations; it shares with Alan Love's notion of "problem agenda" the idea that what happens with various challenges to the MS is less the replacement of an explanation by another, than a reshuffling of the explanatory landscape, even though I use different concepts than the "problem agenda." See Love (this volume).

selection-independent—available for explaining diversity, and it would not have any epistemic privilege. This attention paid to the explanatory structure of the theory casts a light on the current controversies. Recall that in the recent *Nature* paper, challengers (Laland et al., 2014) argued that newly attested facts of nongenetic inheritance, niche construction, or plasticity, among others, are so relevant to evolution that the theory should incorporate new concepts to account for that: This is a qualitative change. Conversely, defenders of the MS (Wray et al., 2014) argue that these facts are fascinating but do not call for entirely new concepts to be accounted for, except if the MS basics are caricatured—that is, some of the novelties brought about by the challengers are not so novel. Granted, the MS has always recognized the fact that evolution and diversity are either mostly explained by selection, or mostly explained by variation, and the notion of constraints faced by selection did not wait for the spandrels paper: Questions of the limits of selection were indeed the matter of crucial controversies between Wright and Fisher, Dobzhansky and others, then Kimura and Gillespie and other population geneticists, and so on. More generally, population genetics itself is familiar with the fact that genetic profiles of population are shaped by many other forces than selection. Wray, Hoekstra, and the conservatives in the dual *Nature* 2014 paper have a point when they emphasize that the MS framework can make room for many putatively novel facts, because it already includes the notion that the explanatory weight of selection is possibly balanced by other explanantia. But what drastically changes, if one takes seriously the claims that novel facts at all levels are indeed challenging for MS, is *the overall structure of explanation*, which connects various explanantia and explananda.

When considering the alternative explanatory framework, in order to take seriously the calls for an alternative theory it is not enough to acknowledge that there are cases where mutations, drift or plasticity play indeed a stronger explanatory role than selection: What should change is the structuring of all these explanantia together and the way explananda are interconnected in an explanatory picture. It may even happen that the very notion of adaptation not only changes its explanatory role regarding commonalities but also receives a new interpretation, since Odling-Smee et al. (2003) claim that niche construction is another pathway to adaptation, alternative to natural selection, and even more radically Walsh (2003) argues on philosophical grounds derived from the so-called statisticalist view of natural selection (which is not under focus here) that natural selection is not even the cause of adaptation. The question of the robustness of the MS as advocated by its defenders is therefore the question of how many shifts can the MS explanatory scheme undergo—for example, combinations of explanantia that mitigate natural selection's explanatory force, adding new explananda such as the "organic form," and so forth—without being turned into a radical alternative explanatory scheme.

The next section therefore questions, among types of empirical facts about variation, which ones would legitimate a switch toward an alternative explanatory scheme, rather than a weak modification of the MS explanatory scheme.

2.4. Random, Directed, and Adaptive Patterns of Variation and Their Implications

Focusing on these new explanantia, the question we face now is the nature of this variation that would compel us to change explanatory schemes, and the possible empirical mechanisms that may instantiate it. Variation as assumed by MS is genotypic (due to recombination and mutation, see, e.g., Huxley discussed earlier) and for this reason often assumed to be "random" or "blind." Challengers of MS have argued that other kinds of variation are met, that to some extent defeat the trade-off adaptationism program as illustrated above, and that may be due to various specific mechanisms—genetic, genomic, organismic, molecular. This section considers these various modes of non-MS variation, but to start with I make clear the role of inheritance in this regard.

The claim that variation is not only caused by mutation challenges gene centrism and not selectionism; the claim that variation is clustered, biased, nonisotropic, and so on, challenges selectionism but not directly gene centrism (it may be still purely genetic). However, if variation involves much more than genes, then it may employ epigenetic changes and other kinds of nongenetic inheritance; as such, to the extent that many epigenetic changes react to the environment and hence are adaptive (Bonduriansky & Day, 2009; Jablonka & Raz, 2009), then we will be committed to a kind of directedness of variation that therefore challenge selectionism. Hence, some of the challenging views of variation embed claims about nongenetic inheritance—and reciprocally, challenges to the MS based on the recognition of the fact of nongenetic inheritance (e.g., Danchin et al., 2011) involve to some extent a commitment to a possibly directed variation and therefore a challenge to selectionism. However such commitment comes in degree, and for instance Danchin et al. (2011) would be less willing than Jablonka and Raz (2009) to downplay the role of natural selection.

In this section I explore the reasons why, in principle, if a significant amount of variation of the kind X is empirically attested, then we should go for the alternative explanatory scheme. I start by abstractly defining types of variation, and examine their in-principle compatibility with the MS explanatory scheme. Then I review cases of allelic mutation, of nongenetic variation, and of directed variation, and question their consequences on adopting either an alternative explanatory scheme or a weakly revised MS scheme.

2.4.1. Characterizing Types of Variation

I am looking for a characterization of variation general enough to decide, then, which kinds of variation would compel us to switch to an alternative explanatory scheme, where the causes of adaptation and diversity have to be looked for directly at the level of variation. The difficulty here is that there are many senses in which people talk about variation, and even if in a given research context they understand each other, at the required level of generality we may find some contradictions.[21]

Notably, variation can concern either the set of variants in one generation or the new extant variants provided at each generation by mutation and recombination, as summarized by Huxley. The first is synchronic variation, the second is diachronic variation, and their relation is complex, entailing that there is no straightforward way to characterize randomness and directedness.

Let us characterize types of variation, starting with isotropy. Prima facie, isotropic synchronic variation is such that for a randomly chosen individual, all values of traits are equiprobable. An equivalent formulation is that a given trait value has the same chances as other trait values to be taken by a given amount of individuals. Nonisotropic variation is such that some intervals of values may be more probable than others for a randomly taken individual at each generation, but all values in these intervals are equally probable, and no interval has probability 0. When variation is isotropic it can often be said to be random; such randomness as equiprobability would mean that two randomly chosen trait values have equal chances to be displayed by one individual in a generation. Interestingly, it might be that allelic variation is not isotropic but that *phenotypic* variation is, for instance because many allelic variants are synonymous due to redundancies in the genetic code.

Concerning diachronic variation, things are mostly affected by heritability, which is (most generally speaking) the pattern of covariance between variation at one generation and variation at the next one (de Visser et al., 2003). Diachronic phenotypic variation is generally not random precisely because of heritability, which means that there is a positive statistical correlation between the values of the offspring's traits and the values of the parents' traits. Going this way, diachronic variation seems reducible to synchronic variation combined with heritability, but "combined" remains problematic, as it will appear now.

When biologists say that variation, including diachronic variation, is *random*, it mostly means that it is not necessarily beneficial, or that the fact that it is

---

21. The whole section on randomness is indebted to Merlin's extensive treatment of the question of randomness of genetic mutations and developmental noise (Merlin, 2014), even though the perspective defended here is quite different and much less complex and encompassing.

beneficial does not cause, or increase the probability, of a given variant—"blind" variation is a better phrase. It is this property of mutations that has been empirically tested and established in the 1950s by especially the experiments of Luria and Delbrück, and that supported definitively the MS framework. This of course does not contradict the fact that synchronic variation could be nonisotropic.

This may not match many technical (algorithmic, mathematical) concepts of randomness, yet it includes the idea that from the perspective of the environment, there is no statistical correlation—hence predictability—between the state of the environment and the likelihood of a given variant. Hence this vernacular use of "randomness" retains the notion of *statistical unpredictability*, which essentially belongs to the notion of randomness. A random sequence of digits is actually a sequence where the knowledge of a digit or a subsequence is not (or only weakly) predictive regarding the next digit or subsequence.[22]

There is a sense in which diachronic variation can be strictly or mathematically *nonrandom*, though still statistically unpredictable *from the environment*— namely, the case of a predictable change across generations of the mean value of the trait in a population. Because of this diachronic predictability, there is no randomness mathematically speaking, though according to the idiosyncratic evolutionists' meaning variation is random (= blind). Here, diachronic variation would be *directed*, in the sense that some trend is visible in the data regarding variation along generations. This directedness can be produced by a bias in mutations: Suppose that, irrespective of the value of the trait of the parents, some mutations in the zygotes are more probable than others; then, from generation 0, all diachronic variation will be biased. This is the basis of "genetic channeling," which is a specific form of developmental constraint. Such a fact challenges a reduction of diachronic variation to synchronic variation plus heritability.

Stoltzfus (2006) and Yampolsky and Stoltzfus (2001) have shown that by reducing to a *synchronic* set of variants all variations that can happen along the evolutionary process through new mutations, MS precisely forgets about how mutations can make a difference onto evolution. More generally they claim that when Fisher (1930) demonstrated that mutations are a force which is negligible compared to selection, he assumed such view of variation, but if you consider that mutants may arise at each point of the evolutionary process, the mutations—by biasing the chances of distinct possible outcomes of such process—massively impinge on evolution by natural selection. Similarly, Beatty

---

22. This property has been shown to derive from computational concepts of randomness as absence of algorithmic shortcut (Chaitin, 1970).

(2006) convincingly argued that the chronological order of chance mutations makes a difference to the course of selection. So variation has to be considered also diachronically, otherwise we neglect properties that are relevant to the evolutionary process.

On this basis, I propose a representation of variation in general in order to specify what would be isotropic/nonisotropic synchronic variation, and biased, directed, and adaptive diachronic variation. I intend to be as general as possible concerning the connection between kinds of variation and explanatory structures, in order to provide a general framework for what can be concluded concretely about cases of actual variation, which can therefore be allelic, phenotypic, genotypic, and so forth. To this end, it is better to simply start from trait values and represent the population as inhabiting a hyperdimensional space where each individual occupies a point defined by the values of its traits.[23] (When there is only one trait considered, this space is a line.) Isotropy, here, means that there are no large empty patches in this space. This representation captures synchronic variation, and the properties of the change between states of the space at each generation capture diachronic variation. Given that there is an application that projects each individual at generation n onto a parent individual at generation n-1 (or the middle individual of the two parents), we can settle a "mean-offspring individual" i(n) which is defined by the average of the coordinates of the offspring of individual i (n-1)[24] and then measure the degree of deformation of the set of individuals at n-1 compared to the set of individuals at n. The degree of heritability (for a given trait) is the degree to which the set is stable between two generations along a given dimension: It can be statistically described and measured, and averaged across several generations.

Now, let us note dx'_i the difference between value x_i for individual i, and x'_i, which is the trait value for its offspring (dx''_i for another offspring, and $d\hat{x}_i$ for the mean of the offsprings' trait values). Considering the global change of values dx and dy of two traits x and y between generations allows one to measure correlated variation between traits, which is well established since Darwin first talked about the "laws of growth,"[25] and which has been massively studied recently (Pavlicev, Cheverud, & Wagner, 2011).

Now, saying that diachronic variation is biased means that intergeneration variation in trait value is more likely to take place in one direction than in another.

---

23. Dimensions are the numbers of traits, or, when considering genetic variation, the loci that underpin one trait.

24. Considering the mean value controls for the effect of selection, since it brackets the variance in offspring number.

25. Many occurrences in chapter 6 of Darwin (1859) (also "complex laws of growth").

Thereby it can be formulated thus: For a randomly taken individual i at generation n-1, and a randomly taken offspring of i noted i', the probability of $dx'_i > 0$ is higher than the probability of $dx'_i < 0$. (Biases can of course affect multiple traits, and therefore are formulated in a multidimensional way.)[26]

Variation being directed means that one trait value Xo seems to be targeted by an intergenerational trend of variation. In the present perspective it is enunciated thus: Picking randomly an individual i at generation n-1, and one of its offspring i', then the probability that $|x_i\text{-Xo}| > |x_{i'}\text{—Xo}|$ is higher than the probability that $|x_i\text{-Xo}| < |x_{i'}\text{—Xo}|$. (See Figure 2.6 for visualizing kinds of variation.)

In this space a manifold is constituted by the sets of trait values that are optimal or highest in fitness in the environment; this defines the optimal phenotypes. Adaptive variation is therefore easily conceived of as a directed variation where the target of the trend is the value of the optimal phenotypes for x. Intuitively, adaptive variation means that diachronic variation tends to approximate the optimal manifold.

How to articulate in principle these types of variation to evolutionary explanations? I consider only one dimension in this space of variations (one trait), for the sake of simplicity. The upshot is that there is no clear-cut one-to-one correspondence between types of variation and explanatory schemes in evolutionary biology as I'll show now.

Suppose a variation is directed toward Xo; is it really the case that in principle evolutionary change is sufficiently explained by variation, with no important role for selection? Consider here the probability distribution defined by synchronic variation in the population at generation n. At generation n+m, if we leave aside selective pressures, the mean of the distribution will be closer to Xo. However, from there nothing is really entailed about the *variance* of this distribution. Thus if this variance is high, there will be room for selection to act on the population—for the following reason.

Suppose indeed now that selection favors trait values close to X1. Then if the variance is large (that is, if variation is not much clustered around Xo, even though diachronic variation is directed toward Xo), in the distribution, at generation n, individuals with trait values closer to X1 will have more offspring than individuals with trait value closer to Xo, even though these offspring will in probability be even closer to Xo than their parents because of directed variation (Figure 2.7)

---

26. Of course, considering the phenotypic traits of adults is problematic because selection may have already acted and some offspring individuals are lost, so that the absence of some trait values is not due to variation but to the low fitness of these traits. Therefore, a space of genetic variation is less prone to this distinction problem since between two generations, what accounts for the distribution of genes of an individual is the variation-producing processes.

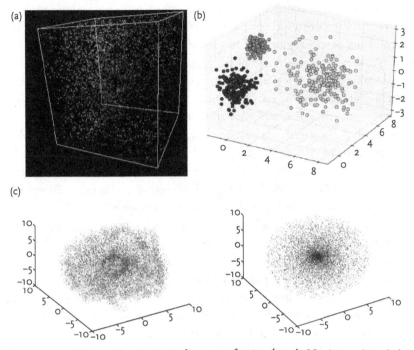

FIGURE 2.6 a. Isotropic variation in the space of trait values. b. Nonisotropic variation. c. Diachronic directed variation (left: generation n; right: generation n+i).

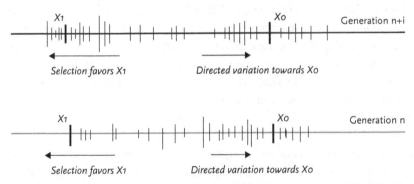

FIGURE 2.7 Directed variation aims at trait value Xo. Each thin vertical black line has a size proportional to the amount of individuals having the trait with the value measured on the line. There is directional selection for X₁. The variance of the distribution at generation n (below) is large. At generation n+i (above), variants closer to X₁ have more offspring because of selection and the population moves towards X₁.

The evolutionary result therefore will be determined by both variation and selection. Heritability is an important factor here because it determines to what extent directionality of variation will affect response to selection: If the trait is poorly heritable, the effect of selection after one generation will be attenuated, so the directionality of variation will tend to overcome selection. But if it is strong, then selection will drive the process. (This very general situation is exemplified in population genetics by studies about mutation-selection balance.)[27]

Since the decisive character of strongly alternative explanatory schemes is that the explanantia of evolutionary change and adaptation have to be sought at the level of variation, this is rather not the case here. Even if the causes of the directionality of variation are elucidated, this will not provide either predictions of the evolutionary trajectory or explanation of the extant adaptations.

However, when variance is very low, selection will be less explanatory because the response to selection will be very weak, given that very few individuals are distant from the X_o trait value and then, being closer to X_1, are likely to outreproduce the individuals displaying a value closer to X_o (because they do reproduce more). Here, the directionality of variation determines intergenerational change; it is plausible that the whole explanatory scheme should be revised, if those cases happen to be rather dominant in our data.

Variance and directionality of variation are thereby explanatory, but only in combination with heritability, and often with selection. Directed variation *as such* does not support, legitimate, or call for a switch in explanatory schemes. However there is one case where variation has such an explanatory weight that it determines the explanation, namely heritable *adaptive* variation. In this case, clearly, explaining why variation directs onto the adaptive value supports a general explanation of the evolutionary change because the value of the trait would be reached even without selection, whereas selection is supposed to depend on variation. This is why some instances of epigenetic inheritance, which seem to include adaptive variation, are so challenging for the MS, at least according to some authors (Jablonka & Lamb, 2005). Yet clearly if some systems exhibit adaptive heritable variation, one might ask a further question about their evolutionary history, and here basic natural selection may provide the answer (see Scott-Phillips, Dickins, & West, 2011).

The next section considers whether some *specific* known cases of variation grant a radically alternative explanatory scheme, or rather, should trigger weak amendments of the explanatory structure.

---

27. If you represented this outcome in terms of the Price equation, all the change in mean trait value would come out in the "selection" term. And in turn selection would lead to a rise in the covariance of parent-offspring trait value (thanks to Denis Walsh for this remark).

2.4.2. Instantiating Patterns of Variation

In this section I consider some known patterns of variation, first at the allelic level, and then at other levels, when there is variation in development or phenotype in general even if the genotype is fixed. I investigate whether specific patterns of variation support specific explanatory schemes.

2.4.2.1. Genetic Variation: Randomness Is Not the Issue

A simple and straightforward approach of variation at the genetic level consists in considering one genotype in relation to the mutations and recombinations it undergoes at the next generation—contextualizing variation in the loci landscape sketched above. A kind of randomness is the fact that no mutation is more likely than another, or that recombination is pairing indifferently the alleles, meaning that chances for allele x to be paired at the next generation with y or with z are equal. However such randomness is not often obtained in biology: On one hand, due to linkage, disequilibrium recombinations may not be random; on the other hand, mutations may actually be biased and nonrandom—especially because given the chemical nature of the nucleotides, some point mutations are more probable than others, and then some mutations from one triplet to another are more probable than others (Yampolski & Stoltzfus, 2001). Hence equiprobability is not the rule for genetic diachronic variation.

Biased mutation is therefore to be taken into account in explaining evolution by natural selection (see Bateson, this volume). It may especially decide between several possible outcomes of selection that have close fitness values, as Stoltzfus and Yampolsky (2009) made clear. However such biased nonadaptive mutation does not require to switch to an alternative explanatory scheme since biased mutation *and selection* are here intertwined to produce the adaptive outcome as well as the evolved diversity. A rather weaker revision of explanatory structure is favored by this case.

While some cases of nonrandom variation at the genetic level are due to the chemical properties of nucleotides, some others may be due to properties of the genetic system, as it has evolved features likely to produce variation for adaptive purposes (Giraud et al., 2001 Moxon et al., 1994).[28] This latter instance of biased allelic variation seems more prone to favor an alternative explanatory scheme.

But while biased mutations may be integrated within a slightly renovated MS schema, inversely random genetic variation does not necessarily support the classical MS scheme, as I show now. Suppose that many mutations are indeed

---

28. Though nothing warrants that it is strictly speaking adaptive variation—see Merlin (2010) on this.

underlying the same phenotype: Then the variants will not be discriminable for selection. Take the extreme case that almost all mutations correlate to a same phenotypic trait value: Then we are exactly in the case presented in section 2.2, Figure 2.3, where constrained variation swamps selection and selection does not play any explanatory role.

The case is not a pure abstraction: It is well known that the relations between genotype and phenotypes, often modeled as genotype-phenotype maps (Wagner & Altenberg, 1996), are not simple, and that one phenotype is generally underpinned by many different genes, even nonsynonymous. So if we characterize variation in this way, the fact that allelic variation is random or isotropic does not ipso facto support the MS explanatory scheme. Current genomic research has found this kind of case. When gene regulatory networks can be built with a sole focus on transcription and binding sites of transcripts, they are called transcriptional networks (see de la Fuente, 2010 for the difference between kinds of networks). Recently, Andreas Wagner and colleagues have considered such networks in cells. Investigation on the transcriptional regulation networks of yeast *Saccharomyses cerevisiae* has shown that they include very few unique regulatory pathways, and a vast majority that are present in two or more occurrences—the mean number of pathways between any source and target pair being 2.01 (Wagner & Wright, 2007, p. 164). This implies that random mutations are not so likely to change the phenotypic outcome, because other pathways can always supplement a pathway damaged by the mutation. A huge set of random mutations map onto very few actual alternative functioning networks: This means that variation is severely constrained, which clearly reduces the scope for selection.

Hence regarding allelic diachronic variation, randomness, isotropy or biases do not, by themselves, favor any decision regarding the explanatory scheme proper to contemporary evolutionary biology. Defining the variations required by MS as mutations and recombination-based variation (as Huxley did) is not the correct way to characterize (negatively) the kinds of variation that would be challenging the MS explanatory scheme. At least, we have shown that randomness or isotropy in mutations is not a sufficient condition for applying the MS explanatory scheme where variation plays no explanatory role (all variants being possibly there). And inversely, cases empirically attested of biased allelic variation do not necessarily support a radical switch in explanatory schemes.

2.4.2.2. Variation with Fixed Genotypes: Why Would Phenotype Matter?

The previous subsection established that randomness of variation is not a general criterion to make sense of the empirical facts that would force one to reconsider the explanatory scheme of the MS. Directed or biased variation does not provide a sufficient reason to downplay the role of selection versus variation, and inversely, there might be cases of random allelic variation, like these analyzed by Wagner and Wright (2007), where in fact phenotypic variation is really clustered in a way that selection faces important constraints, to a point that the mechanisms of variation should play the crucial explanatory role regarding some questions. Adaptive variation is the only clear case where another explanatory scheme of evolution is needed; yet some of the only cases of adaptive variation we know are about phenotypic plasticity (West-Eberhard, 2003), which is not by itself heritable (in the sense that phenotypes are not transmitted).

But precisely, a pervasive claim set by the challengers of MS is that, pace Huxley, there is more to variation than mutation and recombination—supposed to be isotropic and random. And considering these variations will force us to reconsider MS and adopt an alternative explanatory scheme (Kirschner & Gerhart, 2005; West-Eberhard, 2005). Hence it is finally important to consider here the extent to which specific cases of nonallelic variation—hence, variation at the phenotypic level—even nonheritable, could play such an important role in evolution that it forces us to recast our conceptions. This is what West-Eberhard argued regarding phenotypic plasticity—which indeed is pervasive over almost all clades, and is a key feature of development. But West-Eberhard also argued that in many cases a plastic response to environment initiates evolutionary change, and that the genes afterward modify their frequency according to this change. This would define a "genes as followers" scenario of evolution, which contrasts with the usual view of "mutation first" and then selection and evolutionary change (West Eberhard, 2003). It would thereby be an empirical question to decide whether one given trait, or one novelty, or a case of speciation, evolved through one or the other scenarios—and of course another empirical question to determine which scenario has been phylogenetically the most prevalent. This radical view of the evolutionary role of plasticity, emphasized by Walsh (2008) as the highest grade of involvement of development in evolution, fully subscribes to the strong alternative explanatory view since the crucial explanatory role is given to the mechanism of variation, namely plasticity. My argument here is to consider very generally to what extent phenotypic plasticity can make a difference to evolutionary change, and then ask whether it should generally support a rethinking of evolutionary theory, in the way West-Eberhard claims. Then I turn to another kind of nongenetic variation, which attracted attention when evo-devo arose.

2.4.2.2.1. PLASTICITY

Suppose a genotype G with a correlated trait P, which has an average number of offspring w (here, its fitness), and consider an alternative genotype G' which is plastic, so can express either P1 or P2 respectively relative to environments E1 and E2, the two morphs being produced according to environmental cues. P1 and P2 have different chances to have offspring, but the average given the distribution of environments E1 and E2 is also w.

Now consider a mutant genotype G," such that W (G") = w" > w, with its phenotype P". Suppose that, while the fitness of the phenotype P" is higher than the fitness of P, phenotype of G, the fitness of P" has the property that P" is fitter than P1 when it is in an environment mostly made of P1, but less fit than P2 in an environment full of P2.

For example suppose that w = 8, and that there are 3 morphs P1 in E1 and 6 P2 in E2—while W (P1) = 4 and W (P2) = 10.

Then: W (G') = 1/9 (3 × 4 + 6 × 10) = 72/9 = 8 = w.

Now, if we consider the evolutionary dynamics, leaving drift aside (by considering a large enough population), clearly G" will invade a population of Gs, since its fitness is higher.

However, regarding the plastic G', things are less simple. In the E1 environments G" would invade since its fitness is higher than the fitness of P1, which is the fitness of G' in these environments; but in E2, G" won't invade, and G' will still persist. What clearly appears is that even in such a trivial example plasticity makes a crucial difference to the evolutionary dynamics: The genotype fitnesses of G, G', and G" do not accurately predict the dynamics. What explains the evolution of G is what explains phenotypic plasticity, therefore it seems that the cause of phenotypic variation are crucial. But, does it lead to a reconception of the explanatory role of selection?

As such, it does not seem that we should resort to an alternative explanatory framework: Natural selection still drives the dynamics, and the phenotypic traits P2 of G, that defeat the mutant genotype G" in E2, are still here because of natural selection. Hence, plasticity per se does not seem to grant a radical explanatory change, which of course is well illustrated by the fact that various aspects of plasticity—to start with, the "norm of reaction"—have been the topic of intense controversies within the MS framework, at least from Bradshaw (1965) on (Nicoglou, 2015).

What has been emphasized by Walsh (2013, 2015), is that in this case however, phenotypic plasticity as it has been investigated by West-Eberhard (2003) is a feature of organisms rather than genes, and therefore, "adaptation" in the sense of the MS, which is defined by natural selection acting on gene populations, is in turn partially explained by a phenomenon that itself pertains to the organism's activity. This fact should motivate a turn to a radical alternative explanatory scheme: The adaptive capability of organisms explanatorily precedes the adaptation as a trait property, induced by a selective genotypic dynamics.

Yet this conclusion is not so much supported by the sole empirical facts of phenotypic plasticity, than by some conceptual take on plasticity that will simply not be shared by those who stay committed to an MS explanatory framework.

First, this claim about the originality of plasticity faces the usual objection that supporters of the evolutionary relevance of nongenetic adaptive variation meet in general (e.g., Scott-Phillips et al., 2011), which objects that ultimate and proximate causes are mixed in this case. Here, this property of organisms is itself a result of evolution by natural selection, and the MS framework has precisely the resources to ask why plasticity is such a pervasive feature of organisms, and in which environments it has been favored by natural selection. The "evolution of plasticity" is indeed a long-standing program of classical evolutionary biology, as illustrated by the controversy between Via and Schleicher regarding the putative "plasticity genes" (Nicoglou, 2015).

A way for phenotypic plasticity to directly play an evolutionary role would be of course to give rise to heritable variants. This ties the question to nongenetic inheritance, and may by itself justify an extended synthesis or face the above objections. West-Eberhard (2003) conceives of a way for the best phenotypic variants to be passed on to the next generation that does not involve heredity per se, namely "genetic accommodation." Yet she follows a long tradition of concepts in evolutionary thinking aimed at making sense of some intergenerational existence of phenotypic variants with no commitments to nongenetic inheritance: Baldwin effect, genetic assimilation, and so forth. Therefore, it is arguable that many of the plasticity facts were such that the MS had resources to conceive of them.[29]

Second, and even more importantly, the very concept of "plasticity" is controversial, and biologists may be cross-talking when they comment on the findings grouped under such a label. Phenotypic plasticity, as I have reiterated here, can be modeled as reaction norm or even as a plasticity gene in population genetics, which pertains to the MS framework. Unless one has already defended a concept of plasticity that is not the one used in such contexts, it is hard to rely on the empirical facts described as "plasticity" to justify any move away from the classical MS framework. Hence no empirical finding about "plasticity" is likely to settle the issue about a putative explanatory shift, because it already supposes conceptual assumptions about what plasticity is and, before that, about what adaptation is.

So even though the fact of plasticity in principle implies that there is more in evolution than genotypes and their fitnesses—but who really doubted that?—the

_____

29. On genetic accommodation, see Pocheville and Danchin (this volume).

measure to which this should entail an explanatory switch depends on many theoretical assumptions used to describe and model such plasticity.

2.4.2.2.2. HETEROCHRONY AND THE CONCEPTIONS OF VARIATION

One major interest of attention from the side of those who started advocating for a renewal of evolutionary biology is undoubtedly the phenomenon of developmental timing. As Gould (1977) forcefully argued, with a set of coding genes and even a same adult phenotype, any difference in this timing—be it intersection of developmental sequences, deletion of a sequence, or simply changes of speed of some sequences—has induced evolutionary changes. Moreover, some of these changes are major events in phylogenesis, and Gould was not far from argue that we owe them the major novelties in evolutionary history. Those would correspond to important changes in body plans, which define the stages of rapid phylogenetic change and contrast with the phases of evolutionary stasis, which according to a punctuated equilibrium view of phylogenetic patterns make up most of the evolutionary history.

Those variations of developmental timing instantiate the claim that the evolutionary relevant variation is not only—and may be even not principally—the mutation/recombination events mentioned in Huxley's dictum, and this indeed could induce a major rethinking of explanatory architecture of the theory. But why in principle would this variation lead to major epistemic changes?

Suppose that G has two possible developments D', at tempo t', and D at a tempo t much slower.

In case 1, genotype G will compete with other genotypes distributed with a probability distribution X. But in case 2, since the timing is really longer, when the offspring come to reproduction time, they will compete with another probability distribution X' for the other genotypes (Figure 2.8). If there is any frequency dependence of selection, clearly the outcomes of such competition for G will be very different if the timing is t or t'. Therefore considering developmental timing explains important differences in the evolutionary dynamics. And the ultimate cause for such variation is the developmental mechanisms in their relations with environmental conditions.

However, in the case considered here, it is not clear that the whole explanatory weight has to be switched toward variation. It is more plausible that we need to consider at a more fine-grained level the conditions under which natural selection drives evolution. For instance, our understanding of developmental timing has been renewed by the conception and investigation of the role of regulatory genes (Carroll, 2008) and of gene regulatory networks (Davidson, 1986). Heterochrony, in this perspective, may not exactly be a variation with a constant genotype. Regulatory genes govern the expression of the coding genes; each of

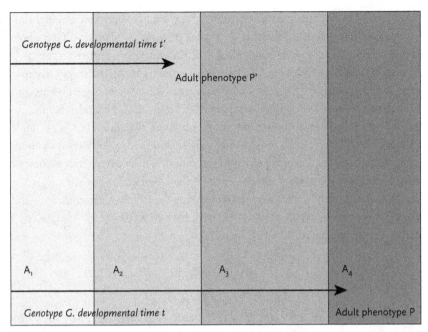

Genotype G. developmental time t'

Adult phenotype P'

A₁   A₂   A₃   A₄

Genotype G. developmental time t   Adult phenotype P

FIGURE 2.8 G genotypes with developmental timings t and t', t > t'. There are four periods A $(1 < i < 4)$, corresponding to 4 generations (or sets of generations) for organisms in the environment. Each color represents one environment characterized by one dominant type of organisms (or traits), which defines a different selective pressure for G.

them may impact at various moments on many genes; and the sequence of expression of various genes is due to the regulatory genes. Hence it is safe to assume that neoteny, paedomorphosis, and heterochrony are related to changes in the regulatory genes. For this reason, regarding the MS explanatory framework one could rather think of this phenomenon as a challenge to integrate a more complex picture of the genome as a sort of complex system, possibly regulative and adaptive.[30] It is not impossible that under this perspective, the MS framework remains plausible, because what is at stake is the evolution of regulatory genes—those being very pleiotropic and displaying different features than coding genes. Going on in this way, one could revisit the case made previously for developmental timing as a motivation for a new synthesis, and instead argue for a weak revision of the explanatory framework. Variation on these genes, and the molecular mechanisms that drive it, are replacing blind allelic mutation and recombination, but this is a move *within* the classical explanatory scheme.

---

30. For a view saying that, one should rethink of evolutionary biology's theoretical structure precisely because genomes are complex adaptive systems, see Walsh (2015, p. 205).

A second perspective on heterochrony and variations in developmental timing consists in seeing this as another instance of phenotypic plasticity. If the whole genome is constant, and the timing can plausibly vary according to some environmental cues, this should indeed be termed phenotypic plasticity. For instance some environments trigger a short development while others trigger a long and slow development. The consequences are what were just sketched in the preceding subsection, and may impinge on the evolutionary dynamics; but in principle this does not differ from the reasons why plasticity, described previously, dramatically changes the evolutionary dynamics. But in this latter case, heterochrony, or developmental timing *in general* can no more than plasticity *in general* trigger any strong revision of the explanatory framework of the MS. More assumptions and more conceptual work, always controversial, have to be done before supporting such a conceptual move.

2.5. Conclusion

Trying to summarize the main radical critiques elaborated against the MS by especially developmental biologists, one quickly has to focus on the compared explanatory roles of variation and selection. A same restructuring of the explanatory scheme of evolutionary theory seems vindicated by many instances of what one uses to call extended synthesis. In order to assess this framework I investigated the kinds of variations that may support the case of such theoretical change. The result is that no clear-cut distinction between cases of variation can be made that would single out which ones are empirically supporting, or infirming, the radical alternative explanatory scheme. Especially the roles of directed and random allelic variation are not as clear as they used to be seen, in the sense that they do not allow one to draw on their sole ground a clear-cut line between the MS and alternative proposals.

The conclusion is that many cases of variation differ from the kind of allelic variation provided through mutation and recombination (appealed to in Huxley's definition of the MS), but many of them seem to call for a diversification of the entanglements between selection and other causal variables in models of evolution that could still be understood within a slightly modified MS explanatory scheme. They are less justifying by themselves a total switch of explanatory schemes, which would downplay the explanatory role of natural selection in favor of an enhanced explanatory role of these patterns of variation and their underlying mechanisms. Such a switch, to be defended, needs more conceptual work regarding the concepts of adaptation, plasticity, and in general genomes as integrated systems.

Acknowledgments

I warmly thank Laura Nuno de la Rosa, Anya Plutynski, Antonine Nicoglou, Francesca Merlin, Arnaud Pocheville, Thibault Racovsky, and Denis Walsh for major criticisms and suggestions that radically improved this chapter, as well as the audiences where it has been first presented, including the workshop "Challenges for Evolutionary Theory" at the University of Toronto in 2010. I'm indebted to Ron Amundson, David Depew, Dick Burian, Etienne Danchin, Jean Gayon, Francesca Merlin, Tobias Uller, and above all Denis Walsh for precious discussions of these ideas. This work is supported by the Grant ANR 13 BSH3 0007 "Explabio".

References

Alberch, P., Gould, S. J., Oster, G. F., & Wake, D. B. (1979). Size and shape in ontogeny and phylogeny. *Paleobiology, 5*, 296–317.

Amundson, R. (1994). Two concepts of constraint: Adaptationism and the challenge from developmental biology. *Philosophy of Science, 61*, 556–578.

Amundson, R. (2001). Homology and homoplasy: A philosophical perspective. *ELS Reviews*. doi:10.1038/npg.els.0003445

Amundson, R. (2005). *The changing role of the embryo in evolutionary thought: Roots of evo-devo.* Cambridge, UK: Cambridge University Press.

Bateson, P. (2005).The return of the whole organism. *Journal of Biosciences, 30*, 31–39.

Beatty, J. (2006). Replaying life's tape. *Journal of Philosophy, 103*, 336–362.

Beatty, J. (2016). The creativity of natural selection? Part I: Darwin, darwinism, and the mutationists. *Journal of the History of Biology, 49*, 659–684.

Bonduriansky, R., & Day, T. (2009). Nongenetic inheritance and its evolutionary implications. *Annual Review of Ecology, Evolution, and Systematics, 40*, 103–125.

Bradshaw, A. D. (1965). Evolutionary significance of phenotypic plasticity in plants. In E. W. Caspari and J. M. Thoday (Eds.) *Advances in genetics* (pp. 115–155).

Brakefield, P. M. (2006). Evo-devo and constraints on selection. *Trends in Ecology & Evolution, 21*, 362–368.

Burian, R. (2005). On conflicts between genetic and developmental viewpoints—and their attempted resolution in molecular biology. In Burian (ed.) *The epistemology of development, evolution and genetics* (pp. 210–233). Cambridge, UK: Cambridge University Press.

Caroll, S. (2005). *Endless forms most beautiful: The new science of evo devo and the making of the animal kingdom.* Norton.

Caroll, S. (2008). Evo-devo and an expanding evolutionary synthesis: A genetic theory of morphological evolution. *Cell, 134*, 25–36.

Carroll, S. B. (2005). Evolution at two levels: On genes and form. *PLoS Biology, 3*, e245.

Chaitin, G. (1970). On the difficulty of computation. *EEE Transactions on Information Theory*, *16*, 5–9.

Cheverud, J. M. (2007). The relationship between development and evolution through heritable variation. *Novartis Foundation Symposia*, *284*, 55–70.

Damuth, J, & Heisler, I. (1988). Alternative formulations of multilevel selection. *Biology and Philosophy*, *3*, 407–430.

Danchin, E., Charmontier, A., Champagne, F. A., Mesoudi, A., Pujol, B., & Blanchet, S. (2011, July). Beyond DNA: Integrating inclusive inheritance into an extendedtheory of evolution. *Nature Reviews Genetics*, *12*, 475–486.

Darwin, C. (1859). *The origin of species.* London: John Murray.

Darwin, C. (1862). *On the various contrivances by which British and foreign orchids are ertilized by insects, and on the good effects of intercrossing.* London, UK: John Murray.

Davidson, E. H. (1986). *Gene activity in early development.* Orlando, FL: Academic Press.

Dawkins, R. (1976). *The selfish gene.* Oxford, UK: Oxford University Press.

De la Fuente. (2010). What are gene regulatory networks ? In D. Das Caragea, M. Welch, & W. Hsu (Eds.), *Handbook of research on computational methodologies in gene regulatory networks* (pp. 1–27). New York, NY: Medical Science Reference.

De Visser, J. A. G. M., Hermisson, J., Wagner, G. P., Ancel Meyers, L., Bagheri-Chaichian, H., Blanchard, J., et al. (2003). Evolution and detection of genetic robustness. *Evolution*, *57*, 1959–1972.

Depew, D. (2011). Adaptation as process: The future of Darwinism and the legacy of Theodosius Dobzhansky. *Studies in History and Philosophy of Science Part C, Studies in History and Philosophy of Biological and Biomedical Sciences*, *42*, 89–98.

Fisher, R. (1930). *The genetical theory of natural selection.* Oxford: Oxford University Press.

Forgacs, G., & Newman, S. (2005). *Biological physics of the developing embryo.* Cambridge, UK: Cambridge University Press.

Gayon, J. (1998). *Darwinism's struggle for survival: Heredity and the hypothesis of natural selection.* Cambridge, UK: Cambridge University Press.

Ghiselin, M. (1980).The failure of morphology to assimilate Darwinism. In E. Mayr & W. B. Provine (Eds.), *The evolutionary synthesis: Perspectives on the unification of biology* (pp. 180–193). Cambridge: Cambridge University Press.

Gilbert, S., Opitz, G., & Raff, R. (1996). Resynthesizing evolutionary and developmental biology. *Development and Evolution*, *173*, 357–372.

Gilbert, S. F., & Sarkar, S. (2000). Embracing complexity: Organicism for the twenty-first century. *Developmental Dynamics*, *219*, 1–9.

Giraud A., Matic I., Tenaillon O., Clara A., Radman M., Fons M., & Taddei, F. (2001). Costs and benefits of high mutation rates: Adaptive evolution of bacteria in the mouse gut. *Science*, *291*, 2606–2608.

Goodnight, C. J. (2005). Multilevel selection: The evolution of cooperation in non kin groups. *Population Ecology*, *47*, 3–12.

Gould, S. J. (1977). *Ontogeny and phylogeny.* Harvard, MA: Belknap.

Gould, S. J. (2002). *The structure of evolutionary theory*. Chicago, IL: University of Chicago Press.

Gould, S. J., & Lewontin, R. (1978). The spandrels of San Marco and the panglossian paradigm: A critique of the adaptationist programme. *Proceedings of the Royal Society of London, B205, 1161*, 581–598.

Gould, S. J., & Lloyd, E. (1999). Individuality and adaptation across levels of selection: How shall we name and generalize the unit of Darwinism? *Proceedings of the National Academy of Science, 11904–11909.*

Griffiths, P., & Gray R. (1994). Developmental systems and evolutionary explanation. *Journal of Philosophy, 91*, 277–304.

Hall, B. (2003). Evo-devo: Evolutionary developmental mechanisms. *The International Journal of Developmental Biology, 47*(7–8), 491–497.

Helantera, H., & Uller, T. (2010). The Price equation and extended inheritance. *Philosophy and Theory in Biology, 2*, 1–17.

Huneman, P. (2010). Assessing the prospects for a return of organisms in evolutionary biology. *History and Philosophy of Life Sciences, 32*, 341–372.

Huneman, P. (2015). Redesigning the argument from design. *Paradigmi, 33*, 105–132.

Huneman, P. (2014). A pluralist framework to address challenges to the modern synthesis in evolutionary theory. *Biological Theory, 9*, 163–177.

Jablonka, E., & Lamb, M. (2005). *Evolution in four dimensions*. Cambridge, MA: MIT Press.

Jablonka, E., & Raz, G. (2009). Transgenerational epigenetic inheritance: Prevalence, mechanisms, and implications for the study of heredity and evolution. *Quarterly Review of Biology, 84*(2), 131–176.

Kauffmann, S. (1993). *Origins of order: Self-organization and selection in evolution*. Oxford, UK: Oxford University Press.

Kirscher, M., & Gerhart J. (2005). *The plausibility of life: Resolving Darwin's dilemma*. New Haven, CT: Yale University Press.

Laland, K., Uller, T., Feldman, M., Sterelny, K., Müller, G. B., Moczek, A., . . . Odling-Smee, J. (2014). Does evolutionary theory need a rethink? Yes, urgently. *Nature, 514*, 161–164.

Lenormand, T., Chevin, L. M. C., & Bataillon, T. (2016). Parallel evolution: What does it (not) tell us and why is it (still) interesting? In G. Ramsey & C. Pence (Eds.), *Chance in evolution*. Chicago, IL: University of Chicago Press.

Lewens, T. (2009). What is wrong with typological thinking? *Philosophy of Science, 76*, 355–371.

Lewontin, R. (2000). *The triple helix: Gene, organism, and environment*. Cambridge, MA: Harvard University Press.

Love, A. (2008). Explaining evolutionary innovation and novelty: Criteria of adequacy and multidisciplinary prerequisites. *Philosophy of Science, 75*, 874–886.

Love, A. C. (2003). Evolutionary morphology, innovation, and the synthesis of evolutionary and developmental biology. *Biology and Philosophy*, *18*, 309–345.

Love, A. C. (2010). Idealization in evolutionary developmental investigation: A tension between phenotypic plasticity and normal stages. *Philosophical Transactions of the Royal Society B*, *365*, 679–690.

Lynch, M. (2007). *The origins of genome architecture*. Sunderland, MA: Sinauer.

Maynard Smith, J., Burian, R., Kauffman, S., Alberch, P., Campbell, J., Goodwin, B., . . . Wolpert, L. (1985). Developmental constraints and evolution. *The Quarterly Review of Biology*, *60*(3), 265–287.

Mayr, E. (1997). The objects of selection. *Proceedings of the National Academy of Sciences*, 2091–2094.

Mayr, E., & Provine, W. (Eds.). (1980). *The evolutionary synthesis: Perspectives on the unification of biology*. Cambridge, MA: Harvard University Press.

Merlin, F. (2010). Evolutionary chance mutation: A defense of the modern synthesis' consensus view. *Philosophy and Theory in Biology*, *2*. doi: http://dx.doi.org/10.3998/ptb.6959004.0002.003

Merlin, F. (2014). *Mutations et aléas*. Paris, France: Hermann.

Moxon, R., Raineyt, B., Nowak, M., & Lenski, R. (1994). Adaptive evolution of highly mutable loci in pathogenic bacteria. *Current Biology*, *4*, 24–33.

Müller, G., & Newman, S. (2005). The innovation triad: An evo-devo agenda. *Journal of Experimental Zoology*, *304*, 487–503.

Newman, S. A., & Bhat, R. (2009). Dynamical patterning modules: A "pattern language" for development and evolution of multicellular form. The *International Journal of Developmental Biology*, *53*, 675–693.

Newman, S. A., & Müller, G. B. (2000). Epigenetic mechanisms of character origination. *Journal of Experimental Zoology Part B: Molecular and Developmental Evolution*, *288*, 304–317.

Nicoglou, A. (2015). The evolution of phenotypic plasticity: genealogy of a debate in genetics. *Studies in the History and Philosophy of Science Part C*, *50*, 67–76.

Odling-Smee, J., Laland, K., & Feldman, M. (2003). *Niche construction: The neglected process in evolution*. Princeton, NJ: Princeton University Press.

Okasha, S. (2006). *Evolution and the levels of selection*. New York, NY: Oxford University Press.

Oliveri, P., Tu, Q., Davidson, E. (2008). Global regulatory logic for specification of an embryonic cell lineage. *Proceedings of the National Academy of Sciences*, *105*, 5955–5962.

Oyama, S., Griffiths, P., & Gray, R. (Eds.). (2001). *Cycles of contingency: Developmental systems and evolution*. Cambridge, MA: MIT Press.

Pavlicev, M., Cheverud, J. M., & Wagner, G. P. (2011). Evolution of adaptive phenotypic variation patterns by direct selection for evolvability. *Proceedings of the Royal Society B: Biological Sciences*, *278*, 1903–1912.

Pearce, T. (2012). Convergence and parallelism in evolution: A neo-Gouldian account. *British Journal for the Philosophy of Science, 63*, 429–448.

Pigliucci, M. (2007). Do we need an extended evolutionary synthesis? *Evolution, 61*, 2743–2749.

Pigliucci, M., & Müller, G. (Eds.). (2011). *Evolution: The extended synthesis.* Cambridge, MA: MIT Press.

Potochnik, A. (2009). Optimality modeling in a suboptimal world. *Biology and Philosophy, 24*(2), 181–197.

Raff, R (1992). Evolution of developmental decisions and morphogenesis: The view from two camps. *Development*, 15–22.

Raff, R. (1996). *The shape of life: Genes, development, and the evolution of animal form.* Chicago, IL: University of Chicago Press.

Raup, D. M. (1966). Geometric analysis of shell coiling: General problems. *Journal of Paleontology, 40*(5), 1178–1190.

Revilla-I-Domingo, R., & Davidson, E. (2003). Developmental gene network analysis. *The International Journal of Developmental Biology, 21*, 695–703.

Rice, S. (2004). *Evolutionary theory.* London, UK: Palgrave.

Richards, R. J. (2012). Darwin's principles of *Divergence* and natural selection: Why *Fodor* was almost right. *Studies in History and Philosophy of Science Part C, 43*, 256–268.

Roughgarden, J. (2009). *The genial gene: Deconstructing Darwinian selfishness.* Berkeley: University of California Press.

Scott-Phillips, T., Dickins, T., & West, S. (2011). Evolutionary theory and the ultimate–proximate distinction in the human behavioral sciences. *Perspectives on Psychological Science, 1*, 38–47.

Smocovitis, V. B. (1996). *Unifying biology: The evolutionary synthesis and evolutionary biology.* Princeton, NJ: Princeton University Press.

Sober, E, & Wilson, D. S. (1998). *Unto others.* Cambridge, MA: Harvard University Press.

Stoltzfus, A. (2006). Mutationism and the dual causation of evolutionary change. *Evolution and Development, 8*, 304–317.

Stoltzfus, A., & Yampolsky, L. Y. (2009). Climbing Mount Probable: Mutation as a cause of nonrandomness in evolution. *Journal of Heredity, 100*, 637–647.

Wade, M. (1977). An experimental study of group selection. *Evolution, 131*, 134–153.

Wagner, G., & Altenberg, L. (1996). Complex adaptations and the evolution of evolvability. *Evolution, 50*, 967–976.

Wagner, A., & Wright, J. (2007). Alternative routes and mutational robustness in complex regulatory networks. *Biosystems, 88*(1–2), 163–172.

Wake, D. B. (1991). Homoplasy: The result of natural selection, or evidence of design limitations? *The American Naturalist, 138*, 543–567.

Walsh, D. M. (2003). Fit and diversity: Explaining adaptive evolution. *Philosophy of Science, 70*, 280–301.

Walsh, D. M. (2008). Development: Three grades of ontogenetic involvement. In M. Matthen & C. Stephens (Eds.), *Handbook for the philosophy of science* (Vol 4, pp. 179–200). Amsterdam: Elsevier.

Walsh, D. M. (2013). Mechanism, emergence, and miscibility. In P. Huneman (Ed.), *Functions: Selection and mechanisms* (pp. 43–65). Dordrecht, Netherlands: Synthese Library.

Walsh, D. M. (2015). *Organisms, agency, and evolution.* Cambridge, UK: Cambridge University Press.

West-Eberhard, M. J. (2003). *Developmental plasticity and evolution.* Oxford, UK: Oxford University Press.

Williams, G. C. (1966). Adaptation and natural selection. Princeton: Princeton University Press.

Wray, G. A., Hoekstra, H. E., Futuyma, D. J., Lenski, R. E., Mackay, T. F. C., Schluter, D., & Strassmann, J. (2014). Does evolutionary theory need a rethink? No, all is well. *Nature, 514*, 161–164.

Yampolsky, L. Y., & Stoltzfus, A. (2001). Bias in the introduction of variation as an orienting factor in evolution. *Evolution & Development, 3*, 73–83.

Zahavi, A., & Zahavi, A. (1997). *The handicap principle: A missing piece of Darwin's puzzle.* Oxford, UK: Oxford University Press.

3 GENETIC ASSIMILATION AND THE PARADOX OF BLIND VARIATION

Arnaud Pocheville and Étienne Danchin

A core tenet of neo-Darwinism is that of blind variation, according to which heritable variations that fuel natural selection do not arise *because* of their adaptive value.[1] Though not always clearly specified in neo-Darwinian writings (Merlin, 2010), such a tenet is central since at least Weismann (1893), not to mention its importance for Darwin (1859, p. 209).[2] The tenet is essential in arguing for the sovereign explanatory power of natural selection as regards design. In the extreme case where variation is always maximally adaptive, there is, roughly speaking, no natural selection at all. The question of blind variation is probably the most salient stumbling block of the old and reviving debates between neo-Darwinians and neo-Lamarckians (e.g., Romanes, 1888; Jablonka & Lamb, 2005, 2010).

Genetic assimilation characterizes a situation where some phenotypic variations, which are initially caused by environmental events,

---

1. By "adaptive value" we mean that which Burian (1983) calls "engineering fitness": how well a systems behaves in its environment. We purposely use the word "blind" and not "random" because, as we develop here, the kind of randomness at stake is not clear. We also preferentially use "variation" instead of "mutation" not to give the flavor of restricting heritable variation to genetic mutation. In this chapter we only consider potentially heritable variation; notice we refer to molecular genes, not to Mendelian genes (Lu & Bourrat, in press). Last, we preferentially refer to neo-Darwinism rather than to the modern synthesis, which is just one version of it (Mayr & Provine, 1998).

2. "As modifications of corporeal structure arise from, and are increased by, use or habit, and are diminished or lost by disuse, so I do not doubt it has been with instincts. But I believe that the effects of habit are of quite subordinate importance to the effects of the natural selection of what may be called *accidental variations* of instincts;—that is of variations produced by the same unknown causes which produce slight deviations of bodily structure" (Darwin, 1859, p. 209, our emphasis). Darwin's later writings came to emphasize more the inheritance of acquired characteristics, so as to cope with fast adaptation (Hoquet, 2009).

eventually get caused by genetic variations.[3] When the phenotypic variations subject to genetic assimilation are adaptive phenotypic responses to environmental challenges, it is tempting to consider that genetic variation is directed, thus contradicting the principle of blind variation. Subsequently, genetic assimilation has given rise to some controversy, especially regarding whether it contradicts, or fits within, neo-Darwinism (Pigliucci et al., 2006).

In this chapter, we wish to clarify several points regarding genetic assimilation and how it connects with the neo-Darwinian tenet of blind variation. First, we briefly sketch the history and current view of genetic assimilation, including recent extensions to this view. We argue that all the mechanisms of genetic assimilation proposed so far rely on blind genetic variation fueling natural selection. Second, we present a new hypothetical mechanism of genetic assimilation, relying on nonblind genetic variation. This model might be thought of as a highly challenging scenario for neo-Darwinism. Yet, we show that it still relies on blind variation of some sort to explain adaptation. Last, we clarify the tenet of blind variation itself. We argue that, though somehow intuitive, the tenet proves remarkably elusive to formal and empirical characterization. Here, we concentrate on the theoretical novelty of the different models of genetic assimilation and on the way they affect the neo-Darwinian framework. We only briefly discuss their biological relevance or extent of applicability (Beatty, 1997), and predictions.

3.1. Genetic Assimilation

The concept of genetic assimilation can be traced back before the word to the second half of the 19th century, when authors were looking for a way to accommodate what seemed to be events of fast adaptation in a neo-Darwinian way (section 3.1.1). As we show, this motivation remains intact for the current models of genetic assimilation (section 3.1.2).

3.1.1. The Century of Genetic Assimilation

That organisms may adapt faster than what would seem permissible by natural selection has puzzled biologists ever since Darwin. By the end of the 19th century, several mechanisms had been proposed to explain fast adaptation, notably involving the inheritance of acquired characteristics (including by Darwin; see Hoquet, 2009). Not all Darwinians were happy with this idea (Romanes, 1888). The controversies between neo-Lamarckians and neo-Darwinians probably

---

3. "Cause" here means a difference maker (Waters, 2007).

reached a climax with Weismann's famous charge against the inheritance of acquired characteristics (Weismann & Spencer, 1893; Baldwin, 1896, p. 446; Simpson, 1953a, p. 110; Depew, 2003).

In this context, several authors independently thought of a way to reconcile neo-Lamarckism and neo-Darwinism (Spalding, 1873, cited by Pigliucci et al., 2006; Baldwin, 1896; Morgan, 1896; Osborn, 1897a, 1897b; Simpson, 1953a, p. 110; Griffiths, 2003). They postulated the existence of a biological mechanism that would mimic the inheritance of acquired characteristics but that would nevertheless rely on natural selection occurring on blind variation. Framed in modern terms, the mechanism postulates a two-step evolutionary process, where organisms, when facing an environmental challenge, can first adapt through adaptive plastic responses, thus allowing time for the species to evolve heritable (nonplastic) traits by natural selection (e.g., Baldwin, 1896, pp. 445–447).

This mechanism, later most famously known as "the Baldwin effect," was "envisioned . . . as a means of facilitating phenotypic evolution" (Pigliucci et al., 2006, p. 2362). Baldwin argued that the mechanism explained what seemed to be occurrences of determinate variation in the fossil record, a fact that otherwise would have seemed in favor of neo-Lamarckism. To Baldwin by contrast, his new mechanism "d[id] away with the need of appealing to the Lamarckian factor" (Baldwin, 1896, p. 446).

By the mid-20th century, Simpson (1953a) had dismissed the Baldwin effect as "a relatively minor outcome of the [neo-Darwinian] theory" (p. 115), which had "seldom [been] discussed in detail" (p. 110) since its inception. He also noticed the paradoxical situation that, "everyone who has discussed [the Baldwin effect] at any length has taken the position that . . . its real importance is in meeting or explaining away the criticisms leveled at natural selection by, especially, the neo-Lamarckians, the Michurinists, and the finalists" (Simpson, 1953a, p. 115).

At that time, the so-called Baldwin effect nevertheless got a new impulse from the experimental and theoretical works of Waddington (1942, 1952, 1953, 1956, p. 10) and Schmalhausen (1949). Fifty years after Baldwin, the question of determinate variation in evolution seemed intact. As Waddington had put it, "it is doubtful . . . whether even the most statistically minded geneticists are entirely satisfied that nothing more is involved than the sorting out of random mutations by the natural selective filter" (Waddington, 1942, p. 563). Waddington's hypothesis was that by looking at how organisms develop, one could explain some of their seemingly adaptive features. He proposed a mechanism similar to Baldwin's, but framed in a genocentrist way, and with an emphasis put on the importance of canalization in development. A phenotype could be formed as a response to an external stimulus, but during the course of evolution the environmental stimulus could then be superseded by an internal genetical factor (Waddington,

1942, pp. 563–564). To test this hypothesis, Waddington performed some of his most famous experiments on Drosophila, showing cases of genetic assimilation (Waddington, 1953, 1959). Using *Drosophila melanogaster*, he showed that certain phenotypes (e.g., crossveinless) can be induced by an environmental stimulus (in this case, a heat stress). He then showed that, after having selected the novel phenotype for relatively few generations (14), the environmental stimulus was not necessary anymore to induce the new phenotype. Waddington interpreted his results in a neo-Darwinian way as an example of selection acting on preexisting multigenic variation (acting on a threshold for environmental induction), explicitly specifying that, in any case, the hypothesis that the treatment had induced new mutations "could bring little comfort to those who wish to believe that environmental influences tend to produce heritable changes in the direction of adaptation. For there is no reason whatever to suppose that the crossveinless phenotype is adaptive to high temperature" (Waddington, 1953, p. 124).

In spite of these historical origins, we still find an ambiguity in the status of the Baldwin effect and genetic assimilation today, with both supporters and critics pretending to defend the modern synthesis (e.g., De Jong, 2005; contra Pigliucci et al., 2006).

3.1.2. Current Models of Genetic Assimilations

In modern terms, the classical mechanism of genetic assimilation consists in three steps (Pigliucci et al., 2006):

1. First, a population occupies a given environment (Figure 3.1, left).
2. Then, the environment changes, revealing capacity for plasticity within the population (Figure 3.1, center). Given the new selective constraints, this plasticity enables the population to persist, initially with no genetic change.

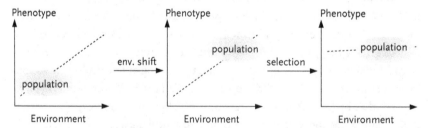

FIGURE 3.1 Mechanism of genetic assimilation (after Pigliucci et al., 2006, fig. 2). First, a population occupies a given environment; there is an unexpressed capacity for plasticity within the population. The environment then changes; the previously unexpressed capacity for plasticity enables the population to persist, initially with no genetic change. Genetic changes can then take place, flattening the plastic reaction norm (because of drift or trade-offs on plasticity).

3. If natural selection persists, genetic changes can make the phenotype constitutive, that is, flatten the plastic reaction norm (Figure 3.1, right). This can occur
because of drift (leading to loss of a previously selected ability to switch phenotypes) and/or trade-offs on plasticity (e.g., a constitutive phenotype might
be less costly than a plastic one).

The model of genetic accommodation of West-Eberhard (2003, p. 140)
generalizes this hypothetical mechanism to cases where the initial phenotypic
change can be either induced by environmental changes, as in step (2) above, or
by mutation.

This mechanism (including West-Eberhard's flavor) is thus one of genetic
assimilation through plasticity. It relies on the assumption of a redundancy
between plastic and nonplastic responses: intragenerational plasticity enables
survival at the populational level, allowing time for blind heritable (i.e., nonplastic) variation to occur and take over (Figure 3.2a).

Possibly because of its neo-Lamarckian flavor, this model of genetic assimilation has been charged with wrongfully attempting to contradict neo-Darwinism
(e.g., De Jong, 2005).[4] However, it is clear that the model solely relies on natural
selection occurring on blind genetic variation in step (3). The blindness of variation was already central to early ancestors of the model (section 3.1.1) and has
been repeatedly emphasized since then (Pigliucci et al., 2006).

Extensions to this model have been proposed. Two of them, which we discuss now, involve nongenetic inheritance (Klironomos, Berg, & Collins, 2013;
Danchin, Pocheville, Rey, Pujol, & Blanchet, 2017).[5]

Klironomos et al. (2013) have proposed a model where the buffering period
enabled by plasticity in step (2) above is now ensured by heritable nongenetic
materials (Figure 3.2b). They consider a population of individuals that can be
adapted by presenting either the right profile of epigenetic marks or the right
profile of genetic variations. They consider a full redundancy between genetic
and epigenetic materials, the only difference being that epigenetic marks are
highly more mutable (to be sure, genetic and epigenetic mutations are blind).
Simulations showed that, when experiencing a selective challenge, a population
first quickly reaches the adaptive peak thanks to variation and selection on epigenetic marks, while genetic variation evolves neutrally. Then, on-peak genetic variants eventually occur by blind mutation and get fixed (being less mutable, on-peak

4. In her paper, De Jong (2005) also argues against slippery uses of the vocable "genetic assimilation" (esp. by Pigliucci & Murren, 2003), which she thinks do not fit with Waddington's use
(Waddington, 1956; De Jong, 2005, p. 115). Being less stringent than she is on the criterium
of family resemblance for the vocable, we follow here the use of Pigliucci and Murren (2003).

5. To be precise, Klironomos et al. (2013) do not call their model a case of genetic assimilation.

genetic variants have a slight advantage over on-peak epigenetic variants). At this point the selective pressures bearing on epigenetic variation are released, due to the redundancy between epigenetic and genetic profiles. Epigenetic variation then evolves neutrally, while on-peak genetic variants are retained by selection. In such an evolutionary process, epigenetic and genetic systems seem tailored for adaptation at different evolutionary timescales. This model represents a mechanism of genetic assimilation through epigenetic inheritance.

Klironomos et al.'s (2013) model surely falls within the frame of neo-Darwinism. The difference with classical genetic assimilation through plasticity is that now intergenerational epigenetic variation, and not intragenerational plasticity, enables survival at the population level while waiting for blind genetic variations to occur (step 2 above). Again, in Klironomos et al.'s model, everything that is heritable and adaptive (be it epigenetic or genetic) comes from blind variation at the individual level, fueling natural selection or drift at the populational level.

We and collaborators have pleaded for extending these models of genetic assimilation (Danchin & Pocheville, 2014; Danchin et al., 2017) (Figure 3.2c). Like previous authors, we assumed redundancy between nongenetic and genetic materials for the adaptation of the individual. In addition, we pleaded for taking into account the fact that nongenetic variation (heritable or not) can both be induced by the environment and impact genetic materials, for example when adaptive regulatory epigenetic marks favor local hypermutability of the genes they regulate (Wright, Longacre, & Reimers, 1999; Wright, 2000; Gorelick, 2003), a mechanism already hypothesized by Davis (1989). Interestingly, epigenetically induced local hypermutability of genes had already been proposed by Jablonka and Lamb in the mid-1990s as a model of genetic assimilation, under the term "mutational assimilation" (Jablonka & Lamb, 1995, pp. 167–171). We thus modified the model of Klironomos et al. (2013) exposed above to incorporate (1) an induction of epigenetic marks by the environment (in particular, toward fit) and (2) a mutagenic effect of epigenetic marks on the corresponding genes. In addition, we assumed that, in some situations, the mutagenicity of epigenetic marks could be context-dependent, that is, that epigenetic marks—or nongenetic materials in general—would be particularly mutagenic only when the individuals would be off-peak. Such context-dependent mutagenicity can occur in situations where a physiological mechanism up-regulates a gene as long as the desired physiological effect is not reached. Then, if the gene is defective and up-regulation is itself slightly mutagenic (Wright, 2000), up-regulation will continue and have more mutagenic effects until the gene becomes functional (or the individual dies). Notice that in this situation, mutagenicity can be context-dependent as an exaptation of plasticity: the context-dependence of regulation (a capacity for plasticity) provides for free the context-dependence of mutagenicity (Danchin & Pocheville, 2014).

Simulations showed that the two-step adaptive process of Klironomos et al. (2013) is considerably accelerated. The population first quickly finds the peak thanks to epigenetic materials being mutable and inducible, and on-peak genetic variants quickly occur and invade the population, thanks to the context-dependent mutagenicity of on-peak epigenetic marks. We also envisaged the hypothesis that plastic or epigenetic responses are not fully redundant with genetic responses (for instance, because of a cost, delay, or unfaithfulness of plastic responses), reaching similar results as with context-dependent mutagenicity.

In short, this model is one of genetic assimilation through epigenetic mutagenicity.

Again, this model is interpreted within the neo-Darwinian frame. Genetic materials are mutable in a context-dependent way, but for a given gene the very variations produced by mutation are supposed to be blind (see section 3.3, Figure 3.4). Epigenetic materials are themselves inducible, possibly toward fit, but such a capacity for induction is supposed to ultimately rely on previous adaptations (or exaptations) provided by blind variation in the past.

3.2. Genetic Memorization

We now argue for the possibility of mechanisms that could accelerate the evolutionary process by several orders of magnitude, which would operate at the intragenerational timescale. Notice that the intragenerational timescale was somehow blackboxed in the models above, which contrasts with the emphasis classically put on development by authors questioning the neo-Darwinian vision of heritable variation (Jablonka & Lamb, 2005; Pigliucci & Müller, 2010; for a historical review, see Loison, 2010).

3.2.1. Heritable Physiological Exploration

Our first speculation concerns what could be called "heritable physiological exploration" at the intragenerational timescale. Authors having expressed views similar to ours that follow include (Kirchner & Gerhart, 2006; Braun & David, 2011, p. 190; Stern et al., 2012; Braun, 2015, p. 45; Soen et al., 2015; see also Lamm & Jablonka, 2008), although we do not mean to imply that they would agree with our exposition here.

Situations may exist where an organism would be able to explore and stabilize variation at the intragenerational timescale and then transmit it to offspring. A putative mechanism of such a pattern would be that when facing an environmental insult, somatic cells in an organism would show: (1) *exploration*, they would try several regulation profiles and/or genetic rearrangements;

(2) *stabilization*, they would retain those regulation profiles that do the job, for example, by epigenetic marks; and (3) *transmission*, they would pass them on to the rest of the soma *and/or* to the germline—something much like Darwin's gemmules (Darwin, 1868, p. 374) (Figure 3.2d).

Hypothesis (1), exploration, is that cells (or living systems in general) are capable of blindly exploring at both the epigenetic and genetic levels, for instance by deregulating gene expression profiles or performing blind genetic rearrangements. (This step can also plausibly involve, when the environmental challenge is not totally new in the lineage, nonblind reaction at the epigenetic level such as adaptive plastic responses. Here we set aside this possibility that relies on past adaptation.)

Hypothesis (2), stabilization, is that cells (or living systems) are not totally blind with respect to their own physiological condition. More precisely, the hypothesis is that if living systems are not able to know how to reach a desired physiological condition (particularly when facing an unknown environmental challenge), at least they know when they have reached it, be it a simply viable or unstressed condition. Notice that this hypothesis departs from previous hypothetical mechanisms of nonblind variation where cells were supposed to select the precise molecular component (protein, RNA, DNA mutation) that would provide a positive fitness effect (Cairns et al., 1988; Cairns & Foster, 1991). That cells be able to distinguish which molecules provided a fitness effect has been deemed implausible (e.g., Stahl, 1988; Foster & Cairns, 1992). In our view by contrast, the whole physiological condition is selected by the system, including, possibly, "superstitious" regulatory features that do not participate in releasing the stress, but happen to do no harm.

Hypothesis (3), transmission, is that cells are able to communicate aspects of their condition across the organism, and to realize epigenetic and/or genetic transmission of acquired traits (both in the soma and to the germen). Though highly controversial in the past, this hypothesis is now experimentally illustrated, at least regarding epigenetic inheritance. For instance, Devanapally et al. (2015) have shown that, in *Caenorhabditis elegans*, neurons can make double-stranded RNA that can enter the germline and cause transgenerational gene silencing. (See also, e.g., Bohacek & Mansuy, 2015; Rodgers et al., 2015; Sharma, 2015; Szyf, 2015).

Exploration, stabilization, and transmission need not happen in a sequential way. Partially releasing a stress can come with concomitant partial stabilization and transmission. Neither do they need to happen independently in every cell, if cells happen to somehow synchronize their explorations. In some cases, natural selection among cells (see, e.g., Pocheville & Montévil, 2014) might also resonate with the exploratory process and amplify the prevalence of successful phenotypes. Our emphasis on regulation resonates with the cis-regulatory hypothesis, according to which an important part of evolutionary change corresponds to changes occurring in regulatory regions rather than in coding regions (King & Wilson, 1975; Wray, 2007). What we argue here is that this change might be initiated and directed by (parts of) the living system at the intragenerational level.

Though we initially thought to propose this model of heritable physiological exploration for theoretical reasons, several recent experimental results suggest that such or similar intragenerational mechanisms are plausible. For instance, Stern et al. (2012) have confronted the development of *Drosophila melanogaster* to an artificial toxic stress. They showed that not only was development modified, but also that the modified development could coincide with increased tolerance to the artificial challenge, and be epigenetically transmitted to the offspring (from 1 to 24 generations). Part of the modification came from the suppression of the Polycomb group genes, which play an important role in stabilizing development (see also Stern et al., 2014). In a similar vein, Braun (2015) has reviewed a set of experiments on *Saccharomyces cerevisiae*, in which populations were inflicted with an "unforeseen challenge" by linking an essential gene from one biochemical pathway (*HIS3* gene) to the regulatory system of another pathway (GAL promoter). The results present a highly nonparadigmatic picture of adaptation:

1. Adaptation is both fast (10–30 generations) and inherited (though not stably inherited at the beginning of the process); adaptation is not due to the selection of rare advantageous phenotypes; many individual cells independently develop the adaptive phenotype as a response to the challenge (50% on average, and up to 80%); last, "when mutation arise, they are *induced* late in the process" (p. 11, Braun's emphasis).
2. Adaptation comes with global gene expression responses; these responses are irreproducible between replicate experiments with no dominating molecular mechanism, which indicates the exploratory nature of the adaptive process and the degeneracy of the gene expression–phenotype relation.
3. The population itself is a relevant level in the dynamics, showing slow relaxation toward a stable state far beyond the single generation timescale (c. 100 generations); the expression level of the rewired gene also exhibits slow collective modes (10–20 generations) that relax after about 100 generations. Interestingly, Braun (2015, p. 20) provides indirect evidence suggesting that the intercellular environment, not cellular signals, might act as a relay to synchronize cells in the population.

Arguably, these experiments are targeted at challenging the cells at the level of their gene regulatory system, precisely to investigate how cells can respond to such rewiring events when they naturally occur in the course of evolution—that is, not so rarely (Braun, 2015, p. 5). The fact that rewiring events are not rare suggests that cells may have physiological responses handling (or taking advantage of) them. This might be why in these experiments genetic mutations seem to play such a minute role. This might also be why adaptation is so fast and frequent in contrast to other experiments with other kinds of severe stresses (e.g., Balaban, Merrin, Chait, Kowalik, & Leibler, 2004). In any case, these experiments open fascinating avenues that strongly challenge the current view of adaptation.

3.2.2. Beyond the Central Dogma?

So far, the mechanisms discussed made no mention of nonblind genetic variation. One classical reason to suppose that genetic variation is blind is that it seems implausible that a living system, when facing a new challenge, is able to compute a gene product that would solve the problem and, from this, to reverse-translate the gene product so as to build a corresponding coding sequence. The impossibility of reverse-translation corresponds to the central dogma (Crick, 1958).[6] The mechanism, however, seems less implausible when one considers variation at the level of regulatory profiles. This is because regulation both offers unbounded variation[7] (different expression profiles can lead to drastically different phenotypes, such as lung or liver cells) and yet relies on somehow simple and quantitative, finitely describable, variation—to caricature, the change in the constitutive expression level of a gene could for instance be described as an increase, decrease, or status quo. Such a simple description of variation suggests that there is no principled reason to suppose that a cell cannot nonblindly modify the genetic sequence in the regulatory regions, for instance under the influence of epigenetic marks that lie close to these regions. Notice however that the central dogma would remain intact, as it concerns *coding* regions. Such nonblind mutation might occur in particular during the stabilization phase, in which case it could be called "genetic memorization."

3.2.3. Embedded Exploration

In the family of models exposed above, our model of heritable physiological exploration consists in an extension of Jablonka and Lamb's (1995) and Danchin et al.'s (2017) models, now including an explicit intragenerational

---

6. Some have argued that reverse-translation is possible (Nashimoto, 2001; but see Cook, 1977), but we set this discussion aside here.

7. Unbounded variation means here a kind of variation that cannot be described in advance in finite terms (Longo, Montévil, & Kauffman, 2012).

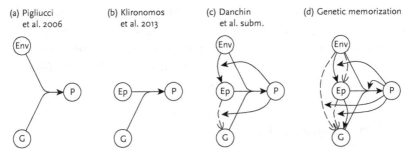

FIGURE 3.2 Models of genetic assimilation at the developmental timescale. G: genetic variation, Env: environment, P: phenotype, Ep: epigenetic variation. Continuous lines: directed effects. Dashed lines: blind effects. (a) Classical model of genetic assimilation tracing back to Baldwin and Waddington (Pigliucci et al., 2006). (b) Genetic assimilation through epigenetic variation (Klironomos et al., 2013). (c) Genetic assimilation through context-dependent epigenetic induction and mutagenicity; notice the context-dependence, represented by the feedbacks from P (Danchin et al., 2017). (d) Genetic memorization, where the phases of exploration (additional blind effects) and stabilization (additional directed effects) are represented. Intraorganismal amplification through intercellular communication and/or somatic selection is not represented.

phase enabling the production of nonblind genetic variation in the offspring (Figure 3.2). Notice, however, that genetic (and possibly epigenetic) variation in the exploratory phase by the cells is still blind, letting, in the end, pure chance fueling the developmental and evolutionary processes with successful variations. (This is the case even in the hypothesis of "genetic memorization"). Blind variation is selected by the cells (or the organism) themselves through their physiological response. This kind of selection is informed, possibly through previous adaptation or exaptation; that is, cells might know how to select their physiological condition because such an ability could have been selected in the past, or cells might have this ability as an exaptation of already existing physiology. It is also possible that blind variation be selected by natural selection occurring at the intraorganismal level, for instance between cells. Thus, "selection" (of now two sorts: *by* the cells and *of* the cells) of blind variation is still the only explanatory resource accounting for adaptation (but now at the individual level of phenotypic exploration and accommodation).

The immediate evolutionary consequence of such a mechanism is that the adaptive process is now embedded at the individual level, considerably accelerating and widening the evolutionary process. We can draw a comparison with learning: a well-known role of learning is to embed the trial-and-error phase at the individual level, enabling adaptation (or accommodation) at the scale of the individual lifetime (Hinton & Nowlan, 1987). Now, the resulting accommodation can sometimes be transmitted to the offspring. This might lead to a nontrivial multiscale view of adaptation, with multiple relevant evolutionary processes occurring at different timescales, in a possibly intricate way. Eventually, our understanding of adaptation in terms of fitness might even be altered, as fitness might vary nontrivially across timescales (Pocheville, 2010; Braun, 2015). We leave these exciting avenues for future work.

3.3. What Blindness Really Is

In the discussion above, we highlighted the central role of blindness in the models of genetic assimilation and heritable physiological exploration. We now wish to give a formal discussion of the concept of blindness to better characterize its epistemic role in evolutionary explanations (see also Huneman, this volume, section 2.4). We argue that blindness constitutes as much an explanatory principle as an empirical claim.

3.3.1. Definitions of Blindness

Neo-Darwinism did not come up with a definitive concept of blind, or "random," variation. Authors of the modern synthesis, for instance, did not define the words "random" or "chance" mutation, and used these terms in a variety of ways, fostering misunderstanding of their theses (Merlin, 2010, p. 3). It is worth recalling here what the term "random mutation" does *not* mean in the modern synthesis (Sober, 1984, p. 104; Merlin, 2010). Random mutations are not mutations that are all equally probable, or inherently unpredictable, or even causally independent from the environment (Sniegowski & Lenski, 1995, p. 572). Nor does randomness mean that mutations will be equally likely to turn out beneficial, deleterious, or neutral. Authors of the modern synthesis, and before them, early neo-Darwinians such as Weissman (1893), generally did not hold such claims. Such claims would have been, in any case, refuted by the (now common) knowledge of the induction of mutation by physical or chemical agents (Benzer & Freese, 1958), the existence of mutation hot-spots (Moxon et al., 1994, 2006), or the common assumption that most mutations are deleterious—this assumption is, of course, context-dependent (e.g., Simpson, 1944, pp. 55–56, quoted by Merlin, 2010, p. 20).

To avoid these ambiguities, mutations are sometimes said to be random with respect to their adaptive value: "A central tenet of evolutionary theory is that mutation is random with respect to its adaptive consequences for individual organisms; that is, the production of variation precedes and does not cause adaptation" (e.g., Sniegowski & Lenski, 1995, p. 553). We take this to mean that there is a statistical and/or causal independence between the probability of occurrence of a mutation and its fitness in a given environment. However, as we will see in what follows, such an independence criterion is not yet precise enough to capture the notion of blind variation. Furthermore, because it is not clear what probability space(s) the term "randomness" is supposed to refer to (i.e., the possible variations and their probability distributions), we prefer to avoid the term and use "blindness" as a default instead.

Merlin (2010) provides a useful review of several concepts of blind variation. They can be classified according to whether they emphasize a statistical or a causal independence criterion between mutation and fitness (Table 3.1). The intuitive idea grounding these criteria is that, for variation to be blind, we either do not want more beneficial variations to be more probable in a given environment, or, if variation presents such a pattern in a given environment, at least we do not want this pattern to be consistently conserved across different environments when the selective challenge changes (Figure 3.3).

In formal terms, we can characterize the statistical criterion of blindness as follows. Let us assume that we are given measures for the heritable variation v and the environment e, and a probability function $p(v,e)$ giving the probabilities of the different variations in the different environments. Let us also assume that a fitness function $w(v,e)$ can be defined. Then the statistical criterion of blindness (in Merlin's version, Table 3.1, with slight modifications) posits that the probability $p(v_i, e_j)$ of occurrence of v_i in e_j is blind to $w(v_i, e_j)$ when at least one of the following criteria is fulfilled:

1. The same variation is not relatively more probable in environments where it is relatively more beneficial: $\mathrm{cov}^{v=v_i}_{e=e_1,\dots,e_m}(w(v,e),p(v|e))=0$. Here the covariance is computed for a given variation in different environments (environments being equally weighted).[8] For nonblindness to be weak or strong, this criterion would have to be refuted for respectively some or most variations.

2. A variation is not more probable than other less beneficial variations, in a given environment: $\mathrm{cov}^{v=v_1,\dots,v_n}_{e=e_j}(w(v,e),p(v|e))=0$. Now the covariance is computed for different variations in a given environment (variations being equally weighted). For nonblindness to be weak or strong, this criterion would have to be refuted in respectively some or most environments.

This formalism is purposely simple for illustration and of course comes with the limitations of the covariance formalism. In particular, such criteria do not reflect

---

8. To avoid capturing nonrelevant effects such as some environments raising the fitness or probability of *all* variations, or increasing the variances of the variables w and p without any skew, w and p in a given environment have to be standardized by their mean and standard deviation across all variations in that environment (variations being equally weighted).

Table 3.1 Criteria of Blind Variation

Statistical criterion	Causal criterion
"[T]he term 'randomness' as applied to mutation often refers to the lack of correspondence of phenotypic effect with the stimulus and with the actual or the adaptive direction of evolution." (Simpson, 1953b, pp. 86–87) A mutation is directed "if it occurs (or occurs more frequently) in the fitness-enhancing or 'selective' environment," i.e., "in an environment where its associated phenotype has an enhanced fitness." (Sarkar, 1991, quoted by Merlin 2010, p. 20n11) "A mutation is 'directed' if and only if it fulfills the two following conditions: (1) It is more probable in an environment where it is beneficial than in another environment where it is deleterious or neutral (2) It is clearly more probable in an environment where it is beneficial than other deleterious or neutral mutations (in the same environment)" (Merlin, 2010, p. 7)*	"The defensible idea in the claim that mutation is random is simply that mutations do not occur *because* they would be beneficial. (. . .) It is a little misleading to summarize this result by saying that 'mutations occur at random.' One might just as well say that the weather occurs at random, since rain doesn't fall because it would be beneficial." (Sober, 1984, p. 105, italics in original) "We define as directed a mutation that occurs at a higher rate specifically when (and even because) it is advantageous to the organism, whereas comparable increases in rate do not occur either (i) in the same environment for similar mutations that are not advantageous or (ii) for the same mutation in similar environments where it is not advantageous." (Lenski & Mittler, 1993, p. 188) "[A] mutation is directed if and only if it is specifically caused by environmental stress in an exclusively adaptive manner." (Millstein, 1997, p. 151)

* Merlin (2010) does provide a statistical criterion but frames her discussion in causal terms: "According to the most influential and persistent meaning in biology, variations are not caused because they can provide adaptation to the individual organisms concerned and their offspring" (p. 20, n4, see also her n11 criticizing Sarkar's [1991] criterion for being statistical). This slight discrepancy probably comes from the difficulty to frame a causal independence criterion in verbal terms when probabilities come into the dance, hence the proposal made here.

statistical independence per se, but the less stringent criterion of an absence of *direction* in the relationship between the variables w and p. In any case, we are confident that other formalisms would not change the rest of the argument.[9]

Arguably, a statistical criterion of blindness is not enough to speak of blindness in causal terms. This is because confounding factors can blur or create any directional relationship between variables. To be able to speak of causation, one has to intervene on the system, so as to break any spurious (non)-directional relationship between variables.[10] A causal equivalent of the formal statistical criterion of blindness can be obtained by intervening on the environment to artificially change the selective pressures at play, while keeping the rest of the system constant. In equations, the fact that the value of the environment is now fixed by an intervention can be written by putting a hat ^ on the variable e (this is a common notation). Then, the two criteria above become:

(1) $\text{cov}_{e=e_1,\dots,e_m}^{v=v_i}(w(v,\hat{e}), p(v|\hat{e})) = 0$; the covariance is computed for a given variation in different environments each fixed by an intervention; variables are standardized as above.

(2) $\text{cov}_{e=e_j}^{v=v_1,\dots,v_n}(w(v,\hat{e}), p(v|\hat{e})) = 0$; the covariance is now computed for different variations in a given environment (fixed by an intervention).

These criteria come close to what would be a causal criterion of blindness, that is, a formal criterion translating the idea that a heritable variation does not occur because of its adaptive value.

3.3.2. Levels of Blindness

The criteria (1) and (2) of blindness can be applied at several levels. For instance, we can speak of genetic variation being blind at the level of the genome or at the level of the locus (up to the level of the nucleotide). Imagine a mutator mechanism that changes the probability of several loci to be mutated in different environments in an adaptive way (as we proposed above), but that leaves the relative probability distribution of mutations unchanged for each

---

9. Notice that these criteria will be violated as well in the case of nonblindness for worse (that is, in short, when more beneficial variations are less probable).

10. We embrace here the interventionist view of causation, which we find the most operative for experimental sciences. See Woodward (2003) for a philosophical introduction; Pearl (2009) for a formal introduction; and Griffiths et al. (2015) and Pocheville (under review) for formal measures of causation applied to biology.

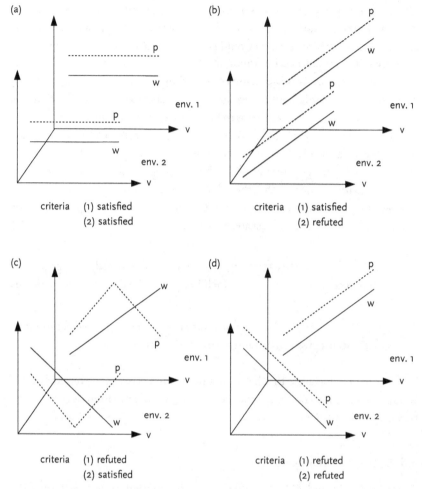

FIGURE 3.3 Four situations illustrating the criteria of nonblind variation. v: variations; p, w: probability and fitness of each variation in a given environment; env: environment. (a) (b) (c) illustrate blind variation (d) illustrates nonblind variation. (a) The probabilities and fitnesses of the variations are not context dependent (both criteria are satisfied). (b) In a given environment beneficial variations are more probable (criterion 2 refuted), but for a given variation its probability of occurrence does not depend on its fitness across environments in which the living system can live, thus the system cannot be said to non-blindly respond to these environments (criterion 1 satisfied). (c) For a given variation, the probability of occurrence does depend on its fitness across environments (criterion 1 refuted), however in a given environment more beneficial variations are not more probable than other variations (criterion 2 satisfied). (d) In a given environment more beneficial variations are more probable (criterion 2 refuted) *and* for a given variation, its probability of occurrence is higher in environments where it is more beneficial (criterion 1 refuted). This last situation illustrates nonblind variation.

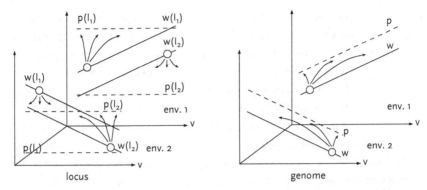

FIGURE 3.4 Blind variation at the locus level leading to nonblind variation at the genomic level. A living system is put in environment 1, where its current variation at locus 1 does worse than average, while its current variation at locus 2 does better than average. By increasing the mutation rate preferentially at locus 1, the living system would increase its chances of doing better while avoiding the cost of mutating locus 2, even when mutation is blind at locus 1. The situation would be similar in environment 2, if the variation at locus 2 were now the one that did worse than average.

locus. Such a mechanism would be nonblind at the level of the genome, but still blind at the level of the locus (Figure 3.4). This kind of blindness can be formally expressed by conditioning on the locus of study: we would simply replace $p(v_i|\widehat{e}_j)$ by $p(v_i|\widehat{e}_j,\widehat{l}_k)$ in the criteria above, where l_k is the locus producing variation v_i, and we would restrict the formula to the variation produced at that given locus. From a mechanistic point of view, such a conditioning enables us to show how, for instance, nonblind variation at the level of the genome can be obtained without refuting the central dogma at the level of the gene (as far as coding regions are concerned). From an evolutionary point of view however, the fact that variation is nonblind at a given level remains significant.

3.3.3. Epistemological Status of Blindness

There is an epistemological difficulty in the causal criteria of blindness stated previously that is worth spelling out. Usually, when we say that a putative cause has no effect on a putative effect, we mean that there is no relation between any intervention on the cause and the value obtained for the effect (this would, for instance, result in a null covariance between the manipulated cause and the effect).

Now, when we say that variations do not occur because of their fitness, the putative cause in question is, supposedly, fitness, the putative effect being the probability of occurrence of variations. However, the putative cause bearing a hat in our formula above is the environment, not fitness. That is, both terms in the covariance (fitness and probability of occurrence) are now effects of a putative intervention on the environment. Ideally, we might have desired to contrast a term reflecting the cause only, bearing a hat (i.e., fitness) and a term being the effect only, with no hat (i.e., the probability distribution of variations).

However, since fitness is a function of the variation and the environment, it is impossible to manipulate the fitness of a given variation independently from the environment, and thus without potentially affecting at the same time the probability of occurrence of that variation in that environment.[11] In formal terms, fitness is not "modular" with respect to variation and the environment.[12] Thus, we cannot check whether fitness is a cause increasing the probability of occurrence of variations. This means that one cannot distinguish the direct effects of the environment (Darwin, 1859, p. 11) on the probability distribution of variations from the effects of an adaptive system that senses the environment and produces variations according to fitnesses in that environment.[13] Thus, we cannot distinguish fitness from the environment as a cause of nonblind variation. Statements such as mutations not occurring "because they would be beneficial" (Sober, 1984, p. 105), or "advantageous to the organism" (Lenski & Mittler, 1993, p. 188) are, taken literally, causal claims where the cause is not open to manipulation. They are useful as catchphrases, but they are empirically empty.[14]

By contrast, the causal criteria of blindness given above constitute an empirically testable claim: variations can surely occur because of the environment, but any directional association (if any) between the fitness and the probability of variations should not be consistently conserved when drastically changing the adaptive challenge.

There is, however, more to the tenet of blind variation than a mere empirical claim. The tenet of blind variation is also an explanatory principle. This principle is that we do not need to suppose that adaptive variations *have a special reason* to

---

11. Paul Griffiths, whom we warmly thank, directed our attention to the fact that Waddington precisely seems to be able to manipulate the fitness independently of the environment. Indeed, Waddington arbitrarily selects the crossveinless phenotype, which he supposes is not adaptive to high temperatures (Waddington, 1953, p. 124). The contradiction stems from an ambiguity in the term "fitness." Waddington, say, is external to his experimental set-up. What he does, then, is not to modify the adaptive value (engineering fitness) of the crossveinless phenotype, but rather the realized reproductive success. The latter is only a proxy for the former and there is no mystery that a proxy can be manipulated independently of its target.

12. On modularity see Woodward (2003, p. 329), Pearl (2009, p. 42), Pocheville and Montévil (in prep.).

13. We thank Philippe Huneman for this formulation.

14. This consideration, of course, holds only insofar as we adopt an interventionist stance on causal claims.

be produced, that is, we do not need to suppose any other explanation of design than natural selection, because natural selection is enough. This is, for instance, how Sniegowski and Lenski (1995, p. 556) read (in part) the wide acceptance of the hypothesis of random mutations among geneticists and evolutionary biologists in the 1930s, before direct experimental evidence. The virtue of such an explanatory principle is that it provides a kind of teleological void from which teleological explanations can be derived. This principle is an explanatory choice, that is, it consists in segregating out of the theory other (compatible) explanatory schemes. It is not an empirical claim, but it is not independent from empirical results either. For, of course, if we were to find consistent nonblind variation (as defined above) across various circumstances, we would have to seek alternative or complementary explanatory schemes for this fact.

3.3.4. Implausibility of Nonblindness

It has been claimed that finding consistent nonblind heritable variation in the living is implausible, particularly as regards genetic variation.

A first reason is the implausibility that a physiological mechanism producing nonblind variation be *generated* in evolutionary history. The problem of finding (in advance) which variation will be beneficial seems an uncomputable problem for a living system (Merlin, 2010, p. 15). Rephrased in our own terms, the objection is that living systems seem to be complex systems in somehow critical situations, where small changes can have drastic effects, immersed in spatially and temporally heterogeneous (and similarly unpredictable) environments. However, strictly speaking our theories of biological functioning are too scarce yet to decide what is or is not computable for living systems. At least, it seems that the recruitment of similar mechanisms for plasticity and adaptive processes may provide exaptative pathways to mechanisms of nonblind variation, in particular as regards the regulatory patterns (Jablonka, 2013; Danchin & Pocheville, 2014).

A second reason to dismiss the possibility of nonblind variation is that it might not be *evolutionarily advantageous*. A mechanism linking heritable variation to the state of the environment could jeopardize the heritable materials. For instance, the sequestration of heritable materials from external influences provides protection against noisy environments (Piraveenan, Polani, & Prokopenko, 2007). This sequestration hypothesis, however, does not preclude the possibility of a beneficial directed influence from the environment in stressful situations. Actually, we would argue that nonblindness should constitute a large evolutionary attractor: mechanisms of heritable nonblind variation, once they have come into existence (for instance as an exaptation of plasticity), should tend to be selected because of their positive effects, which may also result in a virtuous circle if these mechanisms happened to favor the improvement of nonblindness itself. Nonblindness may thus be advantageous at the level of first order (nonblind variation) and second order (nonblind evolvability) evolutionary processes. We are not going to defend such an optimistic view here; suffice it to remark that the sequestration of heritable materials (if any sequestration) may also be discriminatory in a beneficial manner, and that nonblindness may indeed be evolutionarily advantageous. This line of reasoning, when pushed to the limit, could as well lead to interpreting any occurrence of nonblind variation as the result of past selective processes (on blind variation of variation-producing mechanisms).

Yet another argument against nonblind variation is one of epistemological implausibility. There are indeed good reasons to consider that assessing which variations are possible *in advance*, as well as their respective adaptive (or physiological) value, is an uncomputable problem for us (though not necessarily for the living system) in the current state of science (Longo et al., 2012). If listing possible and beneficial variations is uncomputable for us, a handy hypothesis to model evolution is that of blind variation. We think this is why, to our knowledge, all models of heritable nonblind variation in fine rely on blind variation of some sort.

3.5. Conclusion

In this chapter, we emphasized the compatibility of most models of genetic assimilation with the most orthodox neo-Darwinism. We also proposed slightly more challenging views.

Historically, the Baldwin effect and genetic assimilation have been designed as a defense of neo-Darwinism, even if neo-Lamarckians and others may have intended to use it as an argument for their own views. These mechanisms indeed rely on selection of heritable blind variation to produce heritable adaptive change. In the current view, genetic assimilation relies on a hypothetical redundancy between the plastic response and the genetic variation, which enables classical selection (or drift) to take place on the reaction norm (Pigliucci et al., 2006). In Klironomos et al.'s (2013) model, heritable epigenetic variation now plays the role previously played by intragenerational plasticity, but the role of natural selection acting on blind variation (epigenetic or genetic) is intact. In Jablonka and Lamb's (1995) and Danchin et al.'s (2017) models, the case is more subtle. Due to the inducibility of epigenetic variation, which leads to targeted mutagenicity at the locus level, genetic variation is now nonblind at the genomic level. As we have seen, genetic variation is still blind at the locus level, however, and the central dogma is left unaffected.

We proposed a more stringent model of genetic assimilation where genetic variation is not necessarily blind, thanks to exploration and encoding at the individual level. This model relies on a central role of regulation in evolution and development, postulating that regulation enables unbounded variation to emerge from somehow bounded variation. Yet, the role of blind variation remains central: it is selection of blind variation (at the individual level) that explains how physiological exploration can be successful.

This permanence of blind variation in adaptive explanations is not fortuitous. It comes from the fact that blindness is epistemically economical, enabling us to explain adaptations that literally come from nowhere, in that blindness is precisely that which does not seem to require an explanation. This is probably where the core of the neo-Darwinian paradigm lies.

In general, we are confident that the empirical implausibility of nonblindness will be refuted by biology—there might well be some patterns of nonblind variation in some specific, or not-so-specific, situations. But understanding these patterns with mechanisms that do not rely in fine on blind variation will require us to get out of our current theoretical conceptions of the living.

Acknowledgments

We thank Pierrick Bourrat, Erez Braun, Brett Calcott, Sébastien Dutreuil, Paul Griffiths, Francesca Merlin, Maël Montévil, Aurélien Pocheville, Gaëlle Pontarotti, Livio Riboli-Sasco, Isobel Ronai, Karola Stotz, Elena Walsh, and especially Philippe Huneman, for discussions and thorough readings of a previous version that enabled us to greatly improve the manuscript. Aurélien Pocheville attracted our attention early on to the potential virtuous circle of nonblind variation. This work stemmed from a discussion with Jean Deutsch at the ISHPSSB Meeting, Montpellier, July 10, 2013. We benefited from fruitful feedback from the audience at the following meetings: Congrès de la Société de Philosophie des Sciences, Lille, June 26, 2014; Euro Evo-Devo Meeting, University of Vienna, July 24, 2014; Workshop "How Can We Redefine Inheritance Beyond the Gene-Centered Approach?," IHPST, Paris, October 3, 2014; Reading group with the team Theory and Methods in Bioscience, Charles Perkins Centre, Sydney, October 12, 2016. This work was supported by the Laboratoire d'Excellence (LABEX) entitled TULIP (ANR-10-LABX-41), as well as the Soc-H2 ANR project (ANR-13-BSV7-0007-01). We thank Paul Griffiths for supporting AP's participation to the 2014 Euro Evo-Devo Meeting. This publication was made possible through the support of a grant from the Templeton World Charity Foundation. The opinions expressedin this publication are those of the authors and do not necessarily reflect the views of the Templeton World Charity Foundation.

References

Ay, N., & Polani, D. (2008). Information flows in causal networks. *Advances in Complex Systems*, *11*, 17–41.

Balaban, N. Q., Merrin, J., Chait, R., Kowalik, L., & Leibler, S. (2004). Bacterial persistence as a phenotypic switch. *Science*, *305*, 1622–1625.

Baldwin, J. M. (1896). A new factor in evolution. *The American Naturalist*, *30*, 441–451.

Beatty, J. (1997). Why do biologists argue like they do? *Philosophy of Science*, S432–S443.

Benzer, S., & Freese, E. (1958). Induction of specific mutations with 5-bromouracil. *Proceedings of the National Academy of Sciences U. S. A.*, *44*, 112.

Bohacek, J., & Mansuy, I. M. (2015). Molecular insights into transgenerational nongenetic inheritance of acquired behaviours. *Nature Reviews Genetics*, *16*, 641–652. doi:10.1038/nrg3964

Braun, E. (2015). The unforeseen challenge: From genotype-to-phenotype in cell populations. *Reports on Progress in Physics*, *78*, 1–51.

Braun, Erez, & Lior David. (2011). The role of cellular plasticity in the evolution of regulatory novelty. In S. B. Gissis, & E. Jablonka (Eds.), *Transformations of lamarckism: From subtle fluids to molecular biology* (pp. 181–191). Cambridge, Massachusetts: MIT Press.

Burian, R. M. (1983). Adaptation. In M. Grene (Ed.), *Dimensions of Darwinism* (pp. 287–314). New York, NY: Cambridge University Press.

Cairns, J., & Foster, P. L. (1991). Adaptive reversion of a frameshift mutation in Escherichia coli. *Genetics*, *128*, 695–701.

Cairns, J., Julie, O., & Stephan, M. (1988). The origin of mutants. *Nature*, *335*, 142–145.

Cook, N. D. (1977). The case for reverse translation. *Journal of Theoretical Biology*, *64*, 113–135. doi:10.1016/0022-5193(77)90116-3

Crick, F. H. (1958). *On protein synthesis.* Symposium of the Society for Experimental Biology. New York, NY: Academic Press.

Danchin, É., & Pocheville, A. (2014). Inheritance is where physiology meets evolution. *The Journal of Physiology*, *592*, 2307–2317. doi:10.1113/jphysiol.2014.272096

Danchin, E., Pocheville, A., Rey, O., Pujol, B., & Blanchet, S. (2017). Epigenetics as the hub guiding nongenetic inheritance to genetic assimilation. Manuscript submitted for publication.

Darwin, C. R. (1868). *The variation of animals and plants under domestication* (1st ed.). London: John Murray.

Darwin, C. R. (1859). *On the origin of species by means of natural selection, or the preservation of favoured races in the struggle for life* (1st ed.). London: John Murray.

Davis, B. D. (1989). Transcriptional bias: A non-Lamarckian mechanism for substrate-induced mutations. *Proceedings of the National Academy of Sciences*, *86*, 5005–5009.

De Jong, G. (2005). Evolution of phenotypic plasticity: Patterns of plasticity and the emergence of ecotypes. *New Phytologist*, *166*, 101–118. doi:10.1111/j.1469-8137.2005.01322.x

Depew, D. J. (2003). Baldwin and his many effects. In B. H. Weber & D. J. Depew (Eds.), *Evolution and learning: The Baldwin effect reconsidered* (pp. 3–31). Cambridge, MA: MIT Press.

Devanapally, S., Ravikumar, S., & Jose, A. M. (2015). Double-stranded RNA made in *C. elegans* neurons can enter the germline and cause transgenerational gene silencing. *Proceedings of the National Academy of Sciences.*, *112*, 2133–2138.

Foster, P. L., & Cairns, J. (1992). Mechanisms of directed mutation. *Genetics*, *131*, 783–789.

Gorelick, R. (2003). Evolution of dioecy and sex chromosomes via methylation driving Muller's ratchet. *Biological Journal of the Linnean Society*, *80*, 353–368.

Griffiths, P. E. (2003). Beyond the Baldwin effect: James Mark Baldwin's 'social heredity', epigenetic inheritance, and niche construction. In H. W. Bruce, & J. D. David (Eds.), *Evolution and learning: The Baldwin effect reconsidered* (pp. 193–216). Cambridge, Massachusetts, and London, England: MIT Press.

Griffiths, P. E., Pocheville, A., Calcott, B., Stotz, K., Kim, H., & Knight, R. (2015). Measuring Causal Specificity. *Philosophy of Science*, *82*, 529–555.

Hinton, G. E., & Nowlan, S. J. (1987). How learning can guide evolution. *Complex Systems*, *1*, 495–502.

Hoquet, T. (2009). *Darwin contre Darwin: Comment lire L'Origine des espèces?* Paris: Seuil.

Jablonka, E. (2013). Epigenetic inheritance and plasticity: the responsive germline. *Progress in Biophysics and Molecular Biology*, *111*, 99–107. doi:10.1016/j.pbiomolbio.2012.08.014

Jablonka, E., & Marion J. L. (2010). Transgenerational epigenetic inheritance. In M. Pigliucci, & G. B. Müller (Eds.), *Evolution: The extended synthesis* (pp. 137–174). Cambridge, Massachusetts, and London, England: MIT Press.

Jablonka, E., & Lamb, M. J. (2005). *Evolution in four dimensions: Genetic, epigenetic, behavioral, and symbolic variation in the history of life.* Cambridge, MA: MIT Press.

Jablonka, E., & Lamb, M. J. (1995). *Epigenetic inheritance and evolution: The Lamarckian dimension.* Oxford: Oxford University Press.

King, M.-C., & Wilson, A. C. (1975). Evolution at two levels in humans and chimpanzees. *Science*, *188*, 107–116.

Kirschner, M. W., & Gerhart, J. C. (2006). *The plausibility of life: Resolving Darwin's dilemma.* New Haven, CT: Yale University Press.

Klironomos, F. D., Berg, J., & Collins, S. (2013). How epigenetic mutations can affect genetic evolution: Model and mechanism. *BioEssays*, *35*, 571–578.

Lamm, E., & Jablonka, E. (2008). The nurture of nature: Hereditary plasticity in evolution. *Philosophical Psychology*, *21*, 305–319.

Lenski, R. E., & Mittler, J. E. (1993). The directed mutation controversy and neo-Darwinism. *Science*, *259*, 188–194.

Loison, L. (2010). *Qu'est ce que le néolamarckisme?* Paris: Vuibert.

Longo, G., Montévil, M., & Kauffman, S. (2012). No entailing laws, but enablement in the evolution of the biosphere. In *Proceedings of the Fourteenth International Conference on Genetic and Evolutionary Computation Conference Companion* (pp. 1379–1392). New York, NY: GECCO Companion '12. doi:10.1145/2330784.2330946

Lu, Q., & Bourrat, P. (in press). The evolutionary gene and the extended evolutionary synthesis. *The British Journal for the Philosophy of Science*. doi:10.1093/bjps/axw035

Mayr, E., & Provine, W. B. (1998). *The evolutionary synthesis: Perspectives on the unification of biology, with a new preface*. Cambridge, MA: Harvard University Press.

Merlin, F. (2010). Evolutionary chance mutation: a defense of the modern synthesis' consensus view. *Philosophy and Theory in Biology*, 2. doi:http://dx.doi.org/10.3998/ptb.6959004.0002.003

Millstein, R. L. (1997). *The chances of evolution: An analysis of the roles of chance in microevolution and macroevolution*. PhD thesis, University of Minnesota.

Morgan, C. L. (1896). On modification and variation. *Science*, 4, 733–740.

Moxon, E. R., Rainey, P. B., Nowak, M. A., & Lenski, R. E. (1994). Adaptive evolution of highly mutable loci in pathogenic bacteria. *Current Biology*, 4, 24–33.

Moxon, R., Bayliss, C., & Hood, D. (2006). Bacterial contingency loci: The role of simple sequence DNA repeats in bacterial adaptation. *Annual Review of Genetics*, 40, 307–333.

Nashimoto, M. (2001). The RNA/protein symmetry hypothesis: experimental support for reverse translation of primitive proteins. *Journal of Theoretical Biology*, 209, 181–187. doi:10.1006/jtbi.2000.2253

Osborn, H. F. (1897a). Organic selection. *Science*, 15, 583–587.

Osborn, H. F. (1897b). The limits of organic selection. *American Naturalist*, 31, 944–951.

Pearl, J. (2009). *Causality: Models, reasoning and inference* (2nd ed.). New York: Cambridge University Press.

Pigliucci, M., & Müller, G. B. (2010). *Evolution: The extended synthesis*. Cambridge, MA: MIT Press.

Pigliucci, M., & Murren, C. J. (2003). Perspective: Genetic assimilation and a possible evolutionary paradox: Can macroevolution sometimes be so fast as to pass us by? *Evolution*, 57, 1455–1464.

Pigliucci, M., Murren, C. J., & Schlichting, C. D. (2006). Phenotypic plasticity and evolution by genetic assimilation. *Journal of Experimental Biology*, 209, 2362–2367.

Piraveenan, M., Polani, D., Prokopenko, M. (2007). Emergence of genetic coding: an information-theoretic model. In F. Almeida e Costa, L. M. Rocha, E. Costa, I. Harvey, & A. Coutinho (Eds.), *Advances in Artificial Life*. ECAL 2007. Lecture Notes in Computer Science, vol 4648. Springer, Berlin, Heidelberg.

Pocheville, A. (2010). What niche construction is (not). In *La niche ecologique: Concepts, modèles, applications* (pp. 39–124). Paris: Ecole Normale Supérieure Paris. (PhD Thesis).

Pocheville, A., Griffiths, P. E., & Stotz, K. (in press). Comparing causes—an information-theoretic approach to specificity, proportionality and stability. In H. Leitgeb, I. Niiniluoto, E. Sober, & Seppälä, Päivi (Eds.), *Proceedings of the 15th Congress of Logic, Methodology and Philosophy of Science*. London: College Publications.

Pocheville, A., & Montévil, M. (2014). Ecological models for gene therapy: I. Models for intraorganismal ecology. *Biological Theory*, 9, 401–413. doi:10.1007/s13752-014-0191-x

Rodgers, A. B., Morgan, C. P., Leu, N. A., & Bale, T. L. (2015). Transgenerational epigenetic programming via sperm microRNA recapitulates effects of paternal stress. *Proceedings of the National Academy of Sciences*, 112, 13699–13704.

Romanes, G. J. (1888). Lamarckism versus Darwinism. *Nature*, 38, 413–413. doi:10.1038/038413a0

Sarkar, S. (1991). Lamarck contre Darwin, reduction versus statistics: Conceptual issues in the controversy over directed mutagenesis in bacteria. In A. I. Tauber (Ed.), *Organism and the origins of self, Boston studies in the philosophy of science* (pp. 235–271). Boston Studies in the Philosophy of Science 129. Dordrecht, The Netherlands: Kluwer Academic Publishers. http://link.springer.com/chapter/10.1007/978-94-011-3406-4_11

Schmalhausen, I. I. (1949). *Factors of evolution: The theory of stabilizing selection.* Philadelphia: Blakiston.

Sharma, A. (2015). Transgenerational epigenetic inheritance requires a much deeper analysis. *Trends in Molecular Medicine*, 21, 269–270.

Simpson, G. G. (1953a). The Baldwin effect. *Evolution*, 7, 110–117.

Simpson, G. G. (1953b). *The major features of evolution.* New York: Columbia University Press.

Simpson, G. G. (1944), *Tempo and mode in evolution.* New York: Columbia University Press.

Sniegowski, P. D., & Lenski, R. E. (1995). Mutation and adaptation: the directed mutation controversy in evolutionary perspective. *Annual Review of Ecology and Systematics*, 26, 553–578.

Sober, E. (1984). *The nature of selection: Evolutionary theory in philosophical focus.* London: University of Chicago Press.

Soen, Y., Knafo, M., & Elgart, M. (2015). A principle of organization which facilitates broad Lamarckian-like adaptations by improvisation. *Biol. Direct*, 10, 1–17.

Spalding, D. A. (1873). Instinct: with original observations on young animals. *The British Journal of Animal Behaviour*, 2, 2–11.

Stahl, F. W. (1988). Bacterial genetics: A unicorn in the garden. *Nature*, 335, 112.

Stern, S., Fridmann-Sirkis, Y., Braun, E., & Soen, Y. (2012). Epigenetically heritable alteration of fly development in response to toxic challenge. *Cell Reports*, 1, 528–542.

Stern, S., Snir, O., Mizrachi, E., Galili, M., Zaltsman, I., & Soen, Y. (2014). Reduction in maternal polycomb levels contributes to transgenerational inheritance of a response to toxic stress in flies. *The Journal of Physiology, 592*, 2343–2355. doi:10.1113/jphysiol.2014.271445

Szyf, M. (2015). Nongenetic inheritance and transgenerational epigenetics. *Trends in Molecular Medicine, 21*, 134–144.

Waddington, C. H. (1959). Canalization of development and genetic assimilation of acquired characters. *Nature, 183*, 1654–1655.

Waddington, C. H. (1956). Genetic assimilation of the bithorax phenotype. *Evolution, 10*, 1–13. doi:10.2307/2406091

Waddington, C. H. (1953). Genetic assimilation of an acquired character. *Evolution, 7*, 118–126.

Waddington, C. H. (1952). Selection of the genetic basis for an acquired character. *Nature, 169*, 625–626. doi:10.1038/169625b0

Waddington, C. H. (1942). Canalization of development and the inheritance of acquired characters. *Nature, 150*, 563–565.

Waters, C. K. (2007). Causes that make a difference. *The Journal of Philosophy, 104*, 551–579.

Weismann, A. (1893). *The germ-plasm a theory of heredity*. Translated by N. W. Parker, & H. Rönnfeldt. New York: Charles Scribner's Sons. http://www.esp.org/books/weismann/germ-plasm/facsimile/

Weismann, A., & Spencer, H. (1893). *The all-sufficiency of natural selection: A reply to Herbert Spencer*. https://searchworks.stanford.edu/view/8611658

West-Eberhard, M. J. (2003). *Developmental plasticity and evolution*. New York: Oxford University Press.

Woodward, J. (2003). *Making things happen: A theory of causal explanation*. New York: Oxford University Press.

Wray, G. A. (2007). The evolutionary significance of cis-regulatory mutations. *Nature Reviews Genetics, 8*, 206–216.

Wright, B. E. (2000). A biochemical mechanism for nonrandom mutations and evolution. *Journal of Bacteriology, 182*, 2993–3001.

Wright, B. E., Longacre, A., & Reimers, J. M. (1999). Hypermutation in derepressed operons of Escherichia coli K12. *Proceedings of the National Academy of Sciences., 96*, 5089–5094.

4 EVOLUTIONARY THEORY EVOLVING

Patrick Bateson

Charles Darwin and Alfred Russel Wallace proposed a blind process for biological evolution that started with variation in characteristics between competitors, followed by differential survival and reproductive success and finally some means by which descendants of the successful organisms inherited their characteristics (see also Uller & Helanterä, this volume).

Darwin knew nothing of the mechanisms of inheritance. Gregor Mendel's experiments only became well known at the beginning of the 20th century, long after Darwin's death. By the 1930s, it was possible to put together Darwin's evolutionary mechanism with what was by then the understanding of the mechanisms of inheritance. This gave rise to what was known at the time as the "modern synthesis." On this view, differences between organisms in their genes gave rise to differences in their expressed characteristics. The most successful organism then left its genes to its descendants. So, it was argued, biological evolution involves changes in gene frequencies. This form of neo-Darwinism has held sway until the present and only recently has it begun to suffer from not keeping up with modern evidence (see Bateson, 2013; Laland et al., 2015).

The 1930s neo-Darwinist orthodoxy asserts that speciation results from a slow process of natural selection, that mutations in genes drive evolution with the organism passively playing no role, and that developmental processes are irrelevant to an understanding of evolution. In this chapter I examine all three assertions.

4.1. Is Speciation Slow and Gradual?

Even as Darwin's thinking hit the headlines in 1859, many of his friends doubted whether he had produced a theory that explained the transmutation of species. Natural selection provided a wonderful mechanism to explain adaptation and, as it happened, to discredit the natural theology of William Paley, which had been so popular with the intelligentsia since the beginning of the 19th century. But did this slow postulated process of adaptation explain the formation of a new species? Even Darwin's cousin Francis Galton imagined that speciation might be discontinuous. He wrote:

> The mechanical conception would be that of a rough stone, having, in consequence of its roughness, a vast number of natural facets, on any one of which it might rest in "stable" equilibrium. That is to say, when pushed it would somewhat yield, when pushed much harder it would again yield, but in a less degree; in either case, on the pressure being withdrawn it would fall back into its first position. But, if by a powerful effort the stone is compelled to overpass the limits of the facet on which it has hitherto found rest, it will tumble over into a new position of stability, whence just the same proceedings must be gone through as before, before it can be dislodged and rolled another step onwards. (Galton, 1892, pp. 354–355)

When Darwin was an undergraduate at Cambridge University his mentor was John Henslow, who was fascinated by "monstrosities," the appearance of sports markedly different from their lineages. Homoeotic changes in organisms were also the concern of William Bateson, a kinsman of mine. He coined the term "genetics" after the rediscovery of Mendel's papers, but several years before that he had published a book, *Materials for the Study of Variation Treated with Special Regard to Discontinuities in the Origin of Species* (W. Bateson, 1894). The second part of his title was regarded as a direct challenge to Darwinism and was much criticized by Darwin's supporters, though Bateson was not a critic of natural selection as a driver of adaptedness.

Despite the attacks on Bateson, many modern thinkers have been inclined to support his views that discontinuities do arise in evolution. Hosts of examples of big events having no effect and small events leading to big changes are to be found, and many of these are now formalized by the nonlinear mathematical techniques derived from catastrophe theory and chaos. A biography of William Bateson by the late Alan Cock and Donald Forsdyke (2008) used as its title a splendid piece of Bateson's advice to biologists: "Treasure your exceptions."

Discontinuities in evolution have been given especial prominence by some modern paleontologists who have been impressed by periods of stasis and sudden change in the fossil record (Eldredge, 1995; Gould, 2002). They suggested that after periods of stasis in evolution, sudden changes can occur in the fossil record and these may represent the appearance of new species. This idea of discontinuity has recurred periodically and, notably, was foreshadowed in the writings of Richard Goldschmidt (1940), who, in a memorable phrase, referred to a fresh arrival that might give rise to a new species as a "Hopeful Monster."

Even though the discontinuities in natural variation, which Bateson had documented so carefully, no longer pose a problem in developmental biology, hopeful monsters are disparaged to this day on the grounds that even if a big change in the phenotype could occur as a result of mutation, the hopeful monster would be a novelty on its own with no possibility of finding a mate. Without a mate it could never found a new species. However, if somehow or other, enough hopeful monsters sprang into existence simultaneously and thus were able to breed successfully with each other, the possibility exists of competition between the hopeful monsters and the stock from which they sprang. It is not at all difficult to suppose that, by the process of Darwinian evolution, hopeful monsters could quickly replace their competitors if they were better adapted to the environment. No new fancy principles of evolution are involved here. The critical question for an understanding of biological evolution is how could a whole group of individuals, founders of a new species, suddenly arise at the same time?

Suppose that a population splits into two subgroups as the result, say, of migration. The subgroups remain separated for many generations and a different mutation goes to fixation in each of them. Then, the populations merge again. The combination of the mutated genes in the two previously separated subgroups interacts to produce a radically new phenotype, which is sufficiently frequent in the population to allow breeding to occur. Now the conditions are in place for a competition between the phenotypes. If the new phenotype is more successful than the old one, the historical record would show a discontinuity in evolution (P. Bateson, 2002).

Another model for speciation also suggests that postzygotic isolation results from an interaction between two or more genes (Orr & Presgraves, 2000). Suppose the initial genotype is *aabb*, the population splits and in one population an *A* mutation appears and goes to fixation and in the other population a *B* mutation appears and also goes to fixation. If *A* and *B* do not function well together, then hybrids between the two populations will be less viable or infertile. As Orr and Presgraves (2000) pointed out, this model highlights the role of epistasis in evolution. Though credit is usually given to Dobzhansky (1937) and Muller (1940) for the idea, as Orr (1998) noted, the problem was first solved by William Bateson (1909). Population geneticists have typically not liked to contemplate interactions between genes because it complicates their mathematics, but the empirical evidence for such effects is strong. One example is to be found in one of William Bateson's experiments on chickens. When a White Silky hen and a White Dorking cockerel were mated, the offspring were all brown.

These conjectures about the sudden emergence of discontinuous change in evolution do not necessarily imply speciation. However, they would do so if hybrids between old and the new phenotypes were less viable. A plausible case would be a change in chromosome number, which could prohibit the formation of gametes in hybrids. Examples of closely related species with different numbers of chromosomes are well known. In horses the chromosome number ranges from 32 in *Equus zebra hartmannae* and 46 in *Equus grevyi*, to 62 in *Equus assinus* and 66 in *Equus przewalski*; all but two of the horse hybrids are sterile (King, 1993), therefore the differences between the species must have occurred in jumps.

4.2. Active Role of Organisms in Evolution

A second challenge to the orthodoxies of neo-Darwinism comes from the evidence that organisms play important roles in the evolution of their descendants (see P. Bateson, 2005, 2006, 2014). The evidence is of four types: choice, control of the environment, adaptability, and mobility. I consider these in turn.

4.2.1. Choice

Darwin (1871) himself suggested that choice of a mate could drive evolution. He called the evolutionary process "sexual selection." His ideas were couched in the conventional views of late 19th-century England. Joan Roughgarden (2009) was therefore critical and declared Darwin wrong, but she was mistaken in dismissing the idea of active choice. Alfred Russel Wallace, though the coauthor with Darwin of the first clear statement about the role of natural selection, did not like the idea. Indeed, for many years most biologists did not take sexual selection seriously. Erika Lorraine Milan (2010) has given an excellent historical account of the change in thinking that has taken place. I well remember when I was an undergraduate being confidently told that even if sexual selection were possible in theory the process probably played little part in biological evolution. In recent years, however, many experiments have supported Darwin's thinking. A famous experiment by Malte Andersson (1994) involved lengthening the tail of male long-tailed widow birds. He found that the males with extra-long tails attracted more mates. The reason why longer tails are not found in nature is probably because it carries a big cost for the male so that in rainy weather, a bird with an extra long tail cannot drag the great encumbrance off the ground when attacked by a predator.

Another example of choice is that involved in predators' choice of prey. Gazelles that have seen a predator jump into the air. The behavior pattern is called "stotting." A suggested evolutionary mechanism is that first some gazelles jump after noticing cheetah (Fitzgibbon & Fanshawe, 1988). Cheetahs learn not to chase jumping gazelles because their effort would be wasted. The next step is that all gazelles jump after noticing cheetah. Some gazelles gain advantage by giving an exaggerated jump—a stott—after noticing a cheetah. This is because cheetahs learn not to chase stotting gazelles. The interesting feature of this example is that, while the gazelles inherited their behavior in the course of evolving stotting, no change in the inheritance of choice was required in the cheetahs.

4.2.2. Control of the Environment

The environment does not simply set a problem to which the organism has to find a solution. The organism can do a great deal to create an environment to which it is best suited. This should give pause if evolution is considered purely in terms of selection by external forces (Lewontin, 1983). By leaving an impact on their physical and social environment, organisms may affect the evolution of their own descendants, quite apart from changing the conditions for themselves. Some of the impact is subtle, such as when a plant sheds its leaves, which fall to the ground and change the characteristics of the soil in which its own roots and those of its descendants grow. These ideas have been developed extensively and are now referred to as "niche construction" (Odling-Smee, Laland, & Feldman, 2003). One example is provided by beavers, which changed their environment by building dams and creating lakes for themselves. This set up conditions that affected subsequent evolution. The artificially created aquatic environment led the beavers to evolve adaptations such as webbed feet that facilitated swimming. The beavers' lakes were not merely part of what Richard Dawkins (1982) called the "extended phenotype," they were the environment in which webbed feet were advantageous.

The effect of behavioral control on evolutionary change could be especially great when a major component of the environmental conditions with which animals have to cope is provided by their social environment. A similar type of positive feedback to that flowing from the effects of mate choice could operate in such circumstances (Jolly, 1966). If individuals compete with each other within a social group and the outcome of the competition depends in part on each individual's capacity to predict what the other will do, the evolutionary outcome might easily acquire a runaway property. In discussing the social function of intelligence, Nicholas Humphrey expressed the idea as follows:

> An evolutionary "ratchet" has been set up, acting like a self-winding watch to increase the general intellectual standing of the species. In principle the

process might be expected to continue until either the physiological main-spring of intelligence is full-wound or else intelligence itself becomes a burden. The latter seems most likely to be the limiting factor; there must surely come a point where the time required to resolve a social "argument" becomes insupportable. (Humphrey, 1976, p. 311)

As Humphrey noted, such an explanation makes sense of the astonishing rate of increase in the cranial capacity of humans, if it is assumed (reasonably) that cranial capacity and intellectual ability are correlated. Here again the ideas have been developed extensively in subsequent years (e.g., Byrne, 2000). When play generates novel solutions to environmental challenges, the same directional outcome could have occurred in mammalian and avian evolution (P. Bateson & Martin, 2013).

4.2.3. The Adaptability Driver

The adaptability of the organism is likely to play an important role in initiating evolutionary change. This effect is often called the Baldwin effect after James Mark Baldwin (1896). Two others published the same idea in the same year (Lloyd Morgan, 1896; Osborn, 1896). However, 23 years before, Douglas Spalding (1873) had published the same hypothesis in *Macmillan's Magazine*. It was not an obscure publication—it was the predecessor of *Nature*, which continues to be published by Macmillan.

Given Spalding's (1873) precedence and the simultaneous appearance in 1896 of the ideas about "organic selection," it seems inappropriate to term the evolutionary process the "Baldwin effect." The trouble is that calling the proposed process the "Spalding effect" is not descriptive of what initiates the hypothetical evolutionary process. I have a strong preference for a term that captures the active role of the organism in the evolutionary process, namely the "adaptability driver" (P. Bateson, 2005, 2006, 2014). The notion of a behavioral driver in evolution was introduced by Wyles, Kunkel, and Wilson (1983), but they laid primary emphasis on imitation and, of course, other active behavioral processes such as mate choice had already been recognized by Darwin.

Many changes in the environment and/or in the expression of the genome may require adaptability on the part of the organism; if the same phenotypic effect can be generated at lower cost, Darwinian evolution can lead to a change in the underlying developmental mechanism. Adaptability may save the organism from extinction and thereafter promote a new direction in evolution. George Gaylord Simpson (1953) disputed its importance in biological evolution. He believed that if a new phenotype were valuable to the organism, it would evolve along traditional Darwinian lines. Second, he argued that, if plasticity were a prerequisite for the evolutionary process and was generally beneficial, it would be disadvantageous to get rid of it.

On Simpson's first point, if learning (as an example of one form of adaptability in animals) involves several subprocesses or steps, as in operant chaining, then the chances against an unlearned equivalent appearing in one step in the course of evolution are very small. However, with the learned phenotype as the standard, every small step that cuts out some of the plasticity with a simultaneous increase in efficiency is an improvement. As an example, consider the Galapagos woodpecker finch, which pokes sharp cactus spines into holes thereby obtaining insect larvae as food. Suppose it does so without much learning but that an ancestor did so by trial and error. In the first stage of the evolutionary process, a naïve variant of the ancestral finch, when in foraging mode, might, for example, have been more inclined to pick up cactus spines than other birds. This habit spread in the population by Darwinian evolution because those behaving in this fashion obtained food more quickly. At this stage the birds still learned the second part of the sequence. The second step would have been that a naïve new variant, when in foraging mode, was more inclined to poke cactus spines into holes. Again this second habit could spread in the population by Darwinian evolution. The end result is a finch that uses a tool without having to learn how to do so. Simultaneous mutations increasing the probability of two quite distinct acts (taking cactus spines and poking them into holes in the case of the woodpecker finch) would be very unlikely. Learning makes it possible for them to occur at different times and in any order. Without learning processes, having one act but not the other has no value. As a matter of great interest, it seems to be the case that the Woodpecker finch is half down the evolutionary road from fully learned to fully spontaneous, because naïve birds readily pick up small sticks but then have to learn how to use them (Tebbich, Sterelny, & Teschke, 2010).

As far as learning is concerned, Simpson's second point of criticism is based on an inadequate understanding of how behavior is changed and controlled. Learning in complex organisms consists of a series of subprocesses. If, as suggested by P. Bateson and Horn (1994), an array of feature detectors is linked directly to an array of executive mechanisms as well as indirectly through an intermediate layer, and all connections are plastic, then a particular feature detector can become nonplastically linked to an executive system in the course of evolution without any further loss of plasticity. Whether these replies to Simpson's objections can be applied as cogently to plants or less complex animals requires arguments that have, as far as I know, not yet been developed.

An important empirical demonstration of adaptability driving evolutionary change may be that of the house finch. In the middle of the 20th century, the finch was introduced to eastern regions of the United States far from where it was originally found on the West Coast. It was able to adapt to the new and extremely different climate and spread up into Canada. The finch also extended its western range north into Montana, where it has been extensively studied. After a period involving great deal of plasticity, the house finch populations spontaneously expressed the physiological characteristics that best fitted them to their new habitats without the need for developmental plasticity (Badyaev, 2009). Initially the adaptive onset of incubation and the sex bias in the order of ovulation were affected by ambient temperature in the more northerly climes, but as evolution in the population occurred, these behavioral and physiological effects were no longer dependent on the external cues for their expression. After initially using their adaptability to respond to the new environmental conditions, the house finch populations spontaneously expressed the physiological characteristics that best fitted them to their new habitats.

The hypothesis of the adaptability driver has been repeatedly modeled, both analytically (e.g., Ancel, 1999; Paenke, Kawecki, & Sendhoff, 2009) and by simulation (e.g., Hinton & Nowlan, 1987; Red'ko, Mosalov, & Prokhorov, 2005). The outcome of this theoretical activity has been variable, sometimes supporting the adaptability driver hypothesis and sometimes not. Paenke et al. (2009) proposed a general framework that explained both effects. Spontaneously expressing a behavior pattern that had been learned in previous generations could be costly if it meant that the animal lost all of its ability to learn. Some evidence from fruit flies suggests that this might well be the case, at least in simpler organisms (Kawecki, 2010) The benefit of expressing a behavior spontaneously was found to be outweighed by the cost of losing the capacity to learn about other things. The argument is much less cogent when applied to big-brained animals like birds and mammals with multiple parallel pathways involved in learning. In these animals, the loss of capacity to learn in one way has no effect on the capacity to learn in other ways (P. Bateson, 2004).

Inasmuch as it has been taken seriously, the hypothetical process has usually been taken as providing a mechanism for the slow accretion of spontaneously expressed phenotypic elements in the course of evolution. Emphasis has been placed on how particular behavior patterns initially acquired by learning could be expressed spontaneously without learning in the course of subsequent evolution. Recent developments have shifted the focus to other issues, such as the way in which plasticity can accelerate the rate at which challenges set by the environment can be met (e.g., Paenke et al., 2009), the advantages of plasticity in a changing environment (e.g., Pfennig & McGee, 2010), and the conditions in which plasticity might slow down evolution (e.g., Paenke et al., 2009).

The effect of plasticity on evolution may have become increasingly powerful as animals, in particular, become more complex. The general idea is that a system that enhances the fitness of an individual depends on a number of elements such as the capacity to use information contained in the energy impinging on a sense organ, specific biochemical reactions, and particular effectors that respond adaptively to the stimulation (Anokhin, 1974). Elements may be recombined in different ways to perform different functions. Novel challenges create the conditions for the emergence of new functional systems added to the existing ones either by Darwinian evolution or by an individual's plasticity. Possible examples are the addition of a face recognition module in primates (Leopold & Rhodes, 2010) and the evolution of invasiveness in birds (Sol, Duncan, Blackburn, Cassey, & Lefebvre, 2005). This evolutionary process could lead to the establishment of increasingly elaborate organization and patterns of behavior. When such complexity entailed a greater ability to discriminate between different features of the environment or a greater ability to manipulate the environment, the organism would benefit and would be more likely to survive and reproduce in the face of multiple challenges during its lifetime.

A new adaptation emerges in evolution when the accumulated phenotypic effects of genomic reorganization are added to the existing phenotype. Although these phenotypic effects are specific to the new function, existing parts of the phenotype are also recruited for this function. As a result, phenotypic elements established earlier in evolution should be incorporated in more adaptive systems than later evolved elements. Plasticity would promote much more rapid genetic evolution of complex sets of adaptive systems than can be accomplished by mutation alone. This occurs as previously plastic elements are replaced by inherited elements and the organism is able to fill by plasticity missing elements in subsequent systems.

In general, the proposal is that an ability to cope with complex environmental challenges by means of plasticity opens up ecological niches previously unavailable to the organism. This would inevitably lead to the subsequent evolution of morphological, physiological, and biochemical adaptations to those niches and the likelihood of speciation (e.g., Pfennig et al., 2010). Where an environmental challenge involved greater processing capacity by the brain, this organ too would be expected to evolve with greater rapidity.

4.2.4. Mobility

Organisms and proto-organisms were doubtless passive in the initial stages of biological evolution. However, as they evolved, they would soon have become active. This is the key conceptual point in understanding how plasticity and behavior can drive evolutionary change. Development depends on the constancy of many genetic and environmental conditions. If any of these change, as can happen to environmental conditions when organisms are mobile, the characteristics of the organism can also change. High mobility by animals, such as that involved in active exploration or migration, would have frequently placed them in conditions that revealed heritable variation not previously apparent in the population. By their mobility, in the case of animals, or facility to disperse, in the case of plants (Donohue, 2005), organisms would have exposed themselves to new conditions that might reveal heritable variability, thereby opening up possibilities for evolutionary changes that would not otherwise have taken place.

A striking example of what can happen when an animal is mobile has been provided by the work on the three-spined stickleback (Foster & Wund, 2011). This fish moved from a marine environment to freshwater and thence from living near the surface in streams to much deeper water in lakes. As it moved from one environment to the next its morphological and behavioral characteristics changed.

Development depends on the constancy of many genetic and environmental conditions. If any of these change, as can happen to environmental conditions when organisms are mobile, the characteristics of the organism can change. High mobility by animals, such as that involved in active exploration or migration, would have frequently placed them in conditions that revealed heritable variation not previously apparent in the population.

When discussing his results of experiments on what he called "genetic assimilation," Conrad Waddington (1953) suggested that the heat shock, applied to the larvae of fruit flies, led to the expression of genes that were carried in only a part of the population. Waddington bred from the flies that had developed a particular character (lack of a crossvein in the wings) as a result of their larval experience. He continued to apply heat shock in each generation and to breed selectively from the flies with crossveinless wings. After many generations of heat shock and selective breeding, crossveinless wings developed spontaneously in the absence of the external triggering condition of heat shock.

Waddington's finding involved expression of a novel character in a new environment, but the character was not an adaptation to the triggering condition. Because of artificial selection, however, it did confer some advantage on its possessor. Crossveinless wings do not bear any functional relation to the environment that supplied a heat-shock when the flies were larvae. Nor need there be such a relationship under natural conditions. All that is required initially is that the environmental conditions trigger the expression of a phenotype that can be repeated generation after generation so long as the environmental conditions persist.

Waddington's fruit fly experiment is just one illustration of innumerable possible scenarios. The developmental break-out may provide radically new opportunities for those individuals equipped with the new phenotype (West-Eberhard, 2003). For that reason, behavior, along with other forms of dispersion, was likely to be important in initiating evolutionary change. In addition, behavioral adaptability of the animals would have helped buffer them against extinction in new conditions. Mary Jane West-Eberhard argued that after developmental disruption, the reorganization of the genome might have a much broader effect than that envisaged by Waddington. She suggested that major changes might evolve as the character in question became more variable; in other words it became developmentally less robust or less canalized. The umbrella term that she used for all the heritable changes that might occur in the genetic regulation of development in response to environmental influences was "genetic accommodation."

One instance of the broadening effect suggested by West-Eberhard would be the evolution of a variety of distinct developmental trajectories or polyphenisms, each a canalized response and each expressed under specific conditions. A striking example of genetic accommodation was provided by Suzuki and Nijhout (2006), who studied tobacco hornworm caterpillars, the larvae of the moth *Manduca sexta* L. These larvae are normally green, but a naturally occurring mutant form is black; both color forms are heritable, with the black form arising from a sex-linked recessive allele that reduces juvenile hormone secretion, resulting in a black cuticle. When Suzuki and Nijhout heat-shocked the black larvae at the fourth larval instar, which is the sensitive period for the effect of juvenile hormone on cuticle coloration, some of the fifth larval instars were greenish. The authors bred from the greenest larvae, as well as those that did not show color change on heat shock (the persistent black larvae), and monitored in each generation how the larvae appeared after heat shock and after being reared at different temperatures. By the 13th generation, the line selected for blackness developed robustly as black larvae irrespective of the rearing temperature. At high rearing temperatures, the line selected for greenness was more strikingly green than an unselected line that acted as a control group. However, at low rearing temperatures, the green-selected line was mostly black (although not as black as the black-selected line). Thus, the selection process enabled two different phenotypes to evolve at different constant temperatures. The green-to-black color switch could be seen at a lower temperature, and within a smaller temperature range, compared with the control group. The original article did not mention the important detail that the tested larvae were not exposed to heat shock, but Suzuki (pers. com.) has confirmed that this was indeed the case, showing that the observed phenotypes were inherited.

4.3. Development and Evolution

The gene-centered view of the modern synthesis is discussed at length by Uller and Helanterä (this volume). I have also described its shortcomings in my book with Peter Gluckman (P. Bateson & Gluckman, 2011) and we went on to consider how the rapidly expanding field of epigenetics is changing current thinking about evolution.

Unfortunately, semantic and conceptual confusions are rife. Some authors have supposed that the characteristics of an organism are encoded in the genes, in the sense that all the information required for its development is contained in DNA. The notion of genes coding for phenotypic characteristics was always problematic, but its limitations have become increasingly apparent as molecular and biological knowledge has expanded.

Much epidemiological research in recent years has been based on sequencing the entire human genome and looking at mutant alleles that correlate with disease. A surprise of these genomewide association studies has been that even when large populations are studied and the disease of interest is common, such as diabetes, few significant genetic effects are found and the effects of any one specific polymorphism are generally small. Single-gene effects are unusual and largely restricted to relatively rare diseases such as phenylketonuria or hemophilia (Maher, 2008).

Genes have been defined in many different ways: as units of physiological function, units of recombination, units of mutation, or as units of evolutionary process—when they have sometimes been imbued with "selfish" intentions in order to help with understanding (Dawkins, 1976). The problem of definition has been made worse as it has become clear that the same strand of DNA may serve in processes that differ in function. Indeed, the same strand of DNA might be transcribed in one direction to serve one function and in the other direction to serve a different function. Griffiths and Stotz (2006) have emphasized how, in the postgenomic era, the emerging concepts of the gene pose a significant challenge to conventional assumptions about the relationship between genome structure and function, and between genotype and phenotype.

The word "gene" never had a clear unambiguous meaning: for some it meant simply a sequence of DNA, for others it referred specifically to those segments of DNA that are transcribed into ribonucleic acid (RNA) and then translated into a protein. To be set against that, some segments of RNA—the so-called noncoding RNAs—have regulatory functions, and the term "gene" is extended by many molecular geneticists to include the DNA sequences coding for these RNAs. These different meanings of "gene" get conflated, with subsequent confusion of thought (Keller, 2000). The easiest way out of the conceptual thicket is simply to refer to a gene as a particular sequence of DNA that is inherited from one generation to the next and has functional significance.

"Epigenetics" is a term that had multiple meanings since it was first coined by Conrad Waddington (1957). He used the term, in the absence of molecular understanding, to describe processes by which the inherited genotype could be influenced during development to produce a range of phenotypes. He distinguished "epigenetics" from the 18th-century term "epigenesis," which had been used to oppose the preformationist notion that all the characteristics of the adult were preformed in the embryo.

More recently, epigenetics has become defined as the molecular processes by which traits—as defined by a given gene expression profile—can persist across mitotic cell division without involving changes in the nucleotide sequence of the DNA. Epigenetic processes result in the silencing or activation of gene expression through such modification of DNA and its associated RNAs and proteins. A broader definition of epigenetics is to include elements apart from DNA, RNA, and chromatin, such as components of the cytoplasm outside the cell nucleus and the cell membrane that are passed across in mitosis.

While epigenetics usually refers to what happens within an individual developing organism, a growing body of evidence suggests that epigenetic traits established in one generation may be passed directly or indirectly through meiosis to the next, involving a variety of different processes (Jablonka & Raz, 2009).

A central question in considering evolutionary change driven by the environment is whether the transmitted epigenetic markers could facilitate genomic change (Johnson & Tricker, 2010). The answer is that, in principle, they could if (1) they were transmitted from one generation to the next, (2) they increased the fitness of the individual carrying the markers, and (3) genomic reorganization enabled some individuals to develop the same phenotype at lower cost. Epigenetic inheritance would serve to protect the well-adapted phenotypes within the population until spontaneous fixation occurred. That much is exactly the same as I have suggested earlier for the operation of the adaptability driver. However, another process could be at work.

DNA sequences where epigenetic modifications have occurred may be more likely to mutate than other sites. The consequent mutations could then give rise to a range of phenotypes on which Darwinian evolution could act. If epigenetic change could affect and bias mutation rates, such nonrandom mutation would facilitate fixation (P. Bateson & Gluckman, 2011). To support this argument, it is necessary here to delve into some molecular biology.

Methylated sequences of cytosine and guanine in the genome (known as CpG islands) are mutational hotspots due to the established propensity of methylated cytosine to undergo spontaneous chemical conversion to thymine and methylated guanine to convert to uracil (Pfeifer, 2006). As these are functional nucleotides, they are not recognized as damaged DNA and excised or corrected by DNA repair mechanisms. As a result, the mutation becomes incorporated in subsequent DNA replications. DNA mapping shows fewer CpG sequences in the DNA than expected (Schorderet & Gartler, 1992), and CpG hypermutability has led to a decrease in frequency of amino acids coded by CpG dinucleotides in some organisms. Indeed, comparison of the human and chimpanzee genomes has shown that 14% of the single amino acid changes are due to the biased instability of CpG sequences, which can be subject to methylation and thence to mutations (Misawa, Kamatani, & Kikuno, 2008). The methylation of CpGs is a major contributing factor to mutation in RB1, a gene in which allelic inactivation leads to the developmental tumor retinoblastoma (Mancini, Singh, Ainsworth, & Rodenhiser, 1997).

Further evidence in support of the hypothesis that epigenetic change can lead to mutation is found in the analysis of neutrally evolving strands of primate DNA. The evidence indicates that the phylogenetically "younger" sequences have a higher CpG content than the "older" sequences, due to the reduced opportunity for spontaneous mutation. Intriguingly, the CpG content is strongly correlated with a higher rate of neutral mutation at non-CpG sites (Walser, Ponger, & Furano, 2008), which suggests that CpGs play a role in influencing the mutation rate of DNA not containing CpG, perhaps by influencing the chromatin conformation surrounding the CpG and making it more accessible to other modifying processes. Furthermore, CpG content also appears to influence the type of mutation that occurs, with a higher ratio of transition-to-transversion mutations observed in parallel with the non-CpG mutation rate (Walser & Furano, 2010).

If an epigenetic change has long-term heritable consequences in terms of biased mutation, this process would not necessarily produce a phenotype resembling the one that was initially produced by the epigenetic change. This would only happen if the induced mutation resulted in phenotypic variation and one of the variants was favored by a Darwinian process and resembled the original phenotype.

4.4. Conclusion

The "modern synthesis" of the 1930s no longer looks modern, and some of its premises have been seriously challenged by fresh evidence. Species may form suddenly, organisms (particularly animals) play an active role in the evolution of their descendants, and genes often follow rather than lead evolutionary change. The evidence suggests that the conditions of development can radically affect an organism's characteristics, thereby challenging the third prop of neo-Darwinist orthodoxy that development is irrelevant to an understanding of evolution. The orthodoxy was already under threat from the evo-devo movement that shows how developmental toolkits can profoundly influence the course of evolution. Further support for the importance of development in evolution is provided by the rapidly expanding field of epigenetics. Acquired information can be passed to progeny without changing DNA sequences, and information can be inherited for a period in the *absence* of the initial environmental trigger.

In other words, evolutionary theory is evolving. Would Charles Darwin have been concerned if he had lived to see modern developments? I doubt it. At the end of *The Origin of Species* Charles Darwin wrote, "Contemplate an entangled bank, clothed with many plants of many kinds, with birds singing on the bushes, with various insects flitting about, and with worms crawling through the damp earth ... so different from each other, and dependent on each other in so complex a manner, have been produced by laws acting around us." The sense of wonder he expresses here and his strong commitment to evidence would have equipped him to understand the richness of evolutionary processes. One of his great strengths was his willingness to test his ideas by observation and experiment and to change his thinking when the findings told him to do so. We should all continue to revere him and be delighted that the science of evolutionary biology, which he transformed, remains in such a healthy state. At the end of my book with Peter Gluckman, we wrote about the growing synthesis between evolutionary and developmental biology. We suggested that two words should be added to Theodosius Dobzhansky's famous dictum: "Nothing in biology makes sense except in the light of evolution *and development.*"

References

Ancel, L. W. (1999). A quantitative model of the Simpson-Baldwin effect. *Journal of Theoretical Biology, 196,* 197–209.

Andersson, M. (1994). *Sexual selection.* Princeton, NJ: Princeton University Press.

Anokhin, P. K. (1974). *Biology and neurophysiology of the conditioned reflex and its role in adaptive behavior.* Oxford, UK: Pergamon Press.

Badyaev, A. V. (2009). Evolutionary significance of phenotypic accommodation in novel environment: An empirical test of the Baldwin effect. *Philosophical Transactions of the Royal Society of London, B, 364,* 1125–1141.

Baldwin, J. M. (1896). A new factor in evolution. *American Naturalist, 30,* 441–451, 536–553.

Bateson, P. [P. G.] (2002). William Bateson: A biologist ahead of his time. *Journal of Genetics, 81,* 49–58.

Bateson, P. [P. G.] (2004). The active role of behaviour in evolution. *Biology and Philosophy, 19*, 283–298.

Bateson, P. [P. G.] (2005). The return of the whole organism. *Journal of Biosciences, 30*, 31–39.

Bateson, P. [P. G.] (2006). The adaptability driver: Links between behaviour and evolution. *Biological Theory: Integrating Development Evolution and Cognition, 1*, 342–345.

Bateson, P. [P. G.] (2013). New thinking about biological evolution. *Biological Journal of the Linnean Society, 112*, 268–275. doi:10.1111/bij.12125.

Bateson, P. [P. G.] (2014). Why are individuals so different from each other? *Heredity, 115*, 285–292. doi:10.1038/hdy.2014.103

Bateson, P. [P. G.], & Gluckman, P. (2011). *Plasticity, robustness, development and evolution.* Cambridge, UK: Cambridge University Press.

Bateson, P. [P. G.], & Horn, G. (1994). Imprinting and recognition memory: A neural net model. *Animal Behaviour, 48*, 695–715.

Bateson, P. [P. G.], & Martin, P. (2013). *Play, playfulness, creativity and innovation.* Cambridge, UK: Cambridge University Press.

Bateson, W. (1894). *Materials for the study of variation: Treated with especial regard to discontinuity in the origin of species.* London, UK: Macmillan.

Bateson, W. (1909). Heredity and variation in modern lights. In A. C. Seward (Ed.), *Darwin and modern science* (pp. 85–101). Cambridge, UK: Cambridge University Press.

Byrne, R. W. (2000). Evolution of primate cognition. *Cognitive Science, 24*, 543–570.

Cock, A. G., & Forsdyke, D. R. (2008). *Treasure your exceptions: The science and life of William Bateson.* New York, NY: Springer.

Darwin, C. (1871). *The descent of man, and selection in relation to sex.* London, UK: Murray.

Dawkins, R. (1976). *The selfish gene.* Oxford, UK: Oxford University Press.

Dawkins, R. (1982). *The extended phenotype.* Oxford, UK: Freeman.

Dobzhansky, T. (1937). *Genetics and origin of species.* New York, NY: Columbia University Press.

Donohue, K. (2005). Niche construction through phonological plasticity: Life history dynamics and ecological consequences. *New Phytology, 166*, 83–92.

Eldredge, N. (1995). *Reinventing Darwin.* London, UK: Weidenfeld & Nicolson.

Fitzgibbon, C. D., & Fanshawe, J. H. (1988). Stotting in Thomson gazelles: An honest signal of condition. *Behavioral Ecology and Sociobiology, 23*, 69–74.

Foster, S. A., & Wund, M. A. (2011). Epigenetic contributions to adaptive radiation: Insights from threespined stickleback. In B. Hallgrimsson & B. K. Hall (Eds.), *Epigenetics: linking genotype and phenotype* (pp. 317–336). Berkeley: University of California.

Galton, F. (1892). *Hereditary genius: An inquiry into its laws and consequences* (2nd ed.). London, UK: Watts.

Goldschmidt, R. (1940). *The material basis of evolution*. New Haven, CT: Yale University Press.

Gould, S. J. (2002). *The structure of evolutionary theory*. Cambridge, MA: Harvard University Press.

Griffiths, P. E., & Stotz, K. (2006). Genes in the post-genomic era. *Theoretical Medicine and Bioethics, 27*, 499–521.

Hinton, G. E., & Nowlan, S. J. (1987). How learning can guide evolution. *Complex Systems, 1*, 495–502.

Humphrey, N. K. (1976). The social function of intellect. In P. P. G. Bateson & R. A. Hinde (Eds.), *Growing points in ethology* (pp. 303–317). Cambridge, UK: Cambridge University Press.

Jablonka, E., & Raz, G. (2009). Transgenerational epigenetic inheritance: Prevalence, mechanisms, and implications for the study of heredity and evolution. *Quarterly Review of Biology, 84*, 131–176.

Johnson, L. J., & Tricker, P. J. (2010). Epigenomic plasticity within populations: Its evolutionary significance and potential. *Heredity, 195*, 113–121.

Jolly, A. (1966). Lemur social behavior and primate intelligence. *Science, 153*, 501–506.

Kawecki, T. J. (2010). Evolutionary ecology of learning: Insights from fruit flies. *Population Ecology, 52*, 15–25.

Keller, E. F. (2000). *Century of the gene*. Cambridge, MA: Harvard University Press.

King, M. (1993). *Species evolution: The role of chromosome change*. Cambridge, MA: Cambridge University Press.

Laland, K. N., Uller, T., Feldman, M. W., Sterelny, K., Müller, G. B., Moczek, A.,. . . . Odling-Smee, J. (2015). The extended evolutionary synthesis: Its structure, assumptions and predictions. *Proceedings of the Royal Society B, 282*, 20151019. http://dx.doi.org/10.1098/rspb.2015.1019

Leopold, D. A., & Rhodes, G. (2010). A comparative view of face perception. *Journal of Comparative Psychology, 124*, 233–251.

Lewontin, R. C. (1983). Gene, organism and environment. In D. S. Bendall (Ed.), *Evolution from molecules to men* (pp. 273–285). Cambridge, UK: Cambridge University Press.

Lloyd Morgan, C. (1896). On modification and variation. *Science, 4*, 733–740.

Maher, B. (2008). Personal genomes: The case of the missing heritability. *Nature, 456*, 18–21.

Mancini, D., Singh, S., Ainsworth, P., & Rodenhiser, D. (1997). Constitutively methylated CpG dinucleotides as mutation hotspots in the retinoblastoma gene (RB1). *American Journal of Human Genetics, 61*, 80–87.

Milan, E. L. (2010). *Looking for a few good males: Female choice in evolutionary biology*. Baltimore, MD: Johns Hopkins University Press.

Misawa, K., Kamatani, N., & Kikuno, R. F. (2008). The universal trend of amino acid gain–loss is caused by CpG hypermutability. *Journal of Molecular Evolution, 67*, 334–342.

Muller, H. J. (1940). Bearing of the Drosophila work on systematics. In J. S. Huxley (Ed.), *The new systematics* (pp. 125–268). Oxford, UK: Oxford University Press.

Odling-Smee, F. J., Laland, K. N., & Feldman, M. W. (2003). *Niche construction: The neglected process of evolution*. Princeton, NJ: Princeton University Press.

Orr, H. A. (1998). The population genetics of adaptation: The distribution of factors fixed during adaptive evolution. *Evolution, 52*, 935–949.

Orr, H. A., & Presgraves, D. C. (2000). Speciation by postzygotic isolation: Forces, genes and molecules. *BioEssays, 22*, 1085–1094.

Osborn, H. F. (1896). Ontogenic and phylogenic variation. *Science, 4*, 786–789.

Paenke, I., Kawecki, T. J., & Sendhoff, B. (2009). The influence of learning on evolution: A mathematical framework. *Artificial Life, 15*, 227–245.

Pfeifer, G. P. (2006). Mutagenesis at methylated CpG sequences. *Current Topics in Microbiology and Immunology, 301*, 259–281.

Pfennig, D. W., & McGee, M. (2010). Resource polyphenism increases species richness: A test of the hypothesis. *Philosophical Transactions of the Royal Society B, 365*, 577–591.

Pfennig, D. W., Wund, M. A., Snell-Rood, E. C., Cruickshank, T., Schlichting, C. D., & Moczek, A. P. (2010). Phenotypic plasticity's impacts on diversification and speciation. *Trends in Ecology and Evolution, 25*, 459–487.

Red'ko, V. G., Mosalov, O. P., & Prokhorov, D. V. (2005). A model of evolution and learning. *Neural Networks, 18*, 738–745.

Roughgarden, J. (2009). *Evolution's rainbow: Diversity, gender and sexuality in nature and people*. Berkeley: University of California Press.

Schorderet, D. F., & Gartler, S. M. (1992). Analysis of CpG suppression in methylated and nonmethylated species. *Proceedings of the National Academy of Sciences of the United States of America, 89*, 957–961.

Simpson, G. G. (1953). The Baldwin effect. *Evolution, 7*, 110–117.

Sol, D., Duncan, R. P., Blackburn, T. M., Cassey, P., & Lefebvre, L. (2005). Big brains, enhanced cognition, and response of birds to novel environments. *Proceedings of the National Academy of Sciences, 102*, 5460–5465.

Spalding, D. A. (1873). Instinct with original observations on young animals. *Macmillan's Magazine, 27*, 282–293.

Suzuki, Y., & Nijhout, H. F. (2006). Evolution of a polyphenism by genetic accommodation. *Science, 311*, 650–652.

Tebbich, S., Sterelny, K., & Teschke, I. (2010). The tale of the finch: Adaptive radiation and behavioural flexibility. *Philosophical Transactions of the Royal Society B, 365*, 1099–1109.

Waddington, C. H. (1953). Genetic assimilation of an acquired character. *Evolution, 7*, 118–126.

Waddington, C. H. (1957). *The strategy of the genes*. London, UK: Allen & Unwin.

Walser, J.-C., Ponger, L. C., & Furano, A. V. (2008). CpG dinucleotides and the mutation rate of non-CpG content. *Genome Research*, *18*, 1403–1414.

Walser, J. C., & Furano, A. V. (2010). The mutational spectrum of non-CpG DNA varies with CpG content. *Genome Research*, *20*, 875–882.

West-Eberhard, M. J. (2003). *Developmental Plasticity and Evolution*. New York, NY: Oxford University Press.

Wyles, J. S., Kunkel, J. G., & Wilson, A. C. (1983). Birds, behavior, and anatomical evolution. *Proceedings of the National Academy of Sciences of the United States of America*, *80*, 4394–4397.

DEVELOPMENT

5 EVO-DEVO AND THE STRUCTURE(S) OF EVOLUTIONARY THEORY

A DIFFERENT KIND OF CHALLENGE

Alan C. Love

5.1. Challenges to Evolutionary Theory

New "theories" of evolution have been proposed on a regular basis over the past century. The regularity of these offerings and their common characteristics encourage a general and healthy skepticism from biologists and philosophers alike. These characteristics include a faulty understanding of different parts of standard evolutionary theory, a rehearsal of trumped-up empirical difficulties (often warmed over objections from the 19th century), and claims that the new theory (comprehensively) solves all of these difficulties. Call these *false challenges* to evolutionary theory; that is, they are not genuine challenges to the empirical adequacy, systematicity, or fecundity of evolutionary theory. They can be set aside without concern. While tickling the fancy of journalists and tyros, serious students of evolution typically ignore them and move on.

Not all challenges to evolutionary theory exhibit the telltale signature of a false challenge. Over the past decade, evolutionary theory has received a number of challenges from within the scientific community under the rubric of expanding or extending the evolutionary synthesis or standard evolutionary theory (hereafter, SET) (Carroll, 2008; Laland et al., 2015; Müller, 2007; Pigliucci, 2007; Pigliucci & Müller, 2010b; Weber, 2011). The composition of these challenges varies, but one ingredient is commonly found in cases that involve evolutionary developmental biology (evo-devo): the need to explain the nature and origin of "form," which requires a substantive incorporation of embryology and molecular developmental biology.[1]

> Evolutionary theory has shifted from a theory of **form** to a theory of genes, and . . . it is now in need again of a comprehensive and updated theory of **form**. (Pigliucci, 2007, p. 2743)

> Evolutionary developmental biology (evo–devo) emerged as a distinct field of research in the early 1980s to address the profound **neglect of development** in the standard modern synthesis framework of evolutionary theory, a deficiency that had caused difficulties in explaining the origins of **organismal form** in mechanistic terms. (Müller, 2007, p. 943)

> At the time of the "Modern Synthesis" of evolutionary theory that drew together various disciplines including genetics, paleontology, and systematics, very little could be said about the effects of genes on **development**, let alone on the **evolution of form**. . . . [Evo-devo] discoveries forced . . . evolutionary biologists to confront a new source of unforeseen and penetrating genetic insights into the generation and diversification of **animal form**. (Carroll, 2008, p. 25)

> While much evo-devo research is compatible with standard assumptions in evolutionary biology, some findings have generated debate. Of particular interest is the observation that phenotypic variation can be biased by the processes of **development**, with some **forms** more probable than others. . . . Bias is manifest, for example, in the nonrandom numbers of limbs, digits, segments and vertebrae across a variety of taxa . . . inherent **features of development** may have channelled **morphology** along specific pathways . . . the diversity of organismal **form** is only partly a consequence of natural selection—the particular evolutionary trajectories taken also depend on features of **development**. (Laland et al., 2015, p. 3)

These types of claims can be labeled *developmental form challenges* (hereafter, DFCs). The referent of "form" in these claims is structure or morphology—the composition and arrangement, shape or appearance of organic materials. The contrast is with function—activities performed or displayed by organisms. What is meant by "explaining form" or "theory of form," as well as the referent of "evolutionary theory" in these statements, requires further scrutiny.

---

1. My emphasis added in the quotations. It is sometimes claimed that evo-devo is central to expanding SET: "All these proposals for extending the Modern Synthesis are based upon the incorporation of developmental biology with evolutionary theory" (Weber, 2011, p. 76).

It would be hasty to say that these are *genuine* challenges; the jury is still out on whether they merit full engagement, even if it is agreed that they do not fall into the category of false challenges (e.g., Dickins & Rahman, 2012; Hoekstra & Coyne, 2007). One critic—Michael Lynch—puts it bluntly: "There's no general clamoring in the community for a new synthesis. . . . There are more things to explain, but I think a lot of us are happy with the fundamental framework to do that explaining in" (quoted in Grant, 2010, p. 31; see also Lynch, 2007; Minelli, 2010). To recognize this contested but distinctive status, we can refer to DFCs as *serious*. Here my primary goal is to argue that one can interpret these as challenges to how evolutionary theory is structured and not simply whether more content must be included. Serious DFCs are a different *kind* of challenge. Once DFCs are characterized in light of a distinction between content and structure, a more radical prospect emerges. Instead of a single, overarching organization for evolutionary theory, there can be more than one legitimate structure or multiple structures, and therefore we should maintain a pluralist stance about its "anatomy."

In an effort to describe the contours of this pluralism, I use the idea of *theory presentations* (Griesemer, 1984) to demonstrate how different structures are adopted to achieve specific methodological and epistemological goals in evolutionary theory, such as investigating or explaining form. These theory structures are analogous to idealizations, which intentionally depart from features known to be present in nature—knowingly ignoring variation in properties and excluding values for variables (Jones, 2005; Weisberg, 2007). This interpretation of DFCs differs from those previously offered (e.g., Laubichler, 2010) and is anchored by a perspective on how scientific theories can be structured. It also yields unexpected consequences: different DFCs may be genuine challenges to one another as much as they are serious challenges to SET.

5.2. Evolutionary Theory: Expansion and Extension

One of the unstated assumptions of both false and serious challenges is that there is a single evolutionary theory under scrutiny.[2] Unstated is not the same as unjustified, but there are implications. For example, the spatial metaphors used to describe the effect of serious challenges presume a relatively cohesive unit. To *expand* SET is to increase its content; to *extend* SET is to enlarge the boundaries of the theory such that more things are encompassed within it. The argument for expansion or extension as the appropriate modifiers for empirical and theoretical

2. "Darwinian evolutionary theory is generally still discussed today as a single theory" (Bock, 2010, p. 74).

developments that offer serious challenges derives from the perceived need to preserve much of what is found in SET.

> Evolutionary biologists have responded to various crises by augmenting the preexisting framework, building on what was already there, without overthrowing any of the previous foundations. . . . Yet, many . . . insist that—paradigm shift or not—the [modern synthesis], . . . is in need of some significant extension. . . . One can reasonably argue that none of this contradicts any tenet of the [modern synthesis], [but] to go further and state that there is not much new here and that all of this is already part of the [modern synthesis], implicitly or not, would be intellectually disingenuous and historically inaccurate. . . . Once a science is established, conceptual frameworks tend to expand, more than being replaced. (Pigliucci, 2007, pp. 2743, 2748)

This call for an extended or expanded theory of evolution seemingly goes beyond mere augmentation; significant revisions, inclusive of foundational changes in some cases, are being recommended. However, spatial increase (*expand*), rather than organizational rearrangement (*replace* or *reorganize*), is how this set of changes is usually described. This can be seen in the language educed to contrast SET assumptions with those of an extended evolutionary synthesis (EES) (Laland et al., 2015, p. 2; emphasis added).

> **Inheritance extends beyond genes to encompass** (transgenerational) epigenetic inheritance, physiological inheritance, ecological inheritance, social (behavioural) transmission and cultural inheritance
>
> Evolution [is] redefined as a transgenerational change in the distribution of heritable traits of a population. There is a **broadened notion of evolutionary process and inheritance**
>
> **Additional evolutionary processes**, including developmental bias and ecological inheritance, help explain macroevolutionary patterns and contribute to evolvability

The terminological choices of "extending beyond," "encompass," "broadened," and "additional" jointly align with the metaphor of spatial increase. New or revised concepts, such as evolvability, may be a part of this expansion in addition to empirical content. These conceptual additions can represent biological phenomena that fall within the scope of the theory but may not be explained in terms of it: "many of the empirical findings and ideas . . . are simply too recent and distinct from the framework of the Modern Synthesis [i.e., SET] to be

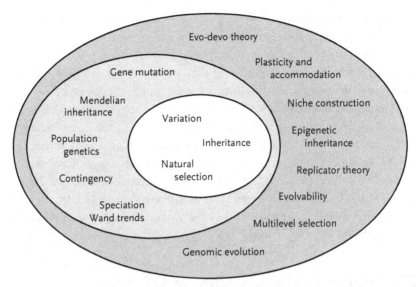

FIGURE 5.1 Key evolutionary concepts organized schematically in terms of Darwinism (center field), the modern synthesis (intermediate field), and the extended synthesis (outer field), representing the trend of a continuous expansion of evolutionary theory (from Pigliucci & Müller, 2010a).

reasonably attributed to it without falling into blatant anachronism" (Pigliucci & Müller, 2010a, p. 4). Diagrammatic representations of this expansion and extension depict clearly that what is in view involves a form of spatial increase rather than organizational rearrangement (Figure 5.1).

A second presumption about evolutionary theory—that it has a relatively uncomplicated structure—is also present in Figure 5.1 and implied by the verbs that are *not* used typically in serious challenges: *reorganize, rearrange, reorient,* or *restructure*. Uncomplicated is not the same as simple, and the expanded content or extended boundaries can be delimited regionally within the EES, such as conceptions of heredity being expanded to include epigenetic inheritance. Another part of this uncomplicated structure is that it has a "core" or foundational structure, displayed in Figure 5.1 by the central container. Often this is assumed to be the apparatus of evolutionary genetics that undergirds an understanding of how genotypic and phenotypic changes occur in populations as a consequence of natural selection, mutation, migration, and genetic drift.

Both of these unstated assumptions—that there is a single evolutionary theory and it has a relatively uncomplicated structure with a central core—contribute to the appeal of adopting a spatial metaphor to argue in favor of theoretical extension and expansion. This emphasizes new empirical findings and new concepts referring to phenomena that demand an explanation beyond the resources of SET while preserving the centrality of particular theoretical elements. And yet a significant extension that has an impact on conceptual foundations may involve the reorganization or rearrangement of theory components, even if it does not "overthrow" existing foundations. The challenge could be about how SET organizes biological knowledge and guides inquiry, not simply about whether it contains all of the relevant parts. Therefore the spatial container metaphor for an EES, with DFCs construed as demanding extension or expansion, might miss something important about these (and other) serious challenges. First we need to clarify the distinction between content and structure because it will help to fill out an interpretation of the referent of evolutionary theory in DFCs (section 5.3). Then we can turn to what is meant by "theory of" and the significance of needing to explain form (section 5.4).

5.3. Evolutionary Theory: Content Versus Structure

When analyzing a scientific theory, we can draw a distinction between its structure and content. The content is what is "in" the theory (hence the attractiveness, though not necessarily the legitimacy, of a spatial container metaphor); the structure is how this content is organized (e.g., a particular explanatory principle may have narrow or broad applicability across different aspects of a theory). How can we understand the content of evolutionary theory? One pedestrian but reliable approach is to compare the contents of textbooks aimed at introducing upper-level undergraduate and graduate students to evolutionary biology (e.g., Futuyma, 2005; Ridley, 2004). Although textbooks simplify actual practices and truncate or compress historical development (Kuhn, 1996), consensus across different authors can signal commitments among evolutionary researchers about "essential" topics that must be addressed.[3] There is relatively strong correspondence across contemporary evolutionary biology textbooks with respect to content, often signaled by a similar grouping of topics into chapter themes, and it is also manifested stably over time in various editions (Love, 2010). These themes include adaptation and fitness, variation and heredity, form and function, the history of life and diversity, systematics and phylogeny, speciation and coevolution, novelty and evolvability, and biogeography. Within these thematic areas, several dimensions of content are recognizable, including empirical findings about

---

3. Textbooks at the level of secondary education often err on the side of inclusiveness, but higher-level textbooks cannot include everything because they treat the subject matter with a higher degree of sophistication (often with an eye to being used in ongoing investigation). Thus, topical consensus can be an informative signal.

these domains, dynamical models of how phenomena in these domains behave, and key concepts that represent diverse instantiations of the phenomena that are being explained or doing the explaining.

Much of the literature discussing the adequacy of SET has concentrated on whether its content is sufficient. Does it capture the different facets of evolutionary processes in the sense of having models for their operation (apart from their empirical adequacy)? Did SET exclude developmental biology, have an overly narrow understanding of inheritance, or overlook particular domains of life that did not fit its explanatory paradigm? Does SET have the resources for dealing with variable rates of evolution, selection processes at the species level, or epigenetic inheritance (Kutschera & Niklas, 2004)? Once again, we see the appeal of the spatial metaphor, which relies on assumptions about the unit integrity of evolutionary theory and its relatively uncomplicated structure. An EES appears to involve "adding more" to SET with concepts that represent phenomena in need of explanation and empirical findings that were previously unavailable (Figure 5.1). Even in challenges that purport to be about structure, claims about content predominate (e.g., Gould, 2002). This intertwining of content and structure is explicable though it is not often considered in treatments of the structure of evolutionary theory.

The issue of evolutionary theory structure involves descriptive questions about how this conceptual content is organized: How is knowledge referred to as "evolutionary theory" ordered and arranged (Bock, 2010; Tuomi, 1981)? Does it have a structure that is similar to or different from other scientific theories (Rosenberg, 1985, ch. 5; Ruse, 1973, ch. 4)? The questions also can be posed prescriptively: *Should* we organize evolutionary theory in a particular way? *Should* it be similar in structure to other theories? Questions of this kind draw us into a philosophical conversation about the nature of scientific theories (Suppe, 1977). These questions are not wholly separated from the content of a theory, but they require exploring evolutionary theory from an angle that is less frequently adopted by working biologists.

The epistemic organization of SET has long worried historians and philosophers of biology (Depew & Weber, 1996). It seems anomalous in comparison with other scientific theories, which raises a question about whether existing philosophical analyses of theory structure are applicable (Caplan, 1978; Shapere, 1980). Because SET appears to include numerous commitments beyond the standard population genetic models of evolution, there might be good reason to distinguish it from more circumscribed domains (Beatty, 1986).[4] Although this maneuver might have warrant, it is common to view it all as a single package:

---

4. "It is very difficult to define the modern synthesis as a 'theory' ... it is indeed questionable whether the modern synthesis should be considered as one single theory" (Gayon, 1990, p. 3).

"the standard modern synthesis framework of evolutionary theory [i.e., SET]" (Müller, 2007, p. 943). Following out the idea of a "framework," this package might be construed as a hypertheory that subsumes subordinate theories of evolutionary mechanisms (Wasserman 1981), as a multifield theory (Darden, 1986), as a bundle of related theories that involve nomological and historical explanations (Bock, 2010), or as a kind of Kuhnian paradigm (Gayon, 1990). However, these suggestions all treat SET as a product rather than as a process (Reif et al., 2000). Contemporary defenders stress that it has changed over time and therefore has a dynamic, responsive—evolving—structure (Grant, 2010).

Philosophers have looked more narrowly at the theory structure associated with evolutionary genetics in terms of two main options: the syntactic view and the semantic view (e.g., Lloyd, 1988; cf. Suppe, 1977).[5] The syntactic view argues that a scientific theory is best reconstructed as an axiom system in first-order predicate logic that includes some (small) set of exceptionless laws as core principles and an empirical interpretation of its terms.[6] The semantic view generally treats scientific theories as families of models, where these models are indispensable to or constitutive of the meaning of the theory. Variations of the semantic view have been preferred for interpreting evolutionary theory, in part because they do not require universal laws, which appear to be absent from biology (e.g., Lloyd, 1988). However, the semantic view does not delimit the boundaries of a scientific theory (Suppe, 2000) and can be used to represent similar content differently. What kinds of internal organization might be relevant? The semantic view also does not indicate how theoretical content is appropriately specified in particular applications of models (Morrison, 2007).[7] How is content parceled out to the relevant models in different areas of evolutionary biology, especially when we take into consideration a variety of domains other than evolutionary genetics, including form and function, the history of life and diversity, systematics and phylogeny, novelty and evolvability, and biogeography? These questions

---

5. "[Philosophers] seem to agree on one thing: if you can characterize formal population genetics, then you have characterized the "guts" or "core" of evolutionary theory" (Lloyd, 1988, p. 8). This is not unjustified because many evolutionary biologists explicitly maintain a view in this vicinity: "population genetics provides the fundamental theory of [evolution]" (Ridley, 1993, p. vii).

6. "A theory is a relatively small body of general laws that work together to explain a large number of empirical generalizations, often by describing an underlying mechanism common to them all" (Rosenberg, 1985, p. 121).

7. Other features might play this kind of role, such as a pragmatic view of explanation, which would make the specification of theoretical content a function of the particular research question being asked. The concept of a theory presentation (Griesemer, 1984) is relevant here, and I return to it below (section 5.5).

about structural organization and arrangement seem more pertinent to the scientific practices associated with using SET than debates *between* syntactic and semantic views. That is, the pertinent questions about structure are not whether one or the other of these two approaches is correct, but rather what further needs to be said about structure to comprehend how biologists use evolutionary theory in practice. Although this is fully consistent with the applicability of general views on theory structure, and much of what follows can be explicitly framed in terms of the semantic approach, my analysis draws attention to the role of structural relationships based on material inferences (Brigandt, 2010). To see this more clearly, we need to focus on a subset of these domains germane to questions surrounding DFCs.

5.4. Developmental Form Challenges and the Structure(s) of Evolutionary Theory

Recall that DFCs involve an appeal to an extended or expanded SET that has the capacity to explain form or morphology. Now we must consider in more detail what is meant by "theory of" and the significance of needing to explain *form* in order to connect them both with the issue of theory structure. A convenient point of entry is Sean Carroll's description of a scientific theory in the context of the DFC he puts forward.

> But do questions posed about evo-devo and evolutionary theory matter to anyone besides the specialists and a few future historians? I think the answers matter very much. By "theory" here, I mean "structures of ideas that explain and interpret facts." . . . Without theories to organize and interpret facts, without the power of general explanations, we are left with just piles of case studies. Moreover, we are without the frameworks that enable us to make predictions about any particular case. Here, I will take the position that we have learned enough about the function, regulation, and history of the genes controlling animal development to formulate a general theory of how form evolves and to make predictions about the genetic path of morphological evolution—predictions that are now being fulfilled for a variety of traits and genes in diverse taxa. (Carroll, 2008, p. 25)

According to Carroll, a theory needs to explain, organize, interpret, and predict. These theory "functions" operate on facts, which I take to include particular phenomena (trichromes in fruit flies) or classes of phenomena falling under a concept at different levels of abstraction (serially repeated elements, color patterns, or vertebrate appendages). Scientific theories are described as "structures of ideas" that have "the power of general explanations." They also are supposed to be predictive, but Carroll does not have in mind the ability to predict the future course of phenotypic evolution, but rather the capacity to predict (or retrodict) what we will find in hitherto unstudied lineages with respect to the pattern of genetic regulation underlying their morphological evolution (cf. Stern & Orgogozo, 2009).

If we concentrate on *structures of ideas* and set to one side the meaning of *general explanations*, then a natural question is, What structures ideas? What provides organization to ideas such that they can have explanatory power? Notice that the syntactic view of theory structure answers this question in terms of a small set of axioms (laws) that, when combined with initial and boundary conditions, will explain the requisite myriad of empirical results (facts or case studies). This kind of theory structure does appear to have advocates among biological practitioners. For example, Sean Rice has argued, "evolutionary theory is not just a collection of separately constructed models, but is a unified subject in which all of the major results are related to a few basic biological and mathematical principles" (Rice, 2004, p. xiii).[8] However, this is unlikely to satisfy many proponents of DFCs because the biological principles invoked by Rice do not address directly the "function, regulation, and history of the genes controlling animal development" that are the foundation for retrodiction from Carroll's standpoint. Other evolutionary biologists might not favor Rice's proposal because it ignores key areas of evolutionary research, such as ecology, systematics, and biogeography. And mathematical formulation might not be the sine qua non for evolutionary theory.[9]

We noted previously that there might be more to theory structure than is contemplated in discussions over the relative merits of the syntactic and semantic views. Carroll's articulation of a genetic theory of morphological evolution (Carroll, 2008) suggests that something beyond these debates is relevant. Carroll

---

8. Although it is not stated explicitly, Rice's rejection of evolutionary theory as "a collection of separately constructed models" might be taken as an implicit criticism of the semantic view of theory structure.

9. "There is no need for evolutionary theory to limit its view to those topics which have been successfully formulated in mathematical terms" (Waddington, 1957, p. 102); "We cannot understand variation and adaptation unless we understand the cellular details through which this variation and adaptation is expressed. Phenotypic variation may seem messy and non-mathematical, but the most general truths in science (including evolution) emerged in these qualitative ways. Messy fields full of details, like chemistry, geology, and medicine, have managed to derive powerful theoretical understandings of complex phenomena. ... In some cases, this was followed by mathematical codification and in other cases not" (Kirschner, 2015, p. 216).

offers eight principles, derived from empirical investigation, that comprise (but do not exhaust) an EES that generally explains animal form.[10]

1. *Mosaic Pleiotropy*: Most proteins regulating development participate in multiple, independent processes that form and pattern morphologically disparate body structures.
2. *Ancestral Genetic Complexity*: Morphologically disparate and long-diverged animal taxa share similar toolkits of genes that build and pattern morphology.
3. *Functional Equivalence of Distant Orthologs and Paralogs*: Animal toolkit proteins often exhibit functionally equivalent activities in vivo when substituted for one another; the biochemical properties of these proteins and their interactions with receptors and cofactors have been largely conserved.
4. *Deep Homology*: The formation and differentiation of many structures such as eyes, limbs, and hearts are governed by similar sets of genes and conserved genetic regulatory circuits.
5. *Infrequent Toolkit Gene Duplication*: Duplications within toolkit gene families have been rare relative to duplications of other gene families because of their negative effects on gene dosage-sensitive developmental processes. Gene duplication is not necessary for morphological novelty.
6. *Heterotopy*: Changes in the spatial regulation of toolkit genes and the genes they regulate are associated with morphological divergence.
7. *Modularity of cis-Regulatory Elements (CREs)*: Large, complex, and modular cis-regulatory regions are a distinctive feature of pleiotropic toolkit loci.
8. *Vast Regulatory Networks*: Individual regulatory proteins control scores to hundreds of target gene CREs.

These principles are not exceptionless generalizations—some describe probabilistic tendencies (*infrequent* toolkit gene duplication). They also are not exhaustive—germane principles seem excluded, such as *heterochrony* (changes in temporal regulation). Some principles have clear empirical violations—the functional equivalence of distant orthologs and paralogs does not hold for all *Hox* genes (Heffer, Shultz, & Pick, 2010; Lynch & Wagner, 2010; Zhao & Potter, 2002). Other principles appear to be vague descriptions (*vast* regulatory networks).

What is the status of these principles? How do they figure in organizing or structuring an EES? They appear to be defeasible empirical generalizations

---

10. I have given paraphrases of these eight principles (and others below) that condense Carroll's articulation at numerous points, though preserving their original meaning.

that combine to yield hypotheses about morphological evolution: "To formu-late general genetic explanations—a predictive theory—about the evolution of form, these facts and inferences have to be distilled into hypotheses and tested against direct empirical data" (Carroll, 2008, p. 30).[11] If these hypotheses hold up to empirical testing (e.g., Fraser et al., 2011; Liao, Weng, & Zhang, 2010; Rebeiz, Patel, & Hinman, 2015), then we have in hand "structures of ideas" with "the power of general explanations." According to Carroll, there is one hypothesis with four components that can be distilled from the principles:

> *The cis-Regulatory Hypothesis*: CRE evolution is the predominant mechanism for the evolution of form; CRE mutations are the primary source of mor-phological variation. *(i)*: CRE sequence changes are sufficient to account for the evolutionary divergence of traits and gene regulation among pop-ulations, species, and higher taxa. *(ii)*: CRE evolution is necessary for the rewiring of regulatory networks. *(iii)*: Gene duplication and coding changes alone are insufficient to rewire regulatory networks. *(iv)*: CRE variation and divergence are detectable over much shorter timescales and taxonomic distances than are functional differences in transcription fac-tors or new toolkit gene duplications.

Each of these components also admits of exceptions, which Carroll thinks leads to the following (also defeasible) rule:

> Regulatory sequence evolution is the likely mode of genetic and morphologi-cal change if: a protein plays multiple roles in development, and mutations in its coding sequence are known or likely to have pleiotropic effects, and the locus contains multiple CREs.

Carroll encapsulates his genetic theory of morphological evolution in two sum-mary statements: (a) form evolves *largely* by altering the expression of function-ally conserved proteins; and (b) such changes *largely* occur through mutations in the cis-regulatory regions of developmental regulatory genes exhibiting mosaic pleiotropy and of target genes within the vast regulatory networks they control. These summarizations require drawing inferences among par-ticular principles (e.g., functional equivalence or mosaic pleiotropy) in order to offer a generalized explanation of how the evolution of form occurs. The

---

11. Other interpretations are possible, such as seeing the principles as guidelines for building evolutionary models, though I think Carroll tends to cast them primarily as defeasible empiri-cal generalizations in his discussion.

explanation appeals to changes in causal processes occurring during ontogeny ("altering the expression of functionally conserved proteins" and "mutations in the *cis*-regulatory regions of mosaically pleiotropic developmental regulatory genes"), not population-level mechanisms such as natural selection or genetic drift (Amundson, 2005).

The reader may find that it is hard to discern the structure of Carroll's explanatory proposal. In between eight principles, one hypothesis with four components, a rule with two clauses, and the entire theory summarized in two statements, the organizational features of Carroll's theory are less than transparent. And yet there is at least one interpretation of this structure that is illuminating. The linkages between the principles can be understood as material inferences—inferential relations based on the specific subject matter in view and the concepts that represent this content. Material inferences are not licensed by their form (as is the case for deductive arguments), and include inductive, explanatory, and investigative modes of scientific reasoning (Brigandt, 2010). For example, that CREs are modular (principle #7) implies that mutations in these elements have the potential to change the spatial regulation of toolkit genes thereby leading to morphological divergence (principle #6—heterotopy). Carroll's distillation of the CRE hypothesis and the resulting rule that he articulates are indicators of material inferences among concepts based on their content that yield general explanatory power. The claim that "regulatory sequence evolution is the likely mode of genetic and morphological change" uses mosaic pleiotropy (principle #1), the modularity of CREs (principle #7), and vast regulatory networks (principle #8) to infer the two clauses—a protein plays multiple roles in development, and mutations in its coding sequence are known or likely to have pleiotropic effects, and the locus contains multiple CREs. These two clauses have generality by virtue of three other principles (ancestral genetic complexity [#2], functional equivalence of distant orthologs and paralogs [#3], deep homology [#4]), and are robust because of principle #5 (infrequent toolkit gene duplication), while taking on evolutionary significance from principle #6 (heterotopy). These inferential connections are licensed because of the conceptual content represented in these principles, not by some logical form or abstract schema. The theory that emerges from this DFC is *structured* by the connections among these claims due to their content; its anatomical organization depends on the material structure found among its components.[12]

---

12. As noted previously, this is consistent with and arguably an application of the semantic view of theories. However, the key point is that it speaks to what is required for structure to comprehend how evo-devo biologists use theory in the particular context of explaining form or morphology.

5.4.1. The Structural Consequences of Developmental Form Challenges

In Carroll's DFC, theory *structure* emerges from different aspects of theory *content*. This is accented by other DFCs that emphasize different kinds of conceptual content and, as a consequence, put forward a different view of evolutionary theory structure. Different material inferences emerge from different sets of (defeasible) empirical and theoretical generalizations. Gerd Müller shares Carroll's complaint that SET does not explain form and requires an evo-devo input: "Evolutionary developmental biology (evo–devo) emerged as a distinct field of research in the early 1980s to address the profound neglect of development in the standard modern synthesis framework of evolutionary theory, a deficiency that had caused difficulties in explaining the origins of organismal form in mechanistic terms" (Müller, 2007, p. 943). But when Müller identifies "major theoretical themes of current evo-devo research and highlights how its results take evolutionary theory beyond the boundaries of the Modern Synthesis" (p. 943), there is no place of preeminence for CREs—modularity, plasticity, and innovation take center stage. Evolutionary developmental genetics, which describes the work of Sean Carroll and colleagues, is one among four different evo-devo research programs, including comparative embryology and morphology, experimental epigenetics, and computational modeling of developmental evolution.[13] Additionally, developmental genetics (inclusive of the CRE hypothesis) occupies only a small portion of the space of explanatory contributions required to understand the origin of phenotypic organization, which "necessarily includes many more factors than the evolution of gene regulation alone, notably the dynamics of epigenetic interactions, the chemicophysical properties of growing cell and tissue masses, and the influences of environmental parameters" (p. 944). This DFC involves material inferences that are absent from Carroll's genetic theory of morphological evolution; Carroll has no place for epigenetic interactions or computational modeling.[14] A different array of concepts representing different conceptual content yields a different species of *structural* challenge.

---

13. In fact, the CRE hypothesis is put in the dock: "Further experimental proof will be necessary to determine the extent to which gene regulatory change has a causal role in evolution" (Müller, 2007, p. 944). How much experimental proof would be sufficient is unclear, though evidence has continued to mount in its favor (Rebeiz et al., 2015).

14. "Many theoreticians sought to explain how periodic patterns could be organized across entire large structures. While the math and models are beautiful, none of this theory has been borne out by the discoveries of the last twenty years. The mathematicians never envisioned that modular genetic switches held the key to pattern formation, or that the periodic patterns we see are actually the composition of numerous individual elements" (Carroll, 2005a, p. 123).

So how can these DFCs be such close kin? One reason is that they share a commitment to a type of explanation not found in SET: "Evo-devo represents a causal-mechanistic approach towards the understanding of phenotypic change in evolution. . . . it seeks to explain phenotypic change through the alterations in developmental mechanisms" (pp. 945–946). This kind of argument has been echoed by other authors (e.g., Amundson, 2005; Laubichler, 2010). Diverse DFCs find a common bond in their requirement that typical population level explanations in evolutionary biology are insufficient. Explaining evolutionary change from one adult phenotype to another via population processes (such as natural selection) is not enough.[15] A weak interpretation of this common bond is in terms of *completeness*: population level explanations ignore the developmental details of how heritable variation is causally produced in each generation without which evolutionary explanations are incomplete. Although an integration of these may be difficult or even impossible to accomplish in practice (Amundson, 2005), the claim accords with the demand to expand or extend SET because the goal is to fill out or add more that is necessary to our theory of evolution. A strong interpretation of this common bond is in terms of *centrality*: "Causal-mechanistic" explanation is superior to explanations citing different forces operating on populations, and therefore evolutionary theory must be restructured so that these developmental considerations take a more central or foundational position (Laubichler, 2010). Both weak and strong interpretations suggest a different structuring of evolutionary theory, but the latter is of a more provocative (and perhaps unconvincing) nature. The proposed structures—however interpreted—differ across various DFCs because the conceptual content captured under the heading "developmental considerations" differs, and it is this content that provides the resources for material inferences that reorganize the various aspects of evolutionary theory.

The other feature that binds DFCs together is the stipulation that these new theory structures are required in order to offer general explanations of *form* or morphology. Whereas many aspects of SET focus on activities or functions (e.g., reproduction or foraging), DFC proponents share a commitment to explaining structure, the morphological aspects of life (e.g., shape, pattern, or organization): "I make the sharp distinction between the evolution of anatomy and the evolution of physiology. Changing the size, shape, number, or color patterns of physical traits is fundamentally different from changing the chemistry of physiological processes" (Carroll, 2005b, p. 1159). The contested axis of form

---

15. The converse is equally plausible: Evolutionary explanations that invoke changes from one ontogeny to another in terms of morphogenetic processes are insufficient apart from explanations of phenotypic change in terms of population processes.

(structure/morphology) versus function (adaptation/physiology) has been with us a long time (Russell, 1982 [1916]). We can characterize this axis as a difference in reasoning styles among biologists (Love, 2011).[16] These differences between methodologies allied to either form or function remain salient today, especially in and around evo-devo, where they account for why researchers adopt particular evidential and inferential practices. Comparative morphology and embryology exemplify form-oriented reasoning with practices that track the organization, structure, and transformation of organic parts on developmental and evolutionary timescales (Winther, 2006, 2011). These reasoning style differences also help us to see why continuing difficulties remain in synthesizing disciplinary approaches in biology (Laubichler & Maienschein, 2009). Many of the communities within evo-devo's interdisciplinary nexus favor a form-oriented reasoning style.[17] Those who foreground organismal performance (function) relative to the environment are at odds with "structuralists" (form) who keep development at the center of evolutionary studies (Amundson 2005, ch. 11). Sean Carroll abjures any responsibility for addressing adaptation.[18]

Thus, DFCs are held together by at least two chords of commonality: the requirement of mechanistic explanations that cite developmental factors involved in producing variation (rather than population level mechanisms), and an emphasis on shape, pattern, and organization (morphology) rather than function. To the degree that a DFC emphasizes only the latter, the challenge to SET from evo-devo (if any) is in terms of including relevant conceptual content that is absent (Minelli, 2010). A richer but cognate result is found in the weak interpretation of theory structure alterations due to developmental explanations (*completeness*), and is consonant with expanding or extending SET through some degree of reorganization, though the container metaphor is relatively unhelpful in showing us how the pertinent structural relationships obtain. A strong interpretation of the significance of causal-mechanistic explanations for evolution demands a radical

---

16. Other compatible characterizations are possible, such as using distinct families of models as explanatory.

17. "Proponents of twentieth-century Evo Devo emphasized the need to study form as a corrective to what they perceived to be an over-reliance on functional (fitness) or optimality (adaptation) type explanations in evolutionary biology.... The emphasis within these kinds of developmental explanations has thus been on form—or rather the (conserved) structure—of the developmental system" (Laubichler, 2009, pp. 11, 37).

18. "This is a specific theory about the evolution of animal form, and not a general statement about evolutionary adaptation.... I don't believe that any comparable genetic theory of adaptation is forthcoming, as all types of molecular changes clearly contribute substantially to organismal adaptation" (Carroll, 2008, p. 34). Many of those invested in this controversy will find this unpalatable because they presume the goal is to have a theory of phenotypic evolution, inclusive of morphological and physiological characteristics (Hoekstra & Coyne, 2007).

type of reorientation or reorganization. On this interpretation, developmental explanations—as species of causal-mechanistic explanations—must become the primary or *central* explanatory resource of evolutionary theory; they somehow displace the evolutionary genetics framework commonly held to constitute the core or guts of the theory. This may not be well motivated or warranted, either for reasons of completeness (cf. footnote 15) or because there is no such centrality.

Regardless of the overall merit of this strong interpretation, the inclusion of developmental content related to morphology in both the weak and strong interpretations alters the material inferential structures that are possible in evolutionary theory. Making this structure explicit is critical, which gives a compelling reason to reject the spatial container metaphor and its associated terminology. The DFCs imply much more than an enlargement of SET. The putative reorganization on a strong interpretation makes shape, pattern, and organization (morphology) the foremost set of *explananda*: "the evolution of form is the main drama of life's story, both as found in the fossil record and in the diversity of living species" (Carroll, 2005a, p. 294). An appropriate structure of ideas with general explanatory power for evolutionary theory has development and morphology at the center, not fitness and evolutionary genetics or the theoretical principles associated with them (Rice, 2004). It is a rejection of the claim that "nothing makes sense in evolution except in the light of population genetics" (Lynch, 2007, p. 8597).[19] However, DFCs are serious even on the weak interpretation because they involve a criticism of the current structuring of SET, both in terms of its relatively exclusive focus on the functional properties of organisms and its reliance on population-level explanations as the primary foundation for understanding evolution.[20]

There is a wrinkle in the fabric of these challenges. Serious DFCs do not yield the *same* structure of ideas for evolutionary theory because they include or foreground different content; this generates different networks of material inference relations. One way to smooth over this wrinkle is to claim that one DFC is correct while another is incorrect; there is one right way to challenge the function-oriented, population-level explanation perspective of standard evolutionary theory (whether in terms of a weak or strong interpretation). Alternatively, it might

---

19. "The litmus test for any evolutionary hypothesis must be its consistency with fundamental population-genetic principles . . . population genetics provides an essential framework for understanding how evolution occurs" (Lynch, 2007, p. 8598).

20. "I am not convinced that what we have learned about the evolution of form is being adequately considered in comparative genomics and population genetics, where the potential role of regulatory sequence evolution appears to be a secondary consideration, or ignored altogether" (Carroll, 2005b, p. 1164).

be that the different DFCs are not in direct competition with one another for knocking SET off its foundations (or merely decentering it). Different DFCs offer distinct structures of ideas with general explanatory power because they have different goals in view, such as understanding the nature of character identity versus understanding how changes in gene regulation produce novel patterns of variation. If this is the case, then the unique value of theory structure derived from evolutionary genetics and an orientation toward function changes to become another structured set of ideas with general explanatory power because different research questions are in view, which leads to distinct explanatory goals and criteria of adequacy (Love, 2010). We have arrived at the prospect that there might be more than one legitimate structure for evolutionary theory. Instead of smoothing over wrinkles, a garment with many folds has emerged.

5.5. Pluralism, Presentations, and Theory Structures as Idealizations

What does it mean to say that evolutionary theory has more than one legitimate structure? What does it mean to adopt a pluralist stance toward the structure of evolutionary theory? By way of contrast, we can begin with what a monist stance involves. Despite differences in how DFCs are formulated or how SET is understood currently, "monism holds that all such . . . accounts can be reconciled into a single unified account or that there is a single perspicuous representation system within which all correct accounts can be expressed" (Kellert, Longino, & Waters, 2006, p. xv). Therefore different DFCs must be reconciled with one another and ultimately reconciled with the principles from evolutionary genetics. The difficult task consists in connecting, with revision and reworking, the incompatible structural joints of the different accounts. Whether or not this can be accomplished, sooner or later, in practice or only in principle, is not as important. But it is crucial from the monistic stance to engage in this work of reconciliation, because the incompatibilities (and even inconsistencies) among the accounts indicate a problem with the theory. These different structures cannot all be true simultaneously, and we typically do not want a theory to contain falsehoods or contradictions.

A pluralist stance on theory structure takes the situation of current practice seriously and involves "a commitment to avoid reliance on monist assumptions in interpretation or evaluation coupled with an openness to the ineliminability of multiplicity in some scientific contexts" (Kellert et al., 2006, p. xiii). In other words, a single perspicuous representation system within which all correct accounts can be expressed may not be forthcoming. One reason to avoid relying on monist assumptions is that they can encourage us to obscure the very scientific contexts under scrutiny. A strategy for reconciling incompatible structure in evolutionary theory involves the postulation of "hidden" theory structure that is not present in scientific discourse. However, this makes scientific reasoning opaque because it obscures how biologists themselves access this structure to evaluate their own reasoning in the context of evolutionary theory. As a consequence, a monist stance can ignore features of scientific practice that deviate from the hidden structure sought, which prevents them from facing real counterexamples derived from the actual scientific context (Love, 2012).[21]

In the scientific context of different DFCs and the principles from evolutionary genetics, reconciling these theory structures with one another is not a criterion of adequacy for a pluralist stance; failed attempts at reconciliation do not represent a deficiency in scientific knowledge, but rather imply that different theory structures are playing distinct roles in evolutionary biology. This gives us an answer to what it means to say that evolutionary theory might have more than one legitimate structure. The legitimacy of a structure for evolutionary theory is licensed by its performance with respect to one or more structural roles. Possibilities for these roles include (Love, 2010): (1) articulating the relationships among different scientific disciplines in the theory (or parts of it); (2) facilitating ongoing research with respect to the theory (or parts of it), including offering explanations to answer specific research questions; (3) presenting the theory (or parts of it) in pedagogical contexts for others to assimilate; (4) comprehending the theory's historical development (or parts of it); and (5) providing a means to respond to internal or external challenges to the theory (or parts of it). For example, to facilitate research with respect to the phenomenon of adaptation in evolutionary theory, I can knowingly ignore aspects of the composition or arrangement of materials for a trait or lineage under scrutiny.

There is no reason to think that success with respect to one or more of these roles will beget success with respect to another—trade-offs are likely. "Scientists and philosophers should recognize that different descriptions and different approaches are sometimes beneficial because some descriptions offer better accounts of some aspects of a complex situation and other descriptions provide better accounts of other aspects" (Kellert et al., 2006, p. xxiv). But these different roles do provide criteria for deciding when and where different theory structures

---

21. Earlier philosophers of science who advocated the syntactic view of theory structure were criticized for just this reason: "The standard construal was never claimed to provide a descriptive account of the actual formulation and use of theories by scientists in the ongoing process of scientific inquiry; it was intended, rather, as a schematic explication that would clearly exhibit certain logical and epistemological characteristics of scientists' theories" (Hempel, 2001/1970, p. 222; cf. Griesemer, 1984).

obtain legitimacy. The main difference for a pluralist stance is that legitimacy is measured in more than one dimension, which assists in achieving a key goal for philosophical accounts of evolutionary theory's structure: "An adequate account of the structure of evolutionary theory will be capable of handling the range of theories already included under the name 'evolutionary theory'" (Lloyd, 1988, p. 8). The different material structures for evolutionary theory found in DFCs or in the principles of evolutionary genetics are responses to different sets of roles for evolutionary theory. A pluralist stance not only leads us to expect diverse structures but also explains why they diverge in the ways that they do.

These different structures for evolutionary theory can be interpreted as theory *presentations* (Griesemer, 1984), which refers to how content is specified in particular applications of a theory to answer research questions.[22] A particular presentation of evolutionary theory is recommended when it produces advantages to some particular end, such as facilitating ongoing investigative and explanatory research on natural phenomena within its domain. How scientists actually use evolutionary theory in practice guides preferences about its structural representation; choices are made in terms of what we do with the theory (Love, 2013). Theory presentations provide "a roadmap showing the topology of links connecting various conceptual items in the presentation of a theory . . . not only a specification of the kinds of system to which the theory applies, but also how the domain may be structured, what counts as an "inference," how measurements are to be made, etc." (Griesemer, 1984, p. 108). As might be expected, this structure exhibits a high degree of complexity, some of which is vague and allows for flexibility in the application of the theory presentation (see section 5.4). There is no essence to evolutionary theory, no canonical presentation that is preferred over all others.[23]

Theory presentations are always (and only) partial specifications that are heuristic in nature and thus prone to systematic biases (Wimsatt, 1980, 2007). The communicative context of a particular theory presentation must be consulted to understand its aims and significance (Griesemer, 1984, pp. 107ff).

---

22. Although I do not dwell on it here, the notion of a theory presentation was articulated explicitly in the context of a semantic view of theories.

23. "Theories are not picked out by presentations of theories. Rather, theories are picked out by a collection of [theory presentations] in a community of interacting authors and readers" (Griesemer, 1984, pp. 108–109). This conception fits with an evolutionary approach to scientific theories that emphasizes variations in presentations as the source of theory change over time (e.g., Hull, 1988). One can also connect these insights with a modified form of the semantic view of theory structure, if so desired. Importantly, this shared commitment to identifying theory presentations in an interactive community plays a key role in orienting researchers to the fact that they are working on the same theory.

By implication, theory presentations are not in constant competition with one another. Whether they are rivals depends on whether they align with respect to their goals. And, in fact, the explanation of form in causal-mechanistic terms is not one of the desiderata of many evolutionary researchers. Thus, a more standard theory presentation focusing on principles from evolutionary genetics that appeals to population-level processes need not be reconciled with each DFC, at least if understood in terms of *completeness*. Even the question of *centrality* might dissolve; references to the core or center of evolutionary theory can be reinterpreted as structural claims with embedded valuations about the particular roles a kind of theory presentation plays. The degree to which centrality remains a genuine issue is the degree to which explanations of function and form are subsumed under a broader category (e.g., phenotypes or traits) that would put theory presentations in competition with one another. Regardless of the issue of centrality, it does appear that there is a potential conflict between Carroll's DFC and that offered by Gerd Müller. These theory presentations share a common goal, the explanation of form, but diverge substantially in their structures (i.e., divergent arrays of conceptual content). Carroll sees the cis-regulatory hypothesis as sufficient for this explanatory goal; Müller and colleagues take this hypothesis as potentially necessary but not sufficient. For the latter, the explanatory goal requires contributions from the physics of development and mathematically modeling morphogenetic processes.

If theory presentations are understood from the stance of pluralism, the one thing we still lack is an account of how they can be incompatible or conflicting structures and yet presentations of the "same" evolutionary theory. Here the concept of idealization largely supplies what is missing. An idealization intentionally departs from features known to be present in nature—knowingly ignoring variation in properties and excluding values for variables (Jones, 2005; Weisberg, 2007). Applied to a theory presentation, an idealization intentionally departs from features known to be present *in the theory*. Researchers in the community of evolutionary biology knowingly ignore aspects of the theory, such as excluding variables that are clearly relevant to understanding evolutionary processes (see, e.g., Lloyd, Lewontin, & Feldman, 2008). A theory presentation focused on the explanation of form using a causal explanation in terms of development with phylogenetic context intentionally ignores natural selection. In many cases, natural selection or other population level processes are known to have operated on the morphology in question. This is knowingly ignored in the theory presentation. A theory presentation focused on allele frequency change in population genetics appealing to mutation, drift, or selection intentionally ignores development. However, development is known to have operated, translating allelic variation into phenotypic variation through ontogeny.

This is knowingly ignored in the theory presentation. Other theory presentations with different idealization choices are not difficult to find; they appear regularly in scientific practice.

One of the historical dangers associated with idealizations is that they can transmute into presumptions over time. Since theory presentations are always selective, they require choosing some subset of principles to use. What was knowingly ignored can become unknowingly ignored or even thought to be insignificant because of the empirical fruitfulness of an idealization.[24] Preferred idealizations (theory presentations) metamorphose to become central or the "core" of a theory. This pitfall of forgetfulness for idealization provides at least part of the historical explanation for the types of structures observed in SET (population-level explanations and function-oriented research), and offers one reason why DFCs have emerged in a particular guise. DFCs are a way of exposing idealization choices with respect to theory presentations that have been forgotten; theoretical amnesia has deleterious consequences with respect to our understanding of the features knowingly ignored in SET.

Elsewhere I have argued that an erotetic structuring or presentation of evolutionary theory has many advantages when thinking about how DFCs fit into the context of an EES (Love, 2010; see also Love, 2014). The relatively stable topics of evolutionary biology noted earlier (section 5.3) correspond to organized arrays of conceptual, theoretical, and empirical research questions (problem *agendas*) about complex phenomena that are concurrently tackled by multiple disciplinary approaches. This theory presentation offers a way to understand historical continuity in evolutionary theory (e.g., specific empirical questions can change while the problem agenda retains its identity), draws attention to how different disciplinary communities make distinct explanatory contributions (in the context of different problem agendas), displays how research is guided in particular directions (by research questions in a problem agenda rather than hypothesis testing), and accounts for methodological choices and the adoption of explanatory standards. It also displays how different topics in evolutionary theory coalesce around major divisions of biological science—genetics, cell, and developmental biology; ecology; systematics—and explains why concentrating on one or more problem agendas in evolutionary theory to the exclusion of others can lead to a devaluing of developmental or ecological factors in formulating an account of evolutionary processes (Figure 5.2).

---

24. "Selective emphasis, choice, is inevitable whenever reflection occurs. . . . Deception comes only when the presence and operation of choice is concealed, disguised, denied. . . . Honest empirical method will state when and where and why the act of selection took place, and thus enable others to repeat it and test its worth" (Dewey, 1981/1925, p. 34).

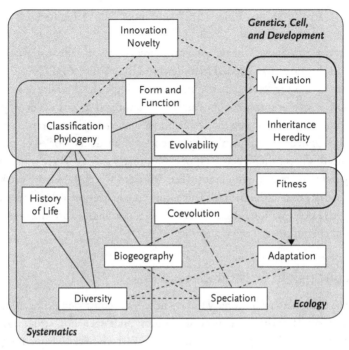

FIGURE 5.2 An idealized picture of an erotetic structure for evolutionary theory that focuses on the main problem agendas and their interrelations, as well their correspondence to primary domains of biological inquiry: systematics, ecology, and genetics, cell, and development (from Love, 2010).

And yet this visualized erotetic presentation of evolutionary theory is not perfect. It does a poor job of recovering the distinction between pattern and process. When evolutionary theory is understood in terms of the *mechanisms* of change, then the inclusion of features such as classification is incongruent. This is exactly what we would expect from distinct choices that motivate different theory presentations. The idealized picture in Figure 5.2 makes it difficult to see how phenotypic plasticity and niche construction fit into evolutionary theory, both of which are important to discussions of an EES (Laland et al., 2015). Phenotypic plasticity is a phenomenon that requires juxtaposing variation and ecology; niche construction is a phenomenon that requires juxtaposing inheritance and ecology. The specific erotetic theory presentation represented in Figure 5.2 makes these connections opaque, if not impossible to recover, but the important point is that it knowingly does so. It is an idealization and the danger lies mainly in forgetting this fact. Disguising or concealing it can lead to the impression that some aspects of evolutionary inquiry are being marginalized, as many have claimed for developmental considerations in SET (Carroll, 2008; Müller, 2007; Pigliucci, 2007; Pigliucci & Müller, 2010a).

We have now achieved the necessary resources to discard the spatial container metaphor; it is no longer needed to understand the nature of DFCs or their significance for ongoing research in evolutionary biology. The diversity of structures observable in evolutionary theory can be reinterpreted as sets of incompatible idealizations that need not be resolvable into a single perspicuous representation system. It is unnecessary to invoke hypertheories, multifield theories, or Kuhnian paradigms to dissect the structure(s) of evolutionary theory. There is no need to identify a core or central foundation. The degree to which DFCs are even competing theory presentations for SET depends on whether the roles for the structure in these presentations align. When there is not alignment, incompatibility and complementarity can travel together.

5.6. Conclusion: Structural Lessons

Serious challenges to evolutionary theory have been construed primarily as content oriented rather than structural in nature. Once seen in a structural light, DFCs that accent the apparent deficit of SET with respect to explaining form and advocate for an EES need not be hostage to the spatial container metaphor. Instead, they can be seen as members in a set that includes a plurality of theory presentations guided by diverse goals in the scientific enterprise of explaining evolution. Just like the choices made for the purposes of idealization, theory presentations knowingly ignore features of evolutionary theory and, in doing so, offer a structure that is incompatible or inconsistent with another theory presentation. If theory presentations align in their goals then they should be seen as potential competitors. Ironically, this means there can be more direct conflict between different DFCs than with the evolutionary genetics perspective of SET. At the same time, we can recognize that another manifestation of conflict comes in forgetting the manifold goals of evolutionary theory and setting up particular presentations as central or making them the "core" of the theory. This is a pseudo-conflict because it relies on forgetting the idealization choices made in articulating a particular theory presentation. Divergent structures serve different sets of methodological and epistemological goals in evolutionary theory, and a pluralist stance can recognize the fruitfulness of maintaining these reasoning strategies: "The explanatory and investigative aims of science [may] be best achieved by sciences that are pluralistic, even in the long run" (Kellert et al., 2006, pp. ix–x).

Philosophers have not fully appreciated serious challenges to scientific theories. They have tended to fixate on issues of content rather than the organization of theoretical knowledge when analyzing how to choose between different theories, evolutionary or otherwise. The question of how to decide between theories in philosophy of science begins with empirical adequacy. Dilemmas emerge when there is more than one empirically adequate theory available (and then epistemic values, such as parsimony or fecundity, are invoked), but little has been said about how there may be more than one structure available within a single empirically adequate theory. The possible relations of complementarity and competition among these structures are recognizable in a pluralist perspective that views these structures as idealizations. More work needs to be done with respect to articulating the diverse roles of evolutionary theory that serve different investigative and explanatory goals; the diversity of these roles and how they are combined is what drives the diversity of theory presentations with distinct structures (Love 2013). That these presentations are incompatible and cannot be reconciled into a single canonical representational structure could be a disturbing conclusion for some. However, the conclusion may be more attractive than it first appears. False theory presentations may be the means to a truer evolutionary theory (Wimsatt, 1987).

Acknowledgments

Thanks to Denis Walsh and Philippe Huneman for the opportunity to participate in "Challenges to Evolutionary Theory II: An Interdisciplinary Workshop" (University of Toronto, February 2010), where an embryonic version of this paper was presented. The feedback from those present was critical in pushing me to think more deeply about the issue of pluralism for theory structure, though clearly more work remains. Additional critical comments from Erik Curiel, James Griesemer, and Elisabeth Lloyd on related material and earlier drafts of this chapter were invaluable in improving its content and structure, though this does not imply their agreement with the theses advanced here.

References

Amundson, R. (2005). *The changing role of the embryo in evolutionary thought: Roots of evo-devo*. New York, NY: Cambridge University Press.

Beatty, J. (1986). The synthesis and the synthetic theory. In W. Bechtel (Ed.), *Integrating scientific disciplines* (pp. 125–135). Dordrecht, Netherlands: M. Nijhoff.

Bock, W. J. (2010). Multiple explanations in Darwinian evolutionary theory. *Acta Biotheoretica, 58*, 65–79.

Brigandt, I. (2010). Scientific reasoning is material inference: Combining confirmation, explanation, and discovery. *International Studies in Philosophy of Science, 24*, 31–43.

Caplan, A. L. (1978). Testability, disreputability, and the structure of the modern synthetic theory of evolution *Erkenntnis, 13*, 261–278.

Carroll, S. B. (2008). Evo-devo and an expanding evolutionary synthesis: A genetic theory of morphological evolution. *Cell, 134,* 25–36.

Carroll, S. B. (2005a). *Endless Forms Most Beautiful: The New Science of Evo-Devo.* New York: W.W. Norton.

Carroll, S. B. (2005b). Evolution at two levels: On genes and form. *PLoS Biology, 3,* e245.

Darden, L. (1986). Relations among fields in the evolutionary synthesis. In *Integrating Scientific Disciplines,* ed. W. Bechtel, 113–123. Dordrecht: M. Nijhoff.

Depew, D. J., & Weber, B. H. (1996). *Darwinism evolving: Systems dynamics and the genealogy of natural selection.* Cambridge, MA: MIT Press.

Dewey, J. (1981/1925). Experience and nature. In *John Dewey: The Later Works, 1925-1953,* ed. J. A. Boydston. Carbondale, IL: Southern Illinois University Press.

Dickins, T. E., & Rahman, Q. (2012). The extended evolutionary synthesis and the role of soft inheritance in evolution. *Proceedings of the Royal Society B: Biological Sciences, 279,* 2913–2921.

Fraser, H. B., Babak, T., Tsang, J., Zhou, Y., Zhang, B., Mehrabian, M., & Schadt, E. E. (2011). Systematic detection of polygenic *cis*-regulatory evolution. *PLoS Genetics, 7,* e1002023.

Futuyma, D. J. (2005). *Evolution* (4th ed.). Sunderland, MA: Sinauer Associates.

Gayon, J. (1990). Critics and criticisms of the modern synthesis: The viewpoint of a philosopher. In M. K. Hecht, B. Wallace, & R. J. Macintyre (Eds.), *Evolutionary Biology* (Vol. 24, pp. 1–49). New York, NY, and London, UK: Plenum Press.

Gould, S. J. (2002). *The structure of evolutionary theory.* Cambridge, MA: Belknap Press of Harvard University Press.

Grant, R. (2010). Should evolutionary theory evolve? *The Scientist, 24,* 24–31.

Griesemer, J. (1984). Presentations and the status of theories. *PSA: Proceedings of the Biennial Meeting of the Philosophy of Science Association, 1,* 102–114.

Heffer, A., Shultz, J. W., & Pick, L. (2010). Surprising flexibility in a conserved Hox transcription factor over 550 million years of evolution. *Proceedings of the National Academy of Sciences USA, 107,* 18040–18045.

Hempel, C. G. (2001/1970). On the "standard conception" of scientific theories. In J. H. Fetzer (Ed.), *The philosophy of Carl G. Hempel: Studies in science, explanation, and rationality* (pp. 218–236). New York, NY: Oxford University Press.

Hoekstra, H. E., & Coyne, J. A. (2007). The locus of evolution: Evo-devo and the genetics of adaptation. *Evolution, 61,* 995–1016.

Hull, D. (1988). *Science as a process: An evolutionary account of the social and conceptual development of science.* Chicago, IL: University of Chicago Press.

Jones, M. R. (2005). Idealization and abstraction: A framework. In M. R. Jones, and N. Cartwright (Eds.), *Idealization XII: Correcting the model; Idealization and abstraction in the sciences* (pp. 173–217). Poznan Studies in the Philosophy of the Sciences and the Humanities (Vol. 86). Amsterdam, Netherlands/New York, NY: Rodopi.

Kellert, S. H., Longino, H. E., & Waters, C. K. (2006). Introduction: The pluralist stance. In S. H. Kellert, H. E. Longino, & C. K. Waters, eds. *Scientific pluralism* (pp. vii–xxix). Minnesota Studies in the Philosophy of Science (Vol. 19). Minneapolis: University of Minnesota Press.

Kirschner, M. W. (2015). The road to facilitated variation. In A. C. Love (Ed.), *Conceptual change in biology: Scientific and philosophical perspectives on evolution and development* (pp. 199–217). Berlin: Springer.

Kuhn, T. S. (1996). *The structure of scientific revolutions* (3rd ed.). Chicago, IL, and London: University of Chicago Press.

Kutschera, U., & Niklas, K. J. (2004). The modern theory of biological evolution: An expanded synthesis. *Naturwissenschaften, 91,* 255–276.

Laland, K. N., Uller, T., Feldman, M. W., Sterelny, K., Müller, G. B., Moczek, A., . . . Odling-Smee, J. (2015). The extended evolutionary synthesis: Its structure, assumptions and predictions. *Proceedings of the Royal Society B: Biological Sciences, 282,* 20151019.

Laubichler, M. D. (2009). Form and function in Evo Devo: Historical and conceptual reflections. In M. D. Laubichler & J. Maienschein (Eds.), *Form and Function in Developmental Evolution* (pp. 10–46). New York, NY: Cambridge University Press.

Laubichler, M. (2010). Evolutionary developmental biology offers a significant challenge to the neo-Darwinian paradigm. In F. J Ayala & R. Arp (Eds.), *Contemporary debates in philosophy of biology* (pp. 199–212). Malden, MA: Wiley-Blackwell.

Laubichler, M. D., & Maienschein, J. (2009). *Form and function in developmental evolution.* New York, NY: Cambridge University Press.

Liao, B., Weng, M. P., & Zhang, J. (2010). Contrasting genetic paths to morphological and physiological evolution. *Proceedings of the National Academy of Sciences USA, 107,* 7353–7358.

Lloyd, E. A. (1988). *The structure and confirmation of evolutionary theory.* Westport, CT: Greenwood Press.

Lloyd, E. A., Lewontin, R. C., & Feldman, M. W. (2008). The generational cycle of state spaces and adequate genetical representation. *Philosophy of Science, 75,* 140–156.

Love, A. C. (2010). Rethinking the structure of evolutionary theory for an extended synthesis. In M. Pigliucci & G. B. Müller (Eds.), *Evolution: The extended synthesis* (pp. 403–441). Cambridge, MA: MIT Press.

Love, A. C. (2011). Darwin's functional reasoning and homology. In M. R. Wheeler (Ed.), *150 years of evolution: Darwin's impact on contemporary thought and culture* (pp. 49–67). San Diego, CA: SDSU Press.

Love, A. C. (2012). Formal and material theories in philosophy of science: A methodological interpretation. In H. W. de Regt, S. Okasha, & S. Hartmann (Eds.), *EPSA philosophy of science: Amsterdam 2009* (pp. 175–185). The European Philosophy of Science Association Proceedings (Vol. 1). Berlin: Springer.

Love, A. C. (2013). Theory is as theory does: Scientific practice and theory structure in biology. *Biological Theory, 7*, 325–337.

Love, A. C. (2014). The erotetic organization of developmental biology. In A. Minelli & T. Pradeu (Eds.), *Towards a theory of development* (pp. 33–55). Oxford, UK: Oxford University Press.

Lynch, M. (2007). The frailty of adaptive hypotheses for the origins of organismal complexity. *Proceedings of the National Academy of Sciences USA, 104*, 8597–8604.

Lynch, V., & Wagner, G. (2010). Revisiting a classic example of transcription factor functional equivalence: Are Eyeless and Pax6 functionally equivalent or divergent? *Journal of Experimental Zoology Part B: Molecular and Developmental Evolution, 316*, 93–98.

Minelli, A. (2010). Evolutionary developmental biology does not offer a significant challenge to the neo-Darwinian paradigm. In F. J. Ayala & R. Arp (Eds.), *Contemporary debates in philosophy of biology* (pp. 213–226). Malden, MA: Wiley-Blackwell.

Morrison, M. (2007). Where have all the theories gone? *Philosophy of Science, 74*, 195–228.

Müller, G. B. (2007). Evo-devo: Extending the evolutionary synthesis. *Nature Reviews Genetics, 8*, 943–949.

Pigliucci, M. (2007). Do we need an extended evolutionary synthesis? *Evolution, 61*, 2743–2749.

Pigliucci, M., & Müller, G. B. (2010a). Elements of an extended evolutionary synthesis. In M. Pigliucci & G. B. Müller (Eds.), *Evolution: The extended synthesis* (pp. 3–17). Cambridge, MA: MIT Press.

Pigliucci, M., &Müller, G. B. (2010b). *Evolution: The extended synthesis*. Cambridge, MA: MIT Press.

Rebeiz, M., Patel, N. H., & Hinman, V. F. (2015). Unraveling the tangled skein: The evolution of transcriptional regulatory networks in development. *Annual Review of Genomics and Human Genetics, 16*, 12.1–12.29.

Reif, W-E., Junker, T., & Hoßfeld, U. (2000). The synthetic theory of evolution: General problems and the German contribution to the synthesis. *Theory in Biosciences, 119*, 41–91.

Rice, S. H. (2004). *Evolutionary theory: Mathematical and conceptual foundations*. Sunderland, MA: Sinauer Associates.

Ridley, M. (1993). *Evolution*. Cambridge, MA: Blackwell Science.

Ridley, M. (2004). *Evolution*. Cambridge, MA: Blackwell Science.

Rosenberg, A. (1985). *The structure of biological science*. Cambridge, MA: Cambridge University Press.

Ruse, M. (1973). *Philosophy of biology*. London, UK: Hutchinson University Library.

Russell, E. S. (1982/1916). *Form and function: A contribution to the history of animal morphology*. Chicago, IL: University of Chicago Press.

Shapere, D. (1980). The meaning of the evolutionary synthesis. In E. Mayr & W. B. Provine (Eds.), *The evolutionary synthesis: Perspectives on the unification of biology* (pp. 388–398). Cambridge, MA: Harvard University Press.

Stern, D., & Orgogozo, V. (2009). Is genetic evolution predictable? *Science, 323,* 746–751.

Suppe, F. (2000). Understanding scientific theories: An assessment of developments, 1969–1998. *Philosophy of Science, 67,* S102–S115.

Suppe, F. (1977). *The structure of scientific theories.* Urbana: University of Illinois Press.

Tuomi, J. (1981). Structure and dynamics of Darwinian evolutionary theory. *Systematic Zoology,* 30, 22–31.

Waddington, C. H. (1957). *The strategy of the genes: A discussion of some aspects of theoretical biology.* London, UK: Allen and Unwin.

Wasserman, G. D. (1981). On the nature of the theory of evolution. *Philosophy of Science, 48,* 416–437.

Weber, B. H. (2011). Extending and expanding the Darwinian synthesis: The role of complex systems dynamics. *Studies in History and Philosophy of Biological and Biomedical Sciences, 42,* 75–81.

Weisberg, M. (2007). Three kinds of idealization. *Journal of Philosophy, 104,* 639–659.

Wimsatt, W. C. (1980). Reductionistic research strategies and their biases in the units of selection controversy. In T. Nickles (Ed.), *Scientific Discovery: Case Studies* (pp. 213–259). Boston Studies in the Philosophy of Science. Boston, MA: Reidel.

Wimsatt, W. C. (1987). False models as means to truer theories. In M. H. Nitecki & A. Hoffman (Eds.), *Neutral models in biology* (pp. 23–55). New York, NY: Oxford University Press.

Wimsatt, W. C. (2007). *Re-engineering philosophy for limited beings: Piecewise approximations to reality.* Cambridge, MA: Harvard University Press.

Winther, R. G. (2006). Parts and theories in compositional biology. *Biology and Philosophy, 21,* 471–499.

Winther, R. G. (2011). Part-whole science. *Synthese, 178,* 397–427.

Zhao, Y., & Potter, S. S. (2002). Functional comparison of the *Hoxa4, Hoxa10,* and *Hoxa11* homeoboxes. *Developmental Biology, 244,* 21–36.

6 TOWARD A NONIDEALIST EVOLUTIONARY SYNTHESIS

Stuart A. Newman

The physical world as conceived by scientists during the years when the theory of evolution by natural selection was taking form was fundamentally inert. Some (Aristotelian) entities only changed their position or state by influence from the outside, while others (Newtonian ones) remained in uniform motion in the absence of such interventions. Anyone with aspirations to a materialist science who studied animals and plants in the 19th century was under pressure not to stray too far beyond the metaphysical framework defined by the era's established physical science (Depew & Weber, 1995; Newman & Linde-Medina, 2013).

But physical science itself was in flux during this period. Beyond mechanics (the best codified physical theory of the time), the early nineteenth century saw the emergence of chemistry, thermodynamics, and electricity and magnetism, which variously dealt with qualitative transformations, irreversible processes, and globally acting fields, concepts more suggestive than those of classical mechanics of biological phenomena.

Some recognition of the protean possibilities inherent in the new chemistry and physics is seen in the work of Jean Baptiste de Lamarck, who described living tissues as being suffused by "contained" and "subtle" fluids (discussed in Newman & Bhat, 2011); rational morphologists like Étienne Geoffroy Saint-Hilaire, Lorenz Oken, and Richard Owen, who postulated "laws of form" (Webster & Goodwin, 1996); and "teleomechanists" such as J. F. Blumenbach, C. F. Kielmeyer, and J. P. Müller, who (as discerned by Timothy Lenoir, who coined this term) adopted a teleological heuristic in their naturalistic anatomical and physiological studies (Lenoir, 1982). But none of these perspectives actually engaged with the new physical sciences, which notwithstanding Friedrich Wohler's stunning 1828 test tube synthesis of urea, had not yet attained the analytical sophistication to bring living systems into their ambit.

Charles Darwin chose to adhere to a conventionally materialistic conception of organismal transformations rather than adopt the more vibrant but speculative one of his Continental contemporaries (and the outlier English rational morphologist, Owen).[1] This was in keeping with the manufacturing wisdom of the burgeoning industrial revolution. Darwin's notion that trial and error acting on incremental change could arrive at all possible endpoints ("If it could be demonstrated that any complex organ existed, which could not possibly have been formed by numerous, successive, slight modifications, my theory would absolutely break down"; Darwin, 1859, p. 158) echoed the experience of his maternal grandfather, the pottery magnate Josiah Wedgwood, who tested approximately 5,000 different earthenware formulations before arriving at the one that made his fortune.[2] Alfred Russel Wallace, the coformulator of the theory of evolution by natural selection, had a somewhat more dynamic concept of living materials, one (as indicated in the following passage) that anticipated what later came to be called "control theory":

> The action of this principle [natural selection] is exactly like that of the centrifugal governor of the steam engine, which checks and corrects any irregularities almost before they become evident; and in like manner no unbalanced deficiency in the animal kingdom can ever reach any conspicuous magnitude, because it would make itself felt at the very first step, by rendering existence difficult and extinction almost sure soon to follow. (Wallace, 1858, p. 61)

But Wallace, like Darwin, also came to advocate a thoroughly gradualistic theory of evolution.

By the mid-20th century physics had undergone additional, dramatic changes. The most famous examples are the emergence of relativity and quantum theory, but these relate to phenomena on larger, smaller, and faster scales than those pertaining to biological systems. More relevant to physiology, development, and evolution was the rise of physics of the middle scale—nonlinear oscillations and multistable dynamical systems, separation and transitions between phases, viscous flow, reaction-diffusion coupling, and the spontaneous

---

1. This was despite Darwin's shared sense of awe with the *Naturphilosophen* at nature's copious creativity (see, e.g., Richards, 2002).

2. See the website of the Wedgwood Museum, Stoke-on-Trent, UK: http://www.wedgwoodmuseum.org.uk/learning/discovery_packs/2179/pack/2184/chapter/2344

breaking of compositional spatial symmetry. Soft, chemically and mechanically excitable matter, even of the nonliving sort, changes in sometimes abrupt ways, partly responsive to environmental influences and typically assuming preferred forms (e.g., waves and vortices in water, wrinkles and cracks in paint) (Forgacs & Newman, 2005). While it is understandable that Darwin and Wallace would have adopted a gradualist perspective, late 20th-century biologists could only do so by ignoring contemporary advances in the sciences of materials.

A nondirectional incrementalism nonetheless remained the ruling doctrine of the modern synthesis (MS), the fusion of Darwinism and population genetics that had come to dominate evolutionary discourse after the 1940s (Huxley, 2010/ 1942; Mayr & Provine, 1998). It was asserted that abrupt phenotypic changes ("saltations") would be exceptionally rare and typically lethal, not contributing significantly to changes in the populational norm of any trait. The idea that certain forms are inherent to living matter and that evolution correspondingly follows preferred pathways ("orthogenesis") was also proscribed in the MS since it undermined the decisive role of adaptation (Simpson, 1944). The most heretical idea of all (although it was increasingly entertained by Darwin himself over the course of his career) was that external influences could have a causal (as opposed to a selective) role in molding propagated phenotypes. According to Ernst Mayr, a founder of the MS, "Nothing strengthened the theory of natural selection as much as the refutation, one by one, of all the competing theories, such as saltationism, orthogenesis [and] inheritance of acquired characters" (Mayr, 1982).

The primary reason that newer knowledge of material processes failed to be incorporated into an evolving theory of evolution was the central assumption of the MS that the genotype is responsible for all the inherited aspects of the phenotype. This view arose from attempts to incorporate the results and interpretations of the plant breeding experiments of Gregor Mendel into the Darwin-Wallace theory. The main problem was that Mendel's findings involved "genes of large effect," the variants of which were associated with dramatic changes in phenotype. A Mendelian school of thought at the turn of the 20th century, "mutationism," placed such genes at the center of a theory of evolution that threatened to marginalize the role of natural selection (Stoltzfus & Cable, 2014). While animals and plants that represented large departures from the phenotypic norm ("sports") were well known to Darwin, this phenomenon was rejected by him and his successors as a basis for evolutionary change, since their occurrence was independent of the allegedly universal functional optimization of traits for which his theory provided a fully gradualistic account.[3]

---

3. In the first edition of *The Origin of Species*, in his discussion of cycles of incremental refinement of form produced by animal breeders, Darwin makes a passing reference to the rise of

Mutationism was laid to rest (prematurely; it now has a body of evidence in favor of it; Chouard, 2010) by the forging of a Darwinian–(weakly) Mendelian consensus that held that all inherited aspects of an organism are determined coordinately by many genes of small effect, and that macroevolutionary changes result solely from accumulated microevolutionary changes. Under this assumption there was no need for a physically based theory of morphological or other aspects of the phenotype, nor, in fact, any requirement to focus (for the purposes of evolutionary accounts) on the phase in the organism's life cycle during which its phenotype was generated, namely, development. The production of organismal form and its transformations over the course of evolution, rather than being based on the dynamics of complex materials (incorporating a genetic component), was conceived of instead as a function of populations of genetically operated machines (later, genetically programmed computers), continually but gradually refined by encounters with external challenges (Linde-Medina, 2010; Newman & Linde-Medina, 2013).

No one has ever come close to delineating the purported rules by which genes direct or program phenotypes without the crucial involvement of material processes. Of course the proponents of the MS recognize that organisms exist in a physical world; they just believe that physics is a constant background condition: necessary, but nonspecific.[4] Since it depends on a hidden genotype-to-phenotype code, the existence of which was guaranteed by the faith that there is no medium other than genes by which specific organismal features could be transmitted from one generation to the next (Moss, 2002; Newman, 1988; Sarkar, 1996), a major tenet of the MS must be counted as a variety of philosophical idealism.[5]

---

unprecedented novelties in both plants and animals (Darwin, 1859). In *Variation of Animals and Plants Under Domestication*), he wrote at greater length about "sudden variations or sports" in bird species, acknowledging that "we are profoundly ignorant of the cause" (Darwin, 1868, vol. 1, p. 213).

4. Richard Dawkins, for example, wrote recently, "Any theory of life has to explain how the laws of physics can give rise to a complex flying machine like a bird or a bat or a pterosaur, a complex swimming machine like a tarpon or a dolphin, a complex burrowing machine like a mole, a complex climbing machine like a monkey, or a complex thinking machine like a person. . . . Darwin explained all of this with one brilliantly simple idea—natural selection, driving gradual evolution over immensities of geological time" (Dawkins, 2011).

5. It should be recognized that modern scientific accounts will necessarily be very different from the mechanistic materialism of the 19th century (see, Moss, 2002), and that the line between materialist and idealist modes of explanation are not as sharp as it was once held to be. It is nonetheless not defensible to adopt a vague idealism (as does Nagel, 2012) in response to the inadequacies of the MS, which are not due to its materialism, but rather to its idealist genetic determinism and disconnection from any notion of developing organisms as mesoscale matter.

An alternative, philosophically materialist, evolutionary synthesis would need to demonstrate the efficacy and specificity of physical determinants of development, at least at the inception of multicellular forms (Newman, 2011a, 2014a). It must describe the means by which this specificity has been propagated and modified over the course of many generations, and the changing relations of physical and genetic mechanisms over the course of evolution (Newman, 2005, 2012, 2014b). It would also need to account for phenomena that appear inconsistent with the MS. These include (1) the striking conservation over hundreds of millions of years and across radical reorganizations of body plans, of the so-called developmental-genetic toolkit (Carroll et al., 2004; Wilkins, 2002); (2) the Avalon and Cambrian explosions and other punctuated episodes in the history of multicellular life (Conway Morris, 2006; Shen et al., 2008); and (3) the recurrent "embryonic hourglass" mode of animal evolution, in which diversification is seen at early and late stages of development, with a largely conserved "phylotypic" stage between the less constrained phases (Newman, 2011b; Sander, 1983). In the following sections I provide the outlines of a materialist synthesis that fulfills the requirements listed above and goes some distance in meeting the challenge of explaining phenomena for which the MS has often needed to resort to unfounded assumptions.

6.1. Dynamical Patterning Modules and the Developmental Genetic Toolkit

The animal or metazoan phyla employ a variety of processes to construct their bodies and organs during embryogenesis. These can be understood in terms of "dynamical patterning modules" (DPMs), associations of specific gene products and physical effects they are capable of mobilizing in the context of cell aggregates (Newman & Bhat, 2008, 2009). In modern animals, a set of ancient proteins, the nontranscription factor subset of products of the developmental-genetic toolkit (termed the "interaction toolkit" in what follows), and other macromolecules (e.g., polysaccharides) they produce, are the molecular components of the DPMs, mediating morphogenesis and cellular pattern formation. Because of their high degree of conservation, the DPM-associated molecules are likely to have served similar functions at the evolutionary origin of these developmental processes (reviewed in Forgacs & Newman, 2005, and Newman & Bhat, 2009).

The DPM that converted unicellular ancestors into cohesive tissues was a *sine qua non* for the emergence of the animals. Specifically, expression of surface molecules (e.g., protocadherins, lectins) in ancestral single-celled organisms called holozoans (Abedin & King, 2008; King et al., 2008; reviewed in Newman,

2016b) provided a basis, under suitable external conditions (e.g., the presence of sufficient calcium ion in the surrounding medium) for mobilization of the force of adhesion between cells (Steinberg, 2007). Simple adhesion was not enough to bring about this new state of matter, however. What was required in addition was a novel protein domain, the transmembrane segment linking the cell-cell adhesion function of cadherins with the cell movement and cell polarity-generating function of the cytoskeleton. Cell clusters mediated by these metazoan-specific "classical" cadherins have the formal properties of liquids, simultaneously exhibiting a high degree of cohesivity and independent mobility of their (cell) subunits (Heisenberg & Bellaiche, 2013). With the emergence of "liquid tissues" an entire suite of novel morphogenetic effects was set into motion.

Like droplets of nonbiological liquids, liquid tissues exhibit *surface tension*. If their subunits are uniform in their surface adhesive properties and isotropic in shape, the principle of *free energy minimization* will thus cause them to lack interior spaces and be spherical by default (reviewed in Forgacs & Newman, 2005; Newman & Forgacs, 2009). If, in contrast, the cells are polarized in either surface composition or shape (each of which is mediated by the animal-exclusive Wnt protein and signaling pathway; Karner, Wharton, & Carroll, 2006a, 2006b; Newman, 2016), the same principle of energy minimization will cause the aggregates to become hollow or elongated. These multicellular-scale effects, physically analogous to the "generic" (i.e., also applicable to nonliving systems; Newman & Comper, 1990; Newman, 1994) processes that lead to micelle and liquid crystal formation in solutions of anisotropic molecules, constitute another set of DPMs (Newman & Bhat, 2008).

Liquid tissues also resemble nonliving liquids in their capacity to undergo phase separation (as in oil and water) when they contain subunits of different types with distinct self- and non-self-affinities. In ordinary liquids such affinities are straightforward properties of molecules and their interactions. Cell-cell affinity strength, in contrast, is not only a function of the number of adhesive molecules on the cell surface but also of the tension exerted on the inner cortex of each cell by its cytoskeleton (Heisenberg & Bellaiche, 2013). In mixed cell type aggregates these two factors, acting in concert, can lead to *tissue multilayering*, manifested as *gastrulation* in early-stage animal embryos and analogous tissue segregation and rearrangement effects during organ formation (Forgacs & Newman, 2005).

All cells potentially exhibit *multistable dynamics* owing to the properties of their internal transcription factor-mediated gene regulatory networks (reviewed in Forgacs & Newman, 2005, and Kaneko, 2011). Such networks are not themselves DPMs, since they operate (at least in animal systems) on the single- rather than multicellular level (Newman et al., 2009). But the physical effect that enables a multicellular cluster to maintain a balance of different cell types—*lateral inhibition*—by which a cell signals adjoining or nearby ones to assume a different state than its own (Meinhardt & Gierer, 2000), is a DPM, since it is manifested only in multicellular aggregates. Lateral inhibition in metazoans is almost invariably mediated by the Notch signaling pathway (Ehebauer, Hayward, & Arias, 2006).

Secreted mobile molecules (*morphogens*, such as members of the hedgehog, bone morphogenetic protein, and fibroblast growth factor families) constitute a DPM in multicellular aggregates by virtue of spreading across them, forming *gradients*. These nonuniformities promote spatially dependent cell differentiation among otherwise equivalent cells. While morphogen transport in embryonic tissues sometimes occurs by free diffusion from a fixed source, a generic physical effect, most often the cells take an active part in secretion, binding, and uptake along the pathway, actively shaping the morphogen's distribution (Lander, 2007).

Aggregated cells can also attain coordinated biochemical states over broad spatial domains by another DPM, based on a ubiquitous single-cell functionality, *oscillation* in internal chemical composition (Reinke & Gatfield, 2006). The synchronization of weakly coupled oscillators is a truly emergent property, since it occurs spontaneously, requiring only weak, relatively nonspecific interactions between the oscillators (Garcia-Ojalvo, Elowitz, & Strogatz, 2004; Strogatz, 2003). Oscillatory or related biochemical or gene regulatory dynamics coupled with an inhibitory morphogen gradient can lead to a variety of tissue patterning effects, some acting as "clock-and-wavefront" generators of segmentation (Cooke & Zeeman, 1976; Dequéant & Pourquié, 2008) and others as reaction-diffusion-based generators of periodic or quasi-periodic skeletal structures or appendages (Meinhardt & Gierer, 2000; Raspopovic et al., 2014; Turing, 1952; Zhu, Zhang, Alber, & Newman, 2010; reviewed in Kondo & Miura, 2010).

Extracellular matrices, secreted materials that remain next to or surrounding the cells that produced them (e.g., collagens, fibronectin, the polysaccharide hyaluronan) can, in the multicellular context, cause epithelial sheets to resist bending deformations and promote invagination, evagination, and branching; turn liquid-like mesenchymal aggregates solid; or cause cohesive tissue masses to disperse into separate cells (Comper, 1996). This combination of toolkit gene products and their derivatives, along with the physical effects they mediate, also constitute a class of DPMs.

The DPMs of developing plants use different toolkit genes and products, which in certain cases (because plant tissues are solid), mobilize different physical effects than those of the liquid tissues of animals (Hernández-Hernández, Niklas, Newman, & Benítez, 2012). The evolutionary-developmental principles are nonetheless the same.

In contrast to the idealistic view implicit in the MS that genes must collectively constitute a program for the generation of morphological phenotypes, the

DPM framework advanced here emphasizes the connection of phenotype generation to the broader material world. Genes act not as conveyers of "information," but as direct and indirect specifiers of building materials, the organizational properties of which inescapably conform to the physical laws appropriate to their composition and scale. What is passed from one generation to the next are not only genes and their variants, but material systems[6], including the specific capacities of their gene products to mobilize physical forces and effects in cell aggregates, and the conditional (environment-dependent) outcomes of those forces and effects.

6.2. Rapid Phylogenesis and the Conservation of the Interaction Toolkit

Although many of the metazoan interaction toolkit molecules and pathways evolved before the Cambrian explosion, probably "for specialized, terminal cell differentiations rather than for the earliest steps in basic patterning" (Duboule & Wilkins, 1998; see also King et al., 2008), they took on their roles as mediators of body plan development and organogenesis coincidentally with the emergence of multicellularity. (This was also the case for molecular novelties like Wnt and the transmembrane portion of the classical cadherins, which appear, enigmatically, with the animals; reviewed in Newman, 2016b). Contrary to the expectations of Darwinian-Wallacean gradualism, many members of both the transcriptional and interaction toolkits perform their functions to surprisingly similar ends across great phylogenetic distances. For example, transcription factors Pax6 and Nkx2.5 act early in the development pathway of eyes and hearts, respectively, in both mice and fruit flies, and Dlx helps specify the distal ends of developing limbs in these same organisms. Furthermore, DPM-associated molecules of the interaction toolkit like Wnt and Notch mediate cell polarity (and its tissue sequelae) and lateral inhibition, respectively, in sponges, cnidarians, chordates, and other phyla. More generally, homologous genes of the transcriptional and interaction toolkits can often be swapped between members of morphologically disparate phyla and maintain their functions (reviewed in Newman, Bhat, & Mezentseva, 2009).

If we assume, as implied by a gradualist scenario invoking genetic changes of small effect, that diversification of the animal phyla took immensely long times, the evolutionary conservation of the toolkit genes and their functions is puzzling. However, we now know that the metazoan phyla appeared in the fossil

---

6. Referred to as "biogeneric," to reflect their commonality with, and distinctiveness from, the organizational properties of non-living materials (Newman, 2016a).

record relatively abruptly during the late Precambrian and early Cambrian periods (Larroux et al., 2008; Rokas, Kruger, & Carroll, 2005). Sheetlike and hollow spherical forms (Yin et al., 2007) and budding and segmented tubes (Droser & Gehling, 2008), possibly representative of the most ancient metazoans, are seen in Precambrian Ediacaran deposits beginning about 630 million years ago. Within a 10-million-year period beginning 575 million years ago (the so-called Avalon explosion; Shen, Dong, Xiao, & Kowalewski, 2008) the two-layered cnidarians (corals, hydroids) and ctenophores (comb jellies) had appeared. Essentially all the three-layered metazoans—arthropods, echinoderms, mollusks, chordates, and so forth, then followed within a space of no more than 20 million years, beginning about 535 million years ago (the Cambrian explosion; Conway Morris, 2006).

The animal phyla are not distinguished from one another by the exclusive presence of any of the standard cell types—muscle, cartilage, nerve, skin, and so on. However, each phylum is characterized by the presence of specific suites of early-developing morphological motifs, constituting its specific range of "body plans." Based on the relation between the body plan characters and differences in the distribution of the interaction toolkit genes across the animal phyla, I have suggested that it is precisely the complement of these genes in a metazoan organism that defines its "phylotype" (Newman, 2011b, 2012). But because there is also a substantial overlap in the genes of the interaction toolkit across wide phylogenetic distances, their conservation presents even more difficulties for the gene centrism of the MS than does that of the cell-type-defining transcriptional toolkit. If the occurrence of rapid phylogenesis is acknowledged however, a causal agency must be identified other than variation of the structural (i.e., protein-specifying regions) of the interaction toolkit genes, and slow selection of the associated incrementally superior phenotypes (Newman, 2006).

One widely discussed proposal suggests that the evolution of regulatory portions of the toolkit genes is more important for the diversification of form than is the evolution of the proteins themselves (Carroll, Grenier, & Weatherbee, 2004). Although there is a debate about this question within the gene-centric framework of the MS (e.g., Hoekstra & Coyne, 2007; Wagner & Lynch, 2008), none of these gene-based perspectives makes a plausible attempt to account for why natural selection would have been so rapid during the period of metazoan phylogenesis, or why the resulting body plans and organs assumed the actual forms they did.

With its focus on the physics mobilized by the products of the interaction toolkit genes, the DPM perspective provides a more plausible account of macroevolution than the exclusively gene-centric MS. One reason for this is that previously irrelevant (in the single-cell world) effects of mesoscale physics are instantly engaged by cell aggregates expressing certain preexisting (or newly acquired) genes, making the rapid diversification of form at the early stages of metazoan evolution mechanistically straightforward. Furthermore, the ancestors of the metazoans are predicted to have had inherently plastic morphologies, since physical determinants are subject to external parameters. While the MS has long dealt with plasticity as an adaptive outcome of selection (e.g., Beldade, Mateus, & Keller, 2011; de Jong, 2005), and selectionist scenarios even inform most analyses that seek to incorporate epigenetic and extraorganismal factors into development (e.g., Gilbert, Sapp, & Tauber, 2012; Jablonka & Lamb, 2005; West-Eberhard, 2003), the plasticity associated with DPMs is a primitive property of tissues as physical materials, not requiring any prior selection. This presents a genuine challenge to the genotype-phenotype mapping assumed in the standard model.

The old question of how complex forms have gradually been built up then shifts to one of understanding the slow process (evidently taking multiple millions of years) of stabilizing and regularizing the multigenerational transmission of morphotypes generated by the DPM-based organizational forces.[7] The DPM framework also provides a set of expectations (based on the predictability of physical determination) concerning the array of forms the semisolid, chemically and mechanically excitable materials constituted by primitive liquid tissue cell clusters and organ primordia would have assumed. This is orthogenesis, but with a material rather than idealist basis.[8]

6.3. The "Embryonic Hourglass" Disconfirms Genetic Determinism

Dynamical patterning modules arose by virtue of the capacity of the ancient gene products of the interaction toolkit to mobilize mesoscale (in this usage, acting on a spatial scale of 0.1–10 mm) physical forces, processes, and effects when they

---

7. I suggest that these transformations, which involve what Waddington (1942) called "canalizing," and Schmalhausen (1949), "stabilizing" evolution, turned families of cellular aggregates with rough-hewn and environmentally conditioned modes of morphological development into populations of recognizable, reliably produced organisms. See also Walsh (2006) and Moss & Newman (2015) for discussions of the related idea that holistically integrated self-organizing physical processes of development are the material basis of Kant's concept, in the *Critique of Judgment*, of organisms as "natural purposes."

8. The notion that physics, as a nonbiological externality, is not involved in the specificity of developmental outcome is contradicted by the DPM framework. The DPMs collectively incorporate a heterogeneous set of mesoscale physical determinants, and the concept entails the recognition that specific categories of molecules (which may not be present in all organismal lineages or at corresponding stages of development), are needed to mobilize the distinct processes and effects (Newman, 2011a).

came to operate in cell clusters. The cells that aggregated to form these ancient clusters, however, were not necessarily genetically identical. Indeed, we know from studies of organisms ranging from social amoeba (Nanjundiah & Sathe, 2011) to mammals (Fehilly, Willadsen, & Tucker, 1984; Meinecke-Tillmann & Meinecke, 1984) that clusters of genetically heterogeneous developmentally competent cells ("chimeras") can undergo development and produce viable outcomes that embody compromises between the characters of their originating groups. This developmental consensus-building within exotic combinations of cells, and the resulting forms, are of course not products of natural selection. Such behavior is much more typical of self-organizing dynamical systems (or concatenations of such systems) than of the sort of deterministic "genomic computers" (Istrail, De-Leon, & Davidson, 2007) envisioned by MS-oriented evolutionary narratives.

Analogously, in some premetazoan cell clusters it is probable that not all the available interaction toolkit genes were present in every cell. In such cases DPMs associated with these genes would have been only partly efficacious. While populations of genetically heterogeneous cells can propagate the capacity to generate approximately defined morphotypes, the origin of individuals capable of passing on to their progeny reliable developmental programs would have depended on mechanisms to enforce genetic uniformity among all cells of an aggregate (Grosberg & Strathmann, 2007; Niklas & Newman, 2013). The emergence of such mechanisms would have also brought into existence populations of organisms characterized by consistent and potentially unique complements of DPMs. According to the hypothesis stated above, these would have been the founding populations of the distinct phyla.

All present-day animal embryos pass through an ontogenetic stage analogous to the primitive cell clusters in which DPMs first appeared. This "morphogenetic stage" (Newman, 2011b; see also Seidel, 1960), which corresponds (depending on the species), to the morula, blastula, blastoderm, or inner cell mass, consists of dozens to hundreds of identically sized cells. This is the precise stage of development at which all the DPMs characteristic of the embryo's phylum become operative.

If the complement of DPMs define phylum-identity, and the DPMs only function developmentally in *medias res*, this suggests that the main evolutionary innovation responsible for the emergence of both organismal individuality and stably distinct phyla among the animals was the *egg* (Newman, 2014c, 2014d). Either by cleavage of a single large cell, or by confinement of the products of blastomere division by a surrounding matrix or shell, the interpolation of an egg stage of development in the animal life cycle ensures that the morphogenetic stage of individual development consists of genetically identical cells rather than the genetically heterogeneous population formed by simple aggregation (Newman, 2011b, 2014c).

Eggs, even of closely related animal groups like mammals and birds, can be morphologically extremely different and still give rise to organisms with similar body plans. This is a straightforward consequence of the fact that it is the DPMs that define the phyla. The size, shape, and internal heterogeneity of the egg should in principle have no role in determining the phylum-specific morphological motifs, and this appears to be the case (reviewed in Newman, 2011b). On the other hand, polarities and heterogeneities present in the egg, or generated post-fertilization, will provide variable (within a phylum) but stereotypical (within a class) alternative boundary and initial conditions for the later operation of the DPMs. This may help provide the classes within phyla with their specific morphological identities (Newman, 2011b).

The convergence of animal embryos of many phyla (such as the chordates, and particularly their vertebrate subphylum) to a mid-embryogenesis conserved morphology, from which they then develop in variable but phylum-associated fashions, has led to the positing of a "phylotypic stage" of development (Sander, 1983) or "zootype" (Slack, Holland, & Graham, 1993). There are a number of interpretations of this effect, which has been termed the "embryonic hourglass" (Duboule, 1994; Hall, 1997; see also Horder, 2008; Raff, 1996). Adaptationist accounts have explained midembryonic conservation by stage-specific selection pressures on various aspects of developmental anatomy, mechanics, or gene expression patterns (often of "master regulatory genes" such as those of the Hox family; Slack et al., 1993; see Galis & Sinervo, 2002; Galis et al., 2002; Raff, 1996), or on the functional requirements of embryonic life, with morphological conservation as a side effect (Wray & Strathmann, 2002).

The DPM framework provides an alternative, materially straightforward interpretation of the transient convergence to a common morphological platform during animal embryogenesis. Trivially, the stage of highest similarity is the morphogenetic stage, which represents a recapitulation of the originating events of metazoan evolution—the formation of clusters of similarly sized cells. In the case of the embryos of present-day animals this stage is achieved by cleavage or blastomere confinement, based on the innovation of a specialized cell type, the egg. The lack of strict constraints on the size and shape of the egg, and the variations of its cytoplasmic patterning, even in morphologically identical or similar species, as well as the extensive tolerance of developmental outcome in the face of perturbations at the earliest stages (reviewed in Newman, 2011b, 2014c), argue strongly against preprogrammed determination of phylotypic body plan formation. The determinants of this process are instead proposed to be the "physico-genetic" (Newman, 2012) DPMs, which only come into play when the morphogenetic stage is reached. During the subsequent embryonic stages, a phylotypic set of DPMs (whose operation is subject to variations in initial and boundary conditions at the egg stage) will generate a correspondingly phylotypic set of morphological motifs with class-specific modifications.

Because of the commonality of DPMs in all members of a given phylum, the early, postmorphogenetic stage development of representatives of different classes within that phylum will necessarily be similar in form and gene expression patterns. While giving the impression of an evolutionarily achieved phylotypic stage, the "embryonic hourglass" is actually evidence of the origination of the phyla in the self-organization mediated by the physical effects harnessed by specific subsets of the interaction toolkit gene products.[9]

6.4. Conclusion

In his book *The Growth of Biological Thought*, Ernst Mayr makes what he considers to be "a rather paradoxical claim" that much of the intellectual structure of *The Origin of Species* can be accounted for by the fact that the leading paleontologists and biologists of Darwin's day were natural theologians whose descriptions routinely invoked "what we would now call adaptations."

He goes on to state that "when 'the hand of the creator' was replaced in the explanatory scheme by 'natural selection,' it permitted incorporating most of the natural theology literature on living organisms almost unchanged into evolutionary biology" (Mayr, 1982, pp. 104–105). For many students of the history and philosophy of science this might have been a tip-off that too much in the way of religious idealism had been carried over into the new theory (see Newman, 1995).

It is perhaps understandable that Mayr, as one of the biologist architects of the MS in the years before the role of development and its physical determinants in the generation of biological form had become evident, would have been captivated by the prospect of a universal explanation based on ordinary gradualist means, that is, the incremental changes that genetic variation imposes on phenotypes. Less excusable were the views of mid-20th-century physicists like Erwin Schrödinger, who in his influential book *What Is Life?* (Schrödinger, 1945) wrote, "The chromosome structures are instrumental in bringing about the development they foreshadow. They are the law-code and executive power—or to use another simile, they are the architect's plan and builder's craft in one," or Max Delbrück,

---

9. Salazar-Ciudad (2010) analyzes the complementary question to the one discussed here (why morphological variation is so prevalent at early embryonic stages despite mid-development conservation) of the mechanistic determinants of morphological variation at later developmental stages across and within phyla.

who wrote three decades later, "It is my contention that Aristotle's principle of the unmoved mover originated in his biological studies. . . . Unmoved mover perfectly describes DNA. DNA acts, creates form in development, and it does not change in the process" (Delbrück, 1976). These views, oblivious of the new physics of the mesocale emerging during this period, which was finding increasing application in physiological studies (Newman, 2014e), were frankly retrograde accounts of development and evolution. Later on, in a period of intellectual ferment signaled by the emergence of the new field of evolutionary developmental biology (Alberch, 1985; Bonner, 1983; Gould, 1977; Gould & Lewontin, 1979; see Müller, 2007; Müller & Newman, 2005; Pigliucci & Müller, 2010, for more recent overviews), a reactionary consolidation of this idealistic ideology among defenders of the Darwinian faith was even more in evidence (Dawkins, 1996; Dennett, 1995).

One thing in favor of the gene-centric gradualism of the MS when other alternatives were not in view was the promise of a way to "get from here to there" in phenotypic terms. All complex systems, if produced in multiple copies, will exhibit variation. For highly integrated systems like present-day animals and plants, small variations will be much more common than large ones. The MS prescribes ignoring the larger changes, which would call for concrete physical explanations, and focusing instead on the minor ones, which in contrast raise no questions about organismal organizational principles.

Admitting the possibility of discontinuous evolution is therefore equivalent to devaluing the decisive role of natural selection. While even Darwin, as noted above, recognized that unprecedented form may suddenly appear in natural populations, such extreme variants—"hopeful monsters"[10]—were dismissed by the MS as being inherently unfit for the ecological niche in which they arose. The notion of an unforgiving world that removes all deviations from contention except those rare ones that are marginally better suited to preexisting slots is essential to the transformation of phenotypes by natural selection.[11]

But if such anomalies had a reasonable chance of surviving and contributing offspring to the next generation, there would be no need to invoke gradualism

10. This term was coined by the geneticist Richard Goldschmidt (Goldschmidt, 1940) to refer to novel phenotypes that might arise in the space of a generation by the mutation of a single broadly acting gene (a "macromutation").

11. The paleontologists Stephen Jay Gould and Niles Eldredge flirted with the idea of nongradual morphological change in some of their earlier writings on the fossil record (e.g., Gould & Eldredge, 1977). But both eventually (and, I suggest, too accommodatingly) settled into an interpretation of "punctuated equilibrium" that was consistent with the population-based incrementalism of the MS (Gould, 2002; Eldredge, 1999). See Wilson (2013) for an evolutionary developmental biology-informed view of the fossil record.

as the exclusive or even primary mode of evolutionary change. In fact, research in the fertile field of "niche construction" has shown that survival of animals and plants does not invariably depend on their suitability to the niches in which they originated (Laland et al., 2008); organisms, far from being passive objects of circumstance are actually active makers of their own ways of life (Lewontin, 2000).

Another theme in Darwin's writings, however, though largely unacknowledged in the MS, points in a different direction, one more compatible with contemporary concepts of the physical determinants of form and niche construction. Frédéric Bouchard, in a recent paper titled "Darwinism Without Populations: A More Inclusive Understanding of the 'Survival of the Fittest'" (Bouchard, 2011), notes, "the notion of fit between an organism and the problems posed by the environment has always been part of Darwinism." Arguing that a focus on the individual organism was eclipsed by the MS's emphasis on allele frequency in populations (see also Ariew & Lewontin, 2004), he calls for moving away from a concept of fitness based exclusively on differential reproductive success (and the resulting changes in the populational distribution of the associated genetic variants) to a more ecological fitness concept.

Although some ecological concepts of fitness (e.g., differential efficacy in resource utilization) can readily be accommodated to classical models within the MS framework, others e.g., behaviors or interactions that enhance the carrying capacity of an ecosystem (Peacock, 2011), or those involving phenotypic plasticity (Peacor, Allesina, Riolo, & Pascual, 2006), open the door to persistence of forms that deviate from the populational norm in ways not contemplated by the MS (Amundson, 2011).

Adopting an ecological notion of fitness and recognizing the reality of niche construction in no way break the connection between genetic and evolutionary change (Kokko et al., 2017). But since under these assumptions discontinuous jumps in morphological phenotype (particularly at early stages of cladogenesis) are neither exceptional nor automatically disqualified from persisting, genes of large effect are no longer fatal to the tenet of descent with modification. Moreover, and also contrary to the standard model, evolutionary change will often have preferred directions dependent on inherent properties of living systems and, insofar as these properties are plastic, register influence from the external environment in the generation of characters. The DPM framework described here provides a material-causal basis for the origin and continued generation of animal form consistent with this perspective, vitiating the notion that each element of the morphological phenotype was built up gradually by successive cycles of past adaptation (discussed in Linde-Medina, 2011). The monocausal genetic idealism of the MS has therefore outlived its usefulness.

Acknowledgments

I thank Philippe Huneman, Marta Linde-Medina, Gerd Müller, and Denis Walsh for helpful comments on earlier drafts of this chapter, and Lenny Moss for his insights on the issues discussed here. This work was supported, in part, by the National Science Foundation program on Frontiers in Integrative Biological Research.

References

Abedin, M., & King, N. (2008). The premetazoan ancestry of cadherins. *Science, 319,* 946–948.

Alberch, P. (1985). Problems with the interpretation of developmental sequences. *Systematic Zoology, 34,* 46–58.

Amundson, R. (2011). *Human relations in view of Evo Devo: Or, revenge of the abnormals.* Conference paper, International Society for History, Philosophy, and Social Studies of Biology, Salt Lake City, Utah.

Ariew, A., & Lewontin, R. C. (2004). The confusion of fitness. *British Journal for the Philosophy of Science, 55,* 365–370.

Beldade, P., Mateus, A. R., & Keller, R. A. (2011). Evolution and molecular mechanisms of adaptive developmental plasticity. *Molecular Ecology, 20,* 1347–1363.

Bonner, J. T. (Ed.). (1983). *Evolution and development: Report of the Dahlem Workshop on Evolution and Development, Berlin 1981, May 10–15.* New York, NY, and Berlin, Germany: Springer Verlag.

Bouchard, F. (2011). Darwinism without populations: A more inclusive understanding of the "survival of the fittest." *Studies in History and Philosophy of Biological and Biomedical Sciences, 42,* 106–114.

Carroll, S. B., Grenier, J. K., & Weatherbee, S. D. (2004). *From DNA to diversity: Molecular genetics and the evolution of animal design.* Malden, MA: Blackwell.

Chouard, T. (2010). Evolution: Revenge of the hopeful monster. *Nature, 463,* 864–867.

Comper, W. D. (Ed.). (1996). *Extracellular matrix.* Amsterdam, Netherlands: Harwood Academic Publishers.

Conway Morris, S. (2006). Darwin's dilemma: The realities of the Cambrian "explosion." *Philosophical Transactions of the Royal Society of London. Series B, Biological Sciences, 361,* 1069–1083.

Cooke, J., & Zeeman, E. C. (1976). A clock and wavefront model for control of the number of repeated structures during animal morphogenesis. *Journal of Theoretical Biology, 58,* 455–476.

Darwin, C. (1859). *On the origin of species by means of natural selection; or, The preservation of favoured races in the struggle for life.* London, UK: Murray.

Darwin, C. (1868). *The variation of animals and plants under domestication.* London, UK: Murray.

Dawkins, R. (1996). *Climbing mount improbable*. New York, NY: Norton.

Dawkins, R. (2011). Attention Governor Perry: Evolution is a fact. In S. Quinn (Ed.), On Faith. *The Washington Post*, Washington, DC. http://www.washingtonpost.com/blogs/on-faith/post/attention-governor-perry-evolution-is-a-fact/2011/08/23/gIQAuIFUYJ_blog.html

de Jong, G. (2005). Evolution of phenotypic plasticity: Patterns of plasticity and the emergence of ecotypes. *New Phytologist, 166*, 101–117.

Delbrück, M. (1976). How Aristotle discovered DNA. In K. Huang (Ed.), *Physics and our world: A symposium in honor of Victor E. Weisskopf* (pp. 123–130). New York: American Institute of Physics.

Dennett, D. C. (1995). *Darwin's dangerous idea: Evolution and the meanings of life*. New York: Simon & Schuster.

Depew, D. J., & Weber, B. H. (1995). *Darwinism evolving: Systems dynamics and the genealogy of natural selection*. Cambridge, MA: MIT Press.

Dequéant, M. L., & Pourquié, O. (2008). Segmental patterning of the vertebrate embryonic axis. *Nature Reviews Genetics, 9*, 370–382.

Droser, M. L., & Gehling, J. G. (2008). Synchronous aggregate growth in an abundant new Ediacaran tubular organism. *Science, 319*, 1660–1662.

Duboule, D. (1994). Temporal colinearity and the phylotypic progression: A basis for the stability of a vertebrate Bauplan and the evolution of morphologies through heterochrony. *Development Supplement*, 135–142. https://www.ncbi.nlm.nih.gov/pubmed/7579514

Duboule, D., & Wilkins, A. S. (1998). The evolution of "bricolage." *Trends in Genetics, 14*, 54–59.

Ehebauer, M., Hayward, P., & Arias, A. M. (2006). Notch, a universal arbiter of cell fate decisions. *Science, 314*, 1414–1415.

Eldredge, N. (1999). *The pattern of evolution*. New York: Freeman.

Fehilly, C. B., Willadsen, S. M., & Tucker, E. M. (1984). Interspecific chimaerism between sheep and goat. *Nature, 307*, 634–636.

Forgacs, G., & Newman, S. A. (2005). *Biological physics of the developing embryo*. Cambridge, UK: Cambridge University Press.

Galis, F., & Sinervo, B. (2002). Divergence and convergence in early embryonic stages of metazoans. *Contributions to Zoology, 71*, http://dpc.uba.uva.nl/ctz/vol71/nr01/a08

Galis, F., van Dooren, T. J., & Metz, J. A. (2002). Conservation of the segmented germband stage: Robustness or pleiotropy? *Trends in Genetics, 18*, 504–509.

Garcia-Ojalvo, J., Elowitz, M. B., & Strogatz, S. H. (2004). Modeling a synthetic multicellular clock: Repressilators coupled by quorum sensing. *Proceedings of the National Academy of Sciences USA, 101*, 10955–10960.

Gilbert, S. F., Sapp, J., & Tauber, A. I. (2012). A symbiotic view of life: We have never been individuals. *The Quarterly Review of Biology, 87*, 325–341.

Goldschmidt, R. B. (1940). *The material basis of evolution*. New Haven, CT: Yale University Press.

Gould, S. J. (1977). *Ontogeny and phylogeny*. Cambridge, MA: Harvard University Press.

Gould, S. J. (2002). *The structure of evolutionary theory*. Cambridge, MA: Belknap Press of Harvard University Press.

Gould, S. J., & Eldredge, N. (1977). Punctuated equilibria: The tempo and mode of evolution reconsidered. *Paleobiology, 3*, 115–151.

Gould, S. J., & Lewontin, R. C. (1979). The spandrels of San Marco and the panglossian paradigm. *Proceedings of the Royal Society of London B, 205*, 581–598.

Grosberg, R. K., & Strathmann, R. (2007). The evolution of multicellularity: A minor major transition? *Annual Review of Ecology, Evolution, and Systematics, 38*, 621–654.

Hall, B. K. (1997). Phylotypic stage or phantom: Is there a highly conserved embryonic stage in vertebrates? *Trends in Ecology & Evolution, 12*, 461–463.

Heisenberg, C. P., & Bellaiche, Y. (2013). Forces in tissue morphogenesis and patterning. *Cell, 153*, 948–962.

Hernández-Hernández, V., Niklas, K. J., Newman, S. A., & Benítez, M. (2012). Dynamical patterning modules in plant development and evolution. *The International Journal of Developmental Biology, 56*, 661–674.

Hoekstra, H. E., & Coyne, J. A. (2007). The locus of evolution: Evo devo and the genetics of adaptation. *Evolution, 61*, 995–1016.

Horder, T. J. (2008). A history of evo-devo in Britain. *Annals of the History and Philosophy of Biology, 13*, 101–174.

Huxley, J. (2010/1942). *Evolution: The modern synthesis*. MIT Press, Cambridge, Mass.

Istrail, S., De-Leon, S. B., & Davidson, E. H. (2007). The regulatory genome and the computer. *Developmental Biology, 310*, 187–195.

Jablonka, E., & Lamb, M. J. (2005). *Evolution in four dimensions: Genetic, epigenetic, behavioral, and symbolic variation in the history of life*. Cambridge, MA: MIT Press.

Kaneko, K. (2011). Characterization of stem cells and cancer cells on the basis of gene expression profile stability, plasticity, and robustness: Dynamical systems theory of gene expressions under cell-cell interaction explains mutational robustness of differentiated cells and suggests how cancer cells emerge. *Bioessays, 33*, 403–413.

Karner, C., Wharton, K. A., & Carroll, T. J. (2006a). Apical-basal polarity, Wnt signaling and vertebrate organogenesis. *Seminars in Cell & Developmental Biology, 17*, 214–222.

Karner, C., Wharton, K. A., & Carroll, T. J. (2006b). Planar cell polarity and vertebrate organogenesis. *Seminars in Cell & Developmental Biology, 17*, 194–203.

King, N., Westbrook, M. J., Young, S. L., Kuo, A., Abedin, M., Chapman, J., . . . Rokhsar, D. (2008). The genome of the choanoflagellate Monosiga brevicollis and the origin of metazoans. *Nature, 451*, 783–788.

Kokko, H., Chaturvedi, A., Croll, D., Fischer, M. C., Guillaume, F., Karrenberg, S., . . . Stapley, J. (2017). Can evolution supply what ecology demands? *Trends in Ecology & Evolution, 32*, 187–197.

Kondo, S., & Miura, T. (2010). Reaction-diffusion model as a framework for understanding biological pattern formation. *Science*, *329*, 1616–1620.

Laland, K. N., Odling-Smee, J., & Gilbert, S. F., (2008). EvoDevo and niche construction: Building bridges. *Journal of Experimental Zoology Part B: Molecular and Developmental Evolution*, *310*, 549–566.

Lander, A. D. (2007). Morpheus unbound: Reimagining the morphogen gradient. *Cell*, *128*, 245–256.

Larroux, C., Luke, G. N., Koopman, P., Rokhsar, D. S., Shimeld, S. M., & Degnan, B. M. (2008). Genesis and expansion of metazoan transcription factor gene classes. *Molecular Biology and Evolution*, *25*, 980–996.

Lenoir, T. (1982). *The strategy of life: Teleology and mechanics in nineteenth century German biology*. Dordrecht, Holland, and Boston, MA: Reidel.

Lewontin, R. C., (2000). *The triple helix: Gene, organism, and environment*. Cambridge, MA: Harvard University Press.

Linde-Medina, M. (2010). Natural selection and self-organization: A deep dichotomy in the study of organic form. *Ludus Vitalis*, *18*, 25–56.

Linde-Medina, M. (2011). Adaptation or exaptation? The case of the human hand. *Journal of Biosciences*, *36*, 575–585.

Mayr, E. (1982). *The growth of biological thought: Diversity, evolution, and inheritance*. Cambridge, MA: Belknap Press.

Mayr, E., & Provine, W. B. (Eds.). (1998). *The evolutionary synthesis: perspectives on the unification of biology*. Cambridge, MA: Harvard University Press.

Meinecke-Tillmann, S., & Meinecke, B. (1984). Experimental chimaeras: Removal of reproductive barrier between sheep and goat. *Nature*, *307*, 637–638.

Meinhardt, H., & Gierer, A. (2000). Pattern formation by local self-activation and lateral inhibition. *Bioessays*, *22*, 753–760.

Moss, L., (2002). *What genes can't do*. Cambridge, MA: MIT Press.

Moss, L., & Newman, S. A. (2015). The grassblade beyond Newton: The pragmatizing of Kant for evolutionary-developmental biology. *Lebenswelt*, *7*, 94–111.

Müller, G. B. (2007). Evo-devo: Extending the evolutionary synthesis. *Nature Reviews Genetics*, *8*, 943–949.

Müller, G. B., & Newman, S. A. (2005). The innovation triad: An EvoDevo agenda. *Journal of Experimental Zoology Part B: Molecular and Developmental Evolution*, *304*, 487–503.

Nagel, T. (2012). *Mind and cosmos: Why the materialist neo-Darwinian conception of nature is almost certainly false*. New York: Oxford University Press.

Nanjundiah, V., & Sathe, S. (2011). Social selection and the evolution of cooperative groups: The example of the cellular slime moulds. *Integrative Biology*, *3*, 329–342.

Newman, S. A. (1988). Idealist biology. *Perspectives in Biology and Medicine.*, *31*, 353–368.

Newman, S. A., & Comper, W. D. (1990). "Generic" physical mechanisms of morphogenesis and pattern formation. *Development, 110,* 1–18.

Newman, S. A. (1994). Generic physical mechanisms of tissue morphogenesis: A common basis for development and evolution. *Journal of Evolutionary Biology, 7,* 467–488.

Newman, S. A. (1995). Carnal boundaries: The commingling of flesh in theory and practice. In L. Birke & R. Hubbard (Eds.), *Reinventing biology* (pp. 191–227). Bloomington: Indiana University Press.

Newman, S. A. (2005). The pre-Mendelian, pre-Darwinian world: Shifting relations between genetic and epigenetic mechanisms in early multicellular evolution. *Journal of Biosciences, 30,* 75–85.

Newman, S. A. (2006). The developmental-genetic toolkit and the molecular homology-analogy paradox. *Biological Theory, 1,* 12–16.

Newman, S. A. (2011a). The developmental specificity of physical mechanisms. *Ludus Vitalis, 19,* 343–351.

Newman, S. A. (2011b). Animal egg as evolutionary innovation: A solution to the "embryonic hourglass" puzzle. *Journal of Experimental Zoology Part B: Molecular and Developmental Evolution, 316,* 467–483.

Newman, S. A. (2012). Physico-genetic determinants in the evolution of development. *Science, 338,* 217–219.

Newman, S. A. (2014a). Physico-genetics of morphogenesis: The hybrid nature of developmental mechanisms. In A. Minelli & T. Pradeu (Eds.), *Towards a theory of development* (pp. 95–113). Oxford, UK: Oxford University Press.

Newman, S. A. (2014b). Development and evolution: The physics connection. In A. Love (Ed.), *Conceptual change in biology: Scientific and philosophical perspectives on evolution and development* (pp. 421–440). Berlin: Springer.

Newman, S. A. (2014c). Why are there eggs? *Biochemical and Biophysical Research Communications, 450,* 1225–1230.

Newman, S. A. (2014d). Excitable media *in medias res*: how physics scaffolds metazoan development and evolution. In L. R. Caporael et al. (Eds.), *Developing scaffolds in evolution, culture, and cognition* (pp. 109–123). Cambridge, MA: MIT Press.

Newman, S. A. (2014e). Form and function remixed: Developmental physiology in the evolution of vertebrate body plans. *The Journal of Physiology, 592,* 2403–2412.

Newman, S. A. (2016a). "Biogeneric" developmental processes: Drivers of major transitions in animal evolution. *Philosophical Transactions of the Royal Society London B, 371,* 20150443.

Newman, S. A. (2016b). Origination, variation, and conservation of animal body plan development. *Reviews in Cell Biology and Molecular Medicine, 2,* 130–162.

Newman, S. A., & Bhat, R. (2008). Dynamical patterning modules: Physico-genetic determinants of morphological development and evolution. *Physical Biology, 5,* 15008.

Newman, S. A., & Bhat, R. (2009). Dynamical patterning modules: A "pattern language" for development and evolution of multicellular form. *The International Journal of Developmental Biology, 53,* 693–705.

Newman, S. A., & Bhat, R. (2011). Lamarck's dangerous idea. In S. Gissis & E. Jablonka (Eds.), *Transformations of Lamarckism: From subtle fluids to molecular biology* (pp. 157–169). Cambridge, MA: MIT Press.

Newman, S. A., Bhat, R., & Mezentseva, N. V. (2009). Cell state switching factors and dynamical patterning modules: Complementary mediators of plasticity in development and evolution. *Journal of Biosciences, 34,* 553–572.

Newman, S. A., & Forgacs, G. (2009). Complexity and self-organization in biological development and evolution. *Encyclopedia of Complexity and Systems Science* (pp. 524–548). New York: Springer-Verlag.

Newman, S. A., & Linde-Medina, M. (2013). Physical determinants in the emergence and inheritance of multicellular form. *Biological Theory, 8,* 274–285.

Niklas, K. J., & Newman, S. A. (2013). The origins of multicellular organisms. *Evolution & Development, 15,* 41–52.

Peacock, K. A. (2011). The three faces of ecological fitness. *Studies in History and Philosophy of Science, 42,* 99–105.

Peacor, S. D., Allesina, S., Riolo, R. L., & Pascual, M. (2006). Phenotypic plasticity opposes species invasions by altering fitness surface. *PLoS Biology, 4,* e372.

Pigliucci, M., & Müller, G. (Eds.). (2010). *Evolution, the extended synthesis.* Cambridge, MA: MIT Press.

Raff, R. A. (1996). *The shape of life: Genes, development, and the evolution of animal form.* Chicago: University of Chicago Press.

Raspopovic, J., Marcon, L., Russo, L., & Sharpe, J. (2014). Modeling digits. Digit patterning is controlled by a Bmp-Sox9-Wnt Turing network modulated by morphogen gradients. *Science, 345,* 566–570.

Reinke, H., & Gatfield, D. (2006). Genome-wide oscillation of transcription in yeast. *Trends in Biochemical Sciences, 31,* 189–191.

Richards, R. J. (2002). *The romantic conception of life: Science and philosophy in the age of Goethe.* Chicago, IL: University of Chicago Press.

Rokas, A., Kruger, D., & Carroll, S. B. (2005). Animal evolution and the molecular signature of radiations compressed in time. *Science, 310,* 1933–1938.

Salazar-Ciudad, I., (2010). Morphological evolution and embryonic developmental diversity in metazoa. *Development, 137,* 531–539.

Sander, K. (1983). The evolution of patterning mechanisms: Gleanings from insect embryogenesis and spermatogenesis. In B. C. Goodwin, N. Holder, C. C. Wylie (Eds.), *Development and evolution.* Cambridge, UK: Cambridge University Press.

Sarkar, S. (1996). Biological information: A skeptical look at some central dogmas of molecular biology. In S. Sarkar (Ed.), *The philosophy and history of molecular biology: New perspectives* (pp. 187–231). Dordrecht: Kluwer.

Schmalhausen, I. I. (1949). *Factors of evolution*. Philadelphia, PA: Blakiston.

Schrödinger, E. (1945). *What is life? The physical aspect of the living cell*. Cambridge, UK: Cambridge University Press; New York, NY: Macmillan.

Seidel, F. (1960). Körpergrundgestalt und Keimstruktur: Eine Erörterung über die Grundlagen der vergleichenden und experimentellen Embryologie und deren Gültigkeit bei phylogenetischen Berlegungen. *Zoologischer Anzeiger, 164*, 245–305.

Shen, B., Dong, L., Xiao, S., & Kowalewski, M. (2008). The Avalon explosion: Evolution of Ediacara morphospace. *Science, 319*, 81–84.

Simpson, G. G. (1944). *Tempo and mode in evolution*. New York: Columbia University Press.

Slack, J. M., Holland, P. W., & Graham, C. F. (1993). The zootype and the phylotypic stage. *Nature, 361*, 490–492.

Steinberg, M. S. (2007). Differential adhesion in morphogenesis: A modern view. *Current Opinion in Genetics & Development, 17*, 281–286.

Stoltzfus, A., & Cable, K., (2014). Mendelian-mutationism: The forgotten evolutionary synthesis. *Journal of the History of Biology, 47*, 501–546.

Strogatz, S. H. (2003). *Sync: The emerging science of spontaneous order*. New York, NY: Theia.

Turing, A. M. (1952). The chemical basis of morphogenesis. *Philosophical Transactions of the Royal Society London B, 237*, 37–72.

Waddington, C. H. (1942). Canalization of development and the inheritance of acquired characters. *Nature, 150*, 563–565.

Wagner, G. P., & Lynch, V. J. (2008). The gene regulatory logic of transcription factor evolution. *Trends in Ecology & Evolution, 23*, 377–385.

Wallace, A. R. (1858). On the tendency of varieties to depart indefinitely from the original type. *Proceedings of the Linnean Society of London, 3*, 53–62.

Walsh, D. M. (2006). Organisms as natural purposes: The contemporary evolutionary perspective. *Studies in History and Philosophy of Science, 37*, 771–791.

Webster, G., & Goodwin, B. C. (1996). *Form and transformation: Generative and relational principles in biology*. Cambridge, UK, and New York, NY: Cambridge University Press.

West-Eberhard, M. J. (2003). *Developmental plasticity and evolution*. New York, NY: Oxford University Press.

Wilkins, A. S. (2002). *The evolution of developmental pathways*. Sunderland, MA: Sinauer.

Wilson, L. A. B. (2013). The contribution of developmental palaeontology to extensions of evolutionary theory. *Acta Zoologica, 94*, 254–260.

Wray, G., & Strathmann, R. (2002). Stasis, change, and functional constraint in the evolution of animal body plans, whatever they may be. *Vie Milieu, 52*, 189–199.

Yin, L., Zhu, M., Knoll, A. H., Yuan, X., Zhang, J., & Hu, J. (2007). Doushantuo embryos preserved inside diapause egg cysts. *Nature, 446*, 661–663.

Zhu, J., Zhang, Y. T., Alber, M. S., & Newman, S. A. (2010). Bare bones pattern formation: A core regulatory network in varying geometries reproduces major features of vertebrate limb development and evolution. *PLoS One, 5*, e10892.

7

EVOLVABILITY AND ITS EVOLVABILITY

Alessandro Minelli

7.1. Defining Evolvability

Eight years ago, Brookfield (2009) reported that a search of the Web of Science for the years from 2004 to 2008 had revealed 402 papers containing the word "evolvability" in either title, abstract, or keywords, compared with 212 papers in the years from 1999 to 2003 and only 33 in the previous 5 years. These figures reveal an obvious trend in popularity of this term, but it may be informative to look even further back in time. Recently (January 2016), a search through Google Scholar for the years 1961–1970 retrieved very few papers mentioning evolvability, and the handful of those actually dealing with biology are limited to issues such as origin of life; things do not change much in the 1970s, with a few titles dealing essentially with artificial systems, as do those (much more numerous, about 180) of the 1980s, expanding on software, complex systems, and artificial life.

7.1.1. One Term, Many Meanings

Eventually, in 1988, Richard Dawkins inaugurated the modern use of the term in biology, introducing "evolvability" to describe the ability of a particular group to spawn an evolutionary radiation.

In practice, however, the picture changed effectively only in 1991, with Pere Alberch's paper on the evolvability of living organisms, which included a lucid analysis of the relationships between genotype and phenotype. In view of the following discussion about the many different perspectives from which evolvability has been defined and studied in the last 2 decades, it is important to remark that for Alberch evolvability was "a property of embryological systems, i.e., certain types of developmental systems are better at evolving" (Alberch, 1991, p. 9).

Significantly, the word "evolvability" appeared in the title of Alberch's paper, and soon thereafter in those of Houle (1992), Messina (1993), and Altenberg (1994), although in Messina's it was given in quotes, testifying a usage that was still far from commonly accepted.

The exponential increase of occurrences of the term in the literature of the following decades witnesses a rapidly growing interest in a series of questions, more or less strictly related to the parallel exponential growth of evolutionary developmental biology (evo-devo). This parallel growth is also reflected in the plurality of explicit or implicit definitions of evolvability given by the different authors (a plurality already noticed by Sniegowski and Murphy (2006), Pigliucci (2008), Brookfield (2009) and others), to a considerable extent matching the plurality of approaches to evo-devo—in particular, the prevalent focus on the genetic side of that genotype→phenotype map (see later, section 7.3.1) to which Alberch first called attention. Let us thus begin with a short survey of definitions given thus far to the term "evolvability."

Dawkins's and Alberch's initial focus on development as the real focus of evolvability, mediating between genetic variation and phenotypic change, has been frequently abandoned by later authors, even by those who maintain an eye open on both genotype and phenotype, for example, Schlichting and Murren (2004, p. 19), who define evolvability as "the tendency of a genotype or lineage to generate genetic variability and produce or maintain phenotypic variation over evolutionary time, enabling it to pursue diverse evolutionary trajectories."

7.1.2. Focus on Genes or Focus on the Phenotype?

For many authors, evolvability is simply the capacity to evolve (Turney, 1999; Yang, 2001), or the evolutionary potential (A. Wagner, 2005; Woods et al., 2011). In a typical neo-Darwinian style, others concentrate on selection and adaptation, thus defining evolvability as the ability of populations to generate heritable phenotypic variation (Brigandt, 2007; Kirschner & Gerhart, 1998; G. P. Wagner & Altenberg, 1996), or to adapt through natural selection (Altenberg, 1994; Draghi & Wagner 2008), or to respond to a selective challenge (Flatt, 2005; Hansen, 2003; Kirschner & Gerhart, 1998). Some authors specifically stress the capacity to evolve new adaptations (Bedau & Packard, 1992). In a more articulated form, Masel and Trotter (2010, p. 406) define evolvability as "the capacity of a population to produce heritable phenotypic variation of a kind that is not unconditionally deleterious. This definition includes both evolution from standing variation and the ability of the population to produce new variants." Reference to selection is rejected instead by Dichtel-Danjoy and Félix (2004), who define evolvability as "the capacity to evolve at the phenotypic level, irrespective of the action of natural selection."

Only Klingenberg (2005) takes into account that evolution is not necessarily due to natural selection, and defines evolvability as "the potential of a population to respond to selection *or to undergo nonadaptive evolution by drift*" (italics mine).

Closer to Alberch's original approach, which treated evolvability in the context of the relationship between genotype and phenotype, West-Eberhard (1998) defines evolvability as "the ability of particular features of systems to facilitate change."

Other authors, however, turn toward the phenotype, and define evolvability as the capacity of a developmental system to generate variation in form (Brakefield, 2011). Strongly focused on the phenotype is Brookfield's (2001, p. R107) notion of evolvability as "the proportion of radically different designs created by mutation that are viable and fertile," where the overall sensible perspective is weakened by the obvious difficulty to agree on how to evaluate whether two "designs" are different enough to be classified as radically different. In principle at least, Brookfield's (2001) definition reveals a quantitative approach to evolvability, but this is more explicit in other formulations: de Visser et al. (2003) and Griswold (2006) treat evolvability as a quantitative measure of the rate of evolution of a given character, and a quantitative aspect is also present in Kirschner's (2013, p. 1) approach, insofar as he defines evolvability as "both the amount of variation that is subject to selection, and the nature of that variation." Operationally, Quayle and Bullock (2006) measure evolvability as the time that it takes for a population to hit a given phenotypic target, although this makes the simplifying assumption that the target itself does not shift over time, as a result of environmental change.

7.1.3. Novelty, Innovation

In more recent papers, some authors make clear that the genetic variation present in a population is not necessarily expressed in full at the phenotypic level, thus Colegrave and Collins (2008) feel obliged to specify that evolvability is "the ability of a population to both generate *and use* [italics mine] genetic variation to respond to natural selection." To quote Brakefield (2011, p. 2072), "The origin of evolutionary novelties is one aspect of evolvability; another is the role of standing genetic variation and mutational input in generating developmental flexibility."

The origin of evolutionary novelties (e.g., Mayr, 1960; Müller, 1990) has been matter of discussion in evolutionary biology long before the birth of evo-devo— witness Darwin's concern for the "abominable mystery" of the origin of flowers.

In a seminal paper on evolutionary novelties, Müller and Wagner (1991, p. 243) defined a morphological novelty as "a structure that is neither homologous to any structure in the ancestral species nor homonomous to any other structure of the same organism." In the same paper, these authors discussed novelties also from a developmental point of view. However, in subsequent studies the focus became more and more strictly limited to morphology, that is, to phenotypic character traits (e.g., Arthur, 2000; Hall, 2005; Leys & Riesgo, 2012; Müller & Newman, 2005; Müller, 2010; West-Eberhard, 2003) and Müller and Wagner themselves (2003, pp. 218–219) redefined evolutionary innovation as "a specific class of phenotypic change that is different from adaptive modification [such as] the origin of new body parts [or] major organizational transitions" and distinguished as novelties those innovations that "introduce new entities, units, or elements into phenotypic organization." Other authors, for example, Love (2003), have used the terms "novelty" and "innovation" interchangeably.

The concept of novelty has been revised by Müller (2010) through the identification of three classes of novelties: type I novelty—primary anatomical architecture of a metazoan body plan; type II novelty—discrete new element added to an existing body plan; type III novelty—major change of an existing body plan character.

Hallgrímsson et al. (2012, p. 502) have recently proposed a definition of novelty that abandons this morphological and hierarchical approach. According to these authors an evolutionary novelty should "involve a transition between adaptive peaks on a fitness landscape [and] a breakdown of ancestral developmental constraints, such that variation is generated in a new direction or dimension."

The contrast between these two approaches to evolutionary novelties parallels the contrast between a phylogenetic notion of homology (thus a hierarchical one, and traditionally based on morphological characters) and a "biological" notion of homology (e.g., G. P. Wagner 1989) based on the modularity of developmental mechanisms. The two concepts, homology and evolvability, are indeed interrelated: Brigandt (2007, p. 710) defines a homologue "as a unit of morphological evolvability, i.e., as a part of an organism that can exhibit heritable phenotypic variation independently of the variation that the organism's other homologues can undergo," or as "a unit of heritable phenotypic variability—a structural unit being able to phenotypically vary in response to genetic variation."

In my opinion, both a morphology-based and a development-based approach to evolutionary novelty (or to homology, by the way) are justified, but their heuristic values are not overlapping. Despite the objections of Peterson and Müller (2013) to Hallgrímsson et al.'s (2012) approach, in discussing evolvability, a development-based approach to evolutionary novelties is more useful (see also Arthur, 2000; Moczek, 2008; Pigliucci, 2008), due to the critical role of development in mediating between genotype and phenotype—an aspect to which I return later.

7.2. A View from the Genes

Most of the literature on evolvability focusing on variation is not simply too narrow because of its essential disregard for the phenotype; it is also inadequate in its view of variation, because it often ignores two important distinctions.

7.2.1. Variation

The first is a distinction between variability (the propensity of characters to vary, through mechanisms such as mutation and recombination) and variation (a measure of the realized differences within a population) (cf. Pigliucci, 2008; Wagner & Altenberg, 1996). It seems reasonable to argue that evolvability revolves around variability as evolution revolves around variation. The second is a distinction between existing genetic variation and expressed variation.

7.2.2. Robustness

So much as the term "evolvability" suggests an intrinsic attitude to change, so does "robustness" suggest an intrinsic resistance to change. However, the robustness of a genetic system does not necessary contrast evolvability. But let us first briefly introduce this unexpected partner of evolvability.

Genetic robustness is "robustness to perturbations both in the form of new mutations and in the form of the creation of new combinations of existing alleles by recombination" (Masel & Trotter, 2010, p. 407).

A. Wagner (2005, 2008, 2011) has convincingly shown that mutations that are not expressed as novel phenotypes nevertheless cause populations to occupy an increasingly larger region of genotype space. As a consequence, the population gains easy access to a greater range of novel genotypes, improving its evolvability, at least in a quantitative sense (see also Draghi et al., 2010).

But there is also a developmental aspect of robustness, that is, the resistance of the processes producing a given phenotype both to genetic changes expressed at the molecular level and to external perturbations. This concept is essentially embodied in Waddington's (1957) notion of canalization, or the reduction of the phenotypic effects of genetic or environmental variation (see also Hendrikse, Parsons, & Hallgrímsson, 2007; Waddington & Robertson, 1966; G. P. Wagner et al., 1997). This also corresponds to the fact that populations harbor amounts of unexpressed variation (*cryptic genetic variation*), that represent a standing potential for evolvability, either by genetic assimilation or by small additional mutational change (e.g., Badyaev, 2005; Flatt, 2005; Gibson & Dworkin, 2004; Moczek, 2007; Rieseberg et al., 2003; Schlichting, 2008).

From this perspective, we must admit, perhaps counterintuitively, robustness may promote evolvability, rather than resisting it.

7.3. Developmental Systems and Evolvability

To understand evolvability, we must keep distance from a simplistic and strictly deterministic view of the relationships between genotype and phenotype.

7.3.1. The Genotype→Phenotype Map

On the one hand, as mentioned before, we have clear evidence that populations harbor smaller or larger amounts of cryptic variation, that is, of variation that is not expressed in the phenotype. On the other hand, the path leading from genotype to phenotype (the genotype→phenotype map; cf. Alberch, 1991; Draghi & Wagner, 2008; Pigliucci, 2001; Wagner & Altenberg, 1996; West-Eberhard, 2003) is complex and not necessarily predictable: The expression of variation depends on the structure (the "logics") of gene networks and especially on the structure and robustness of the developing system (Hansen, 2006; Kirschner & Gerhart, 1998; Wagner, 2005; Wagner & Altenberg, 1996).

One of the few generalizations we are allowed to make about the genotype→phenotype map is the pervasiveness of *pleiotropy*, that is, of the multiple phenotypic effects of mutations affecting a single locus. To some extent, what we document as pleiotropy is in fact a consequence of the way we articulate the phenotype into a plurality of characters: This partitioning into smaller units is perhaps reasonable in terms of morphology, but not necessarily so in terms of developmental processes, or of gene expression patterns and functions. On the other hand, one may argue that the only uniquely controlled phenotype corresponding to a given gene is perhaps its primary mRNA transcript, previous to any possible post-transcriptional editing.

Pleiotropic effects of genes are, however, far from random, and their clustering around different "hubs" translates into *developmental modularity*, "a genotype–phenotype map in which there are few pleiotropic effects among characters serving different functions" (Brandon, 1999; see also Nagy, 1998; Raff, 1996; Raff & Raff, 2000; von Dassow & Munro, 1999; Wagner & Altenberg, 1996).

Like evolvability and robustness, "modularity" is another term quite popular in the evo-devo literature, but—once more—one with multiple and only partially overlapping definitions. I adopt here Klingenberg's (2005, p. 6) definition: "Modules are assemblages of parts that are tightly integrated internally by relatively many and strong interactions but relatively independent of one another because there are only relatively few or weak interactions between modules."

As noted by Pigliucci (2008), most authors agree on recognizing a dependence of evolvability on the modularity of the genetic architecture, but disagree on the nature of this dependence, largely because they move from different concepts of evolvability. In defending a dependence of evolvability on modularity, G. P. Wagner (1996) and Schlosser and Wagner (2003) discuss the issue in terms of developmental modularity, others, for example, Gerhart and Kirschner (1997), Kirschner and Gerhart (1998), von Dassow and Munro (1999) and Griswold (2006), in terms of a modular genetic architecture. But according to Budd (2006), modularity has only a minor role: Evolvability evolves anyway because it allows the evolution of whatever developmental system is suitable to a specific environment.

Summing up, we might expect that a higher degree of independence in respect to other characters should increase the freedom of a character to evolve (Hendrikse et al., 2007), or, that the antagonistic effect of pleiotropy on evolvability will grow with the number of correlated characters (Dickerson, 1955; Kirkpatrick, 2009; Walsh & Blows, 2009). This view is supported by Fisher's (1930) classic demonstration that a steeply decreasing function links the probability of a random mutation being favorable to the number of traits affected by it. But things are more complex: Carter, Hermisson, and Hansen (2005) have shown that negative epistasis reduces evolvability, but positive epistasis increases evolvability. Moreover, Hansen's (2003) computer simulations suggest that genetic architectures with the highest degree of modularity are not necessarily those with the highest evolvability. Effects will depend on the selective regime to which a character is subject and on the possible reciprocal compensation of the constraints that selection of a given character may impose on the evolution of other characters to which it is linked because of pleiotropic effects.

7.3.2. Developmental Modularity and Evolvability

As extensively discussed by Arthur (e.g., 2004a, 2004b), the fact that developmental systems produce *biased embryos* (Arthur, 2004b), translating variation in favor of some alternatives, is probably a universal feature.

The most naïve way to express the impact of development on evolvability and evolution is arguably in terms of *developmental constraints* (e.g., Alberch, 1982; Cheverud, 1984; Maynard Smith et al., 1985; Lande, 1986; von Dassow & Munro, 1999). As defined by Maynard Smith et al. (1985, p. 266), a developmental constraint is "a bias on the production of variant phenotypes or a limitation on phenotypic variability caused by the structure, character, composition, or dynamics of the developmental system."

Rupert Riedl (1977, 1978; see also Wagner & Laubichler, 2004) thought that too many degrees of freedom are likely to limit evolvability, whereas correlations between characters increase it, provided these characters are also functionally coupled, so that a change in one of them can be positively selected only if the other character changes in a way that preserves their overall functionality.

Natural selection operates on organisms as wholes (Kemp, 2007). This suggests that the evolution of the different parts of the body is subjected to *correlated progression* (Budd, 1998; Kemp, 1999, 2007; Lee, 1996). A nice example is the covariation of hands and feet in our species and in the chimpanzee: Rolian, Lieberman, and Hallgrímsson (2010) have shown that the phenotypic covariation between serially homologous fingers and toes in the two species causes these digits to evolve along highly parallel trajectories, even when selection pressures push their means in divergent directions.

A different problem is caused by the lack of a simple correspondence between developmental modules and phenotypic units potentially subjected to independent selection (Kampourakis & Minelli, 2014; Minelli, 2009). This has been elegantly demonstrated by a study of the evolvability of the eyespots on the wings of the butterfly *Bicyclus anynana* (Monteiro, Brakefield, & French, 1997; Beldade & Brakefield, 2002; Brakefield, 2008). To a potential predator, a wing eyespot appears as a unitary body part, but it is composed of several rings of different colors subjected to developmental constraints that cause them to respond to selection somehow independently.

7.3.3. Modularity in the Real World

The effects of modularity are clearly visible in some unusual phenotypes that depart dramatically from those of closely related taxa, but only in a strictly circumscribed body part, whereas the rest of the body conforms to the architecture of its "normal" relatives. In other instances, however, evolution has affected the whole architecture of the animal, or plant, in which case no developmental or evolutionary modularity is observed (Minelli, 2015).

An example of evolutionary trend based on developmental modularity is provided by the gonopods of male millipedes. In principle, we would not expect that just one pair of appendages within a series of otherwise identical limbs, such as the legs of a millipede, would evolve in a strongly diverging way, without correlative change in any of the other limb pairs, those in front of it and those behind it. However, this is exactly what we observe in the males of most millipede species, those belonging to the subclass Helminthomorpha.

The number of walking legs in these millipedes varies from 32 to 375 pairs. Except for the smaller size of the first pair, or the first few pairs, all leg pairs are morphologically identical in the females and also in male juveniles. At each molt, new pairs of legs are added to those already present in the previous stage. At later stages, however, the eighth pair of legs, and often also the ninth, undergoes a unique metamorphosis and is replaced in the adult male by a pair of gonopods, specialized sexual appendages used as claspers or to transfer sperm (Drago, Fusco, Garollo, & Minelli, 2011; Drago, Fusco, & Minelli, 2008). These appendages differ dramatically from walking legs, and present a fantastic diversity of shapes, often too complex to be adequately described in words.

Examples of systemic changes in body architecture are provided instead by the riverweeds and the duckweeds. The riverweeds (family Podostemaceae) are aquatic plants of tropical regions, whose vegetative structures deviate not only in relation to their closest relatives, but also in comparison with all other flowering plants. In the most derived forms, none of the conventional vegetative parts of angiosperms is recognizable and the whole takes a form more reminiscent of an alga than of a plant articulated into roots, stem, branches, and leaves.

In the case of duckweeds, currently regarded as a basally splitting lineage (Lemnoideae) within the arum family (Araceae) (Henriquez, Arias, Pires, Croat, & Schaal, 2014), the whole photosynthetic structure is reduced to an irregular lens- or disk-shaped body, or a small cluster of similar floating bodies, out of which one or more roots sprout inferiorly. The final stage of this trend is *Wolffia arrhiza*, a subspherical grain of green matter without any root or other projection, perhaps one millimeter in diameter.

A modular versus systemic nature of change is not simply a property of morphology, but also one of an organism's ontogeny. In other terms, evolvability can have phenotypic consequences affecting the whole life cycle, or ones limited to a single developmental stage. Here again, some "morphological misfits" (Bell, 2008) strongly deviating from the organization of their closest relatives can offer examples of either systemic or modular deviating phenotypes.

An example where the whole life cycle is represented by oddly shaped stages and the sequence of stages also strongly departs from the norm is provided by the Cycliophora, minuscule animals that live on the appendages of the Norwegian lobster, for which zoologists were forced to introduce new terms such as the Pandora larva and the Prometheus larva, because the terminology available for the other animals failed to offer adequate labels for their unique stages (Obst & Funch, 2003).

In other cases, strong morphological singularities are limited to one stage only within the life cycle. For example in cirripeds like true barnacles (*Balanus*) and goose barnacles (*Lepas*), the larval stages have the general morphology of conventional crustacean larvae, whereas the adult deviates so conspicuously from the crustacean body plan, that the real affinities of these animals were discovered only when their metamorphosis was first observed by Thompson (1830, 1835).

7.3.4. Evolvability Through Plasticity

Phenotypic plasticity (reviewed, e.g., in DeWitt & Scheiner, 2004; Fusco & Minelli, 2010; Greene, 1999; Nicoglu, 2015; Pigliucci, 2001; Schlichting & Pigliucci, 1998; West-Eberhard, 2003, 2005) is "a property of individual genotypes to produce different phenotypes when exposed to different environmental conditions" (Pigliucci, Murren, & Schlichting, 2006, p. 2363). It is through phenotypic plasticity that the different castes are generally produced among social insects (e.g., Miura, 2005), but this applies also to the environmental determination of sex in reptiles such as the alligator (e.g., Janzen & Phillips, 2006) or the origination of predator-induced morphs in water fleas (e.g., Laforsch & Tollrian, 2004) and in frogs (e.g., Relyea, 2004).

It has been suggested (Nijhout, 2003) that nonadaptive or just incidentally adaptive phenotypic plasticity is likely the primitive character state for most if not all traits. Eventually, depending on the effect of phenotypic plasticity on fitness, the ensuing evolutionary change can proceed either by progressive reduction of plasticity, thus buffering the production of a stable phenotype against environmental variation, or by genetic assimilation of multiple phenotypes, thus leading to a genetic polymorphism.

How easily environmentally controlled alternative phenotypes can eventually fall under genetic control has been shown by recent studies on the pea aphid (*Acyrthosiphon pisum*) (Brisson, 2010). In this species, individuals of either sex are either winged or wingless, but in the male this difference is controlled by a single-gene polymorphism (Braendle, Caillaud, & Stern, 2005), whereas the females exhibit a polyphenic response to varying environmental cues. However, the product of the gene (*aphicarus*) responsible for wing development in the male is also involved in the developmental response of the female to the environmental stimulus (Braendle, Friebe, Caillaud, & Stern, 2005).

This example suggests how novel phenotypes can evolve by *genetic accommodation*, to use a term introduced by West-Eberhard (2003). More specifically, such an evolutionary path involving a decrease in environmental sensitivity (environmental canalization) corresponds to Waddington's (1953) genetic assimilation. In other cases, genetic accommodation will result instead in an increase in environmental sensitivity. According to Braendle and Flatt (2006), the process of genetic accommodation begins when previously cryptic, heritable genetic or epigenetic variation is uncovered, caused by a genetic or environmental change that triggers the expression of a novel phenotype, thus exposing it to natural selection. Eventually, selection on existing variation for the expression of that trait may cause it to become more strictly controlled by genes.

According to the *theory of facilitated variation* (Gerhart & Kirschner, 2007, 2010; Kirschner & Gerhart, 2005; Moczek, 2014), plasticity combined with an exploratory behavior is the condition that allows developmental systems to adaptively adjust to new developmental contexts. As a consequence, evolvability is increased because initially random genetic variation becomes available to selection.

The interrelationships between the developing organism and its environment must also be construed as flexible, especially because the organism itself eventually modifies its environment adaptively, as expressed by the *niche construction theory* (Lewontin, 1983; Odling-Smee, 2010; Odling-Smee, Laland, & Feldman, 2003).

7.4. Tempo and Mode

Several authors, especially among the paleontologists interested in macroevolution (e.g., Erwin, 2012), have emphasized the long-term effect that evolutionary novelties can have on the subsequent evolution of a lineage. Maynard Smith and Szathmáry (1995) even defined some evolutionary events as special because they change what is subsequently possible to evolve.

7.4.1. Founder Evolvability

One may be suspicious of this relation, attractive as it may appear, because of an inherent risk of circularity, following our arbitrary choice of considering a novelty the first evidence of a trait that will eventually mark a subsequent radiation. However, experiments on bacterial populations have demonstrated that the specialization of founder individuals can increase the likelihood of diversification of the offspring (Spencer, Tyerman, Bertrand, & Doebeli, 2008). In subsequent experiments, also on bacteria, Flohr, Blom, Rainey, and Beaumont (2013) have demonstrated how the dynamics of adaptive radiation are constrained by the niche occupied by the founder. Based on these results, these authors introduced the new concept of *founder evolvability*, defined (p. 20663) "as the maximal range of derived organisms with different niches that can be accessed from the founding ancestor by mutation and recombination over an interval of evolutionary time."

In the case of evolutionary lineages displaying a trend toward decreasing body complexity, the founder effect may simply follow the economic principle of diminishing returns: The more you cut away, the less remains to be

additionally deleted further ahead. This principle offers a reasonable explanation for the trend observed by Brooks and McLennan (1993) among the parasitic flatworms. These authors reported that of 1,882 character transformations
implied by their phylogeny for these worms, only 10.8% could be described as
a loss. They interpreted this result as evidence against the popular view that
parasites are secondarily simplified. However, a higher prevalence of transitions
representing simplification in the tree nodes closest to the root of the tree can
explain, on the principle of diminishing returns, the scarcity of further losses
along the tree branches: We cannot expect much further reduction from an
already simplified ancestor.

7.4.2. Convergent Evolution

Long ago, Vavilov (1922) proposed a "law of homologous variation," according to
which the similarity of developmental pathways in related species causes similar
variants to appear.

Studies of the last quarter of century have shown indeed that convergent
phenotypes are sometimes dependent on similarly convergent changes in gene
expression, especially in genes controlling early developmental phases. This is the
case of a number of sea urchin species where the typical planktotrophic larva (the
echinopluteus) has been replaced by a lecithotrophic larva (Wray, 2002).

Morphologically similar lecithotrophic larvae have also evolved in other animal phyla (enteropneust hemichordates, nemerteans, and polychaete annelids
including sipunculids), but in this case the underlying genetic and developmental
basis is different, although convergence in the expression of regulatory genes is
probably more common than generally acknowledged (Wray, 2002).

However, convergent phenotype can also result from different genetic change.
In plants, at least four distinct developmental pathways result in strictly unisexual flowers, that is, flowers without any vestigial rudiment of sexual organs of the
opposite sex (Mitchell & Diggle, 2005). In mammals, a reduction in digit number can be caused by extensive cell death in the tissue surrounding the remaining
toe(s), as in the jerboa, the horse, and the camel, or be a consequence of earlier
limb patterning mechanisms including down-regulation of the expression of the
gene *Patched1*, as in the pig (Cooper et al., 2014).

Mutations in many different genes can alter the phenotype in similar ways
(Levy & Dean, 1998; Nüsslein-Volhard & Wieschaus, 1980). Examples of convergent phenotypic evolution are also found within one species, dependent on
different genetic changes, as between different weedy forms of rice, all conver-
gently adapted in morphology to the same agricultural environment (Thurber,
Jia, Jia, & Caicedo, 2013).

7.4.3. Breaking Constraints: Examples of Large Phenotypic Leaps

Natural history offers good empirical tests of our ideas about the strength of putative evolutionary constraints. In nature there are indeed a number of "morphological misfits" (Bell 1991), animals and plants that deviate, often dramatically, from the body architecture of their ascertained, or putative, relatives. Of course, as made clear by Bell (2008, p. 247), morphological misfits are such "for the moment, to botanical [or zoological, by the way] discipline, not misfits for a successful existence." In a recent review (Minelli, 2015) I suggested a classification of morphological misfits into three groups.

A first group is represented by *misfits by reduction*. These animals or plants lack many of the parts or organs usually found in the members of the group to which they belong. In evolutionary perspective, morphological reduction is often the effect of progenesis (Westheide, 1987): Reproductive maturity is reached at a stage corresponding in morphology to an embryonic or larval stage of their relatives. Reduction, however, does not rule out the expression of novel traits, as demonstrated by two different groups of miniaturized misfits evolved among the Cnidaria. All these forms deviate very strongly from the typical morphological organization of these animals, usually represented by polyps or medusae. One of these cnidarian misfits is *Buddenbrockia*, a wormlike parasite of freshwater bryozoans (Jiménez-Guri, Okamura, & Holland, 2007), another is *Polypodium*, a parasite of sturgeon's eggs, in the form of an irregular mass of jelly with finger-like projections (Siddall & Whiting, 1999).

In the second group of misfits the main novelty is found in the structure of the individual cells of which they are formed. The best example is the Loricifera, minuscule but anatomically complex marine animals (about a quarter of a millimeter when adult), formed by a high number of cells of extremely small size that have lost all the organelles of which they could dispose, including mitochondria (Danovaro et al., 2010).

Members of the third group are the misfits by *synorganization*, where body parts that are well recognizable in their "normal" relatives are indistinctly separated, either because of a primary lack of separation, or because of secondary fusion. This is the case of the paussine beetles, whose extraordinary antennae have evolved into a uniquely shaped leaf-like blade, composed of many fused articles.

7.4.4. The Temporal Scale of Change and the Case for Saltational Evolution

Smaller changes and larger transitions are likely mixed together along the evolutionary history of life (Orr, 1998). In other terms, evolution offers examples of discontinuities (cf. Frazzetta, 2012) that defy explanation in terms of small adaptive changes whose effects have been added through time.

These morphological discontinuities have been often exploited by taxonomists to characterize higher taxa, such as the gastropods, the root of which is identified with the advent of the torsion of the visceral sac; or the flatfishes, with their apparently sudden transition from bilateral symmetry to directional asymmetry. Quite popular, in the evo-devo literature, is the case of the turtles, unique among the tetrapods in having the pectoral girdle encased within the ribs rather than external to it. Gilbert, Loredo, Brukman, and Burke (2001) and Rieppel (2001) did not hesitate to describe this change as saltational.

Theißen (2009) has explicitly defended saltational evolution as a concept necessary to describe a number of key innovations and changes in body plan that have not possibly evolved in the more common, but far from universal gradualistic mode. Before the advent of evo-devo, the very idea of saltational evolution as an explanation for macroevolutionary transitions (as suggested, for example, by Goldschmidt, 1940), was strictly banned as heretical. However, an appreciation of the nonlinear character of the genotype→phenotype map is enough to realize how major phenotypic changes can be accomplished in a leap. For example, a single-gene mutation was probably responsible for the evolution of the bilaterally symmetrical orchid flowers from an ancestor with radially symmetrical ones (Theißen, 2009). Similarly, a small genetic change may explain the sudden duplication of the number of leg pairs observed in a lineage of scolopenders (Minelli, Chagas, & Edgecombe, 2009).

7.4.5. Developmental Inertia and Paramorphism

To study evolvability through the glasses of developmental biology does not simply require understanding the complexity of the genotype→phenotype map, it also requires adopting an adequate view of development. This is not necessarily an easy task. At present, there is not even a consensus view, among biologists and philosophers of biology, on how to define development (Pradeu et al., 2016), not to mention on the possible paths toward a theory of development, a target that some regard as a valid and heuristically important task and others dismiss as worthless or even impossible (an articulated sample of alternative perspectives on this question is available in the recent book edited by Minelli and Pradeu, 2014). These issues are not discussed in depth in this chapter. What mainly matters here is to abandon adultocentrism, the common, somehow finalistic view according to which development is the sequence of processes through which a zygote, a spore, or a bud is eventually turned into an adult (Minelli, 2014). I suggested elsewhere (Minelli, 2011a) that we should instead adopt a *principle of developmental inertia*, according to which development is basically an iteration of elementary dynamics such as cell proliferation, onto which more restricted and eventually canalized processes are grafted, in a never-ending progression of change. Spatial and temporal boundaries of developmental systems are thus arbitrary, to a large extent (Minelli, 2011b).

On the other hand, the self-perpetuating nature of development may help explaining the evolutionary origin of many features of bodily organization, such as symmetry and serial repetition of parts (e.g., arthropod segments and leaf + node + internode units in plants) (cf. Schmitt, this book), but also phenomena like regeneration, vegetative reproduction through buds or body fragmentation, and also polyembryony, the production from a single zygote of two or more genetically identical twins.

Finally, *paramorphism* (Minelli, 2000, 2003), that is, the production of new patterned axes by the inertial iteration of existing developmental dynamics already responsible for the production of the main body axis—possibly followed by divergence and specialization further ahead in evolution—offers a plausible parsimonious scenario for the origin of body appendages.

Iteration of existing developmental dynamics from new "worksites" is possibly a key feature of evolvability. Once more, natural history offers many examples, a sample of which, at least, would deserve closer study, also in terms of the underlying patterns of gene expression. A nice example is provided by the accessory contractile vessels repeatedly evolved in insects in addition to the dorsal heart. Topographical distribution and function of these accessory pulsatile organs are diverse (Pass, 1998): The list includes antennal circulatory organs (Pass, 1991) and accessory pulsatile organs associated with the ovipositor valves (Hustert, Frisch, Böhm, & Pass, 2014) as well as wing hearts (Tögel, Pass, & Paululat, 2008). Immediately after the molt to adult, the activity of the wing hearts is required to clear the wing from cell debris that fill the space between the upper and the lower cuticular blade, a cleaning process necessary to allow the wing to get its mature stiffness required to fly.

7.4.6. Macroevolutionary "Laws" and the Predictability of Evolution

Is evolution predictable? This is a question many evolutionary biologists would never ask, perhaps by mistrust that it might be ever answered scientifically, perhaps only to avoid reviving even the slightest bit of late 19th-century orthogenetic views of evolution. Not all of the old macroevolutionary literature, however, has been definitively left in the wastebasket; think of Cope's rule of the unspecialized (Cope, 1887, 1896), Dollo's "law" of the irreversibility of evolution (Dollo, 1893), and especially the so-called Williston's "law" (Williston, 1914) of the progressive reduction and fixation in number and increase in specialization of serially repeated body parts (e.g., teeth, or vertebrae), three controversial principles more than occasionally mentioned in the recent evo-devo literature.

Experimental evidence suggesting a degree of predictability of evolution is found in studies of the genetic changes underpinning convergent evolution. In this context, Stern (2013) has introduced the useful distinction between *parallel genetic evolution* and *collateral genetic evolution*. The first is caused by the independent origin and spread of the same mutations in different populations or species. The second is obtained either by evolution of an allele that was polymorphic in a shared ancestral population, or by introgression, the introduction by hybridization of an allele into a population where originally it was not present. Martin and Orgogozo (2013) have provided a list of more than 100 cases of parallel and collateral genetic evolution. According to Stern (2013), this abundance implies that genome evolution is not random. The origin or the selective consequences of genetic variants, or both, might instead be somehow predictable (see also Christin, Weinreich, & Besnard, 2010; Conte et al., 2012; Gompel & Prud'homme, 2009; Stern, 2010; Stern & Orgogozo, 2008, 2009; Wood, Burke, & Rieseberg, 2005). Into the same direction point recent studies of polygenic traits, at least in comparisons between different populations or strains of the same species, such as Chan et al.'s (2012) work, that suggests widespread parallelism between different mouse strains evolving under the same selection regime.

7.4.7. Evolving Evolvability

The vagaries of tempo and mode of evolution suggest that evolvability is very far from being uniformly distributed across lineages and across temporal segments of the evolutionary history. It is thus sensible to ask whether evolvability is subject itself to evolution and, if so, whether this is an adaptive one, caused by positive selection for enhanced evolvability, or for evolvability biased in some specific direction.

There seems to be large agreement on the fact the evolvability evolves; according to Pigliucci (2008), this can be accepted, no matter how evolvability is defined. Contentious instead is the issue of the existence of selection for evolvability. This has been suggested, perhaps with cautionary comments, by Riedl (1978), Wagner and Altenberg (1996), Earl and Deem (2004), denied instead by Turney (1999). Of course, selection for evolvability, which is a property of the population rather than one of the individual, must be intended as higher-level selection (Alberch, 1991), because it is hard to see why selection on individuals should maximize it (Brookfield, 2009). Kirschner (2013) has floated the idea that evolvability, although not necessarily always so, can simply evolve as a byproduct of the evolution of physiological adaptability, but he favors a different hypothesis, based on a positive selection for the capacity to tolerate nonlethal phenotypic changes exposed to selection suppressing mutants expressed as lethal phenotypes at some stages of development or in particular circumstances, but without eliminating the variant genotype. This causes increased accumulation of genetic variation in a population, thus increasing evolvability.

7.5. Summary and Open Questions

As in the case of "species," "gene," and "homology," "evolvability" is one of those terms in the life sciences whose increasingly popular use is accompanied by a multiplication of definitions (often only implicitly assumed, rather than overtly stated) that often reveal as much about the scientist who adopts them as about the biological systems to whose description the term is used. To many, evolvability is just a property of the genetic system, to others, a property of phenotypes. However, the real heuristic power of evolvability is only found in the evo-devo dimension, that is, when we take into full account the complexities of the genotype→phenotype map. As forcefully argued by Hendrikse et al. (2007), evolvability is the proper focus of evolutionary developmental biology. We can hardly avoid addressing aspects of evolvability, whenever we study cryptic variation and its role in evolution, or the often pervious boundary between phenotypic plasticity and the expression of a genetic polymorphism, or the major phenotypic leaps that the mechanisms of development can produce based on point mutations, or the morphological stasis that reveals how robust a developmental process can be in front of a diversity of genetic changes and environmental stresses.

I agree with Pigliucci (2008), however, in suggesting that different notions of evolvability can have their merit and should be conserved, although perhaps under different, or differently qualified, names. What matters, of course, is to make clear how "evolvability" is used in any specific study.

While stressing the preferred link of the notion of evolvability to evolutionary developmental biology—thus aligning it back to its original meaning as in Dawkins (1989) and Alberch (1991)—I recommend that those who instead prefer to treat evolvability as a concept of either genetics or morphology eventually revisit it in the light of the novel perspectives emerging in the other camp.

Studying evolvability as a property of heredity systems, it may prove worthwhile to consider that heredity is not limited to the transmission (and expression) of genes, and to take a broader approach, in terms of heredity as phenotypic covariance between parent and offspring (e.g., Uller & Helanterä, this volume).

It is quite possible that nongenetic mechanisms of heredity contribute to the origin of novelties (Badyaev, 2008, 2009; Rice, 2012), either through maternal effect genes (Gilbert, 2010) or, as mentioned earlier, through the stabilization of environmentally induced phenotypes (e.g., West-Eberhard, 2003). It has been even suggested that heredity itself evolves following increasing stabilization of life cycles under natural selection (Badyaev, 2009).

There are also potentially far-reaching possibilities for an extension of a phenotype-centered view of evolvability. The phenotype, indeed, is not just the sum of the morphological characters of an organism. Beyond morphology, we can consider indeed the temporal aspect of the deployment of the phenotype (Minelli & Fusco, 2012), the nature of the reproduction modes (Minelli, 2009), or the structure and complexity of the life cycle (Minelli & Fusco, 2010). All of these dimensions of the phenotype are open for a thorough investigation in terms of evolvability, as a precondition to a revisitation of their evolutionary history.

References

Alberch, P. (1982). Developmental constraints in evolutionary processes. In J. T. Bonner (Ed.), *Development in evolution* (pp. 313–332). Berlin: Springer.

Alberch, P. (1991). From genes to phenotype: Dynamical systems and evolvability. *Genetica, 84*, 5–11.

Altenberg, L. (1994). The evolution of evolvability in genetic programming. In J. K. E. Kinnear (Ed.) *Advances in genetic programming* (pp. 47–74). Cambridge, MA: MIT Press.

Arthur, W. (2000). Intraspecific variation in developmental characters: The origin of evolutionary novelties. *American Zoologist, 40*, 811–818.

Arthur, W. (2004a). The effect of development on the direction of evolution: Toward a twenty–first century consensus. *Evolution and Development, 6*, 282–288.

Arthur, W. (2004b). *Biased embryos and evolution*. Cambridge, UK: Cambridge University Press.

Badyaev, A. V. (2005). Stress-induced variation in evolution: From behavioural plasticity to genetic assimilation. *Proceedings of the Royal Society B: Biological Sciences, 272*, 877–886.

Badyaev, A. V. (2008). Maternal effects as generators of evolutionary change: A reassessment. *Annals of the New York Academy of Sciences, 1133*, 151–161.

Badyaev, A. V. (2009). Evolutionary significance of phenotypic accommodation in novel environments: An empirical test of the Baldwin effect. *Philosophical Transactions of the Royal Society B: Biological Sciences, 364*, 1125–1141.

Bedau, M., & Packard, N. (1992). Measurement of evolutionary activity, teleology, and life. In C. Langton, C. Taylor, J. D. Farmer, & S. Rasmussen (Eds.), *Artificial life II: Proceedings of the Workshop on Artificial Life* (pp. 431–462). Redwood City, CA: Addison-Wesley.

Beldade, P., & Brakefield, P. M. (2002). The genetics and evo-devo of butterfly wing patterns. *Nature Reviews Genetics, 4,* 442–452.

Bell, A. (1991). *Plant form: An illustrated guide to flowering plant morphology.* Oxford, UK: Oxford University Press.

Bell, A. (2008). *Plant form: An illustrated guide to flowering plant morphology* (new edition). Portland, OR, and London, UK: Timber Press.

Braendle, C., Caillaud, M. C., & Stern, D. L. (2005). Genetic mapping of *aphicarus*: A sex-linked locus controlling a wing polymorphism in the pea aphid (*Acyrthosiphon pisum*). *Heredity, 94,* 435–442.

Braendle, C., & Flatt, T. (2006). A role for genetic accommodation in evolution? *Bioessays, 28,* 868–873.

Braendle, C., Friebe, I., Caillaud, M. C., & Stern, D. L. (2005). Genetic variation for an aphid wing polyphenism is genetically linked to a naturally occurring wing polymorphism. *Proceedings of the Royal Society B: Biological Sciences, 272,* 657–664.

Brakefield, P. M. (2008). Prospects of evo-devo for linking pattern and process in the evolution of morphospace. In A. Minelli & G. Fusco (Eds.), *Evolving pathways: Key themes in evolutionary developmental biology* (pp. 62–79). Cambridge, UK: Cambridge University Press.

Brakefield, P. M. (2011). Evo-devo and accounting for Darwin's endless forms. *Philosophical Transactions of the Royal Society B: Biological Sciences, 366,* 2069–2075.

Brandon, R. N. (1999). The units of selection revisited: The modules of selection. *Biology and Philosophy, 14,* 167–180.

Brigandt, I. (2007). Typology now: Homology and developmental constraints explain evolvability. *Biology and Philosophy, 22,* 709–725.

Brisson, J. A. (2010). Aphid wing dimorphisms: Linking environmental and genetic control of trait variation. *Philosophical Transactions of the Royal Society B: Biological Sciences, 365,* 605–616.

Brookfield, J. F. Y. (2001). The evolvability enigma. *Current Biology, 11,* R106–R108.

Brookfield, J. F. Y. (2009). Evolution and evolvability: Celebrating Darwin 200. *Biology Letters, 5,* 44–46.

Brooks, D. R., & McLennan, D. A. (1993). Macroevolutionary patterns of morphological diversification among parasitic flatworms (Platyhelminthes: Cercomeria). *Evolution, 47,* 495–509.

Budd, G. E. (1998). Arthropod body-plan evolution in the Cambrian with an example from anomalocaridid muscle. *Lethaia, 31,* 197–210.

Budd, G. E. (2006). On the origin and evolution of major morphological characters. *Biological Reviews, 81,* 609–628.

Carter, A. J. R., Hermisson, J., & Hansen, T. F. (2005). The role of epistatic gene interactions in the response to selection and the evolution of evolvability. *Theoretical Population Biology, 68,* 179–196.

Chan, Y. F., Jones, F. C., McConnell, E., Bryk, J., Bünger, L., & Tautz, D. (2012). Parallel selection mapping using artificially selected mice reveals body weight control loci. *Current Biology, 22,* 794–800.

Cheverud, J. M. (1984). Quantitative genetics and developmental constraints on evolution by selection. *Journal of Theoretical Biology, 110,* 155–171.

Christin, P. A., Weinreich, D. M., & Besnard, G. (2010). Causes and evolutionary significance of genetic convergence. *Trends in Genetics, 26,* 400–405.

Colegrave, N., & Collins, S. (2008). Experimental evolution: Experimental evolution and evolvability. *Heredity, 100,* 464–470.

Conte, G. L., Arnegard, M. E., Peichel, C. L., & Schluter, D. (2012). The probability of genetic parallelism and convergence in natural populations. *Proceedings of the Royal Society B: Biological Sciences, 279,* 5039–5047.

Cooper, K. L., Sears, K. E., Uygur, A., Maier, J., Baczkowski, K.-S., Brosnahan, M., . . . Tabin, C. J. (2014). Patterning and post-patterning modes of evolutionary digit loss in mammals. *Nature, 511,* 41–45.

Cope, E. D. (1887). *The origin of the fittest.* New York, NY: Appleton.

Cope, E. D. (1896). *The primary factors of organic evolution.* Chicago, IL: Open Court.

Danovaro, R., Dell'Anno, A., Pusceddu, A., Gambi, C., Heiner, I., & Kristensen, R. M. (2010). The first metazoa living in permanently anoxic conditions. *BMC Biology, 8,* 30.

Dawkins, R. (1989). The evolution of evolvability. In C. G. Langton (Ed.), *Artificial life: SFI studies in the sciences of complexity* (pp. 201–220). Redwood City, CA: Addison-Wesley.

de Visser, J. A. G. M., Hermisson, J., Wagner, G. P., Meyers, L. A., Bagheri-Chaichian, H., Blanchard, J. L., . . . Whitlock, M. C. (2003). Evolution and detection of genetic robustness. *Evolution, 57,* 1959–1972.

DeWitt, T. J., & Scheiner, S. M. (Eds.). (2004). *Phenotypic plasticity: Functional and conceptual approaches.* New York, NY: Oxford University Press.

Dichtel-Danjoy, M.-L., & Félix, M.-A. (2004). Phenotypic neighborhood and microevolvability. *Trends in Genetics, 20,* 268–276.

Dickerson, G. E. (1955). Genetic slippage in response to selection for multiple objectives. *Cold Spring Harbor Symposia on Quantitative Biology, 20,* 213–224.

Dollo, L. (1893). Les lois de l'évolution. *Bulletin de la Société Belge de Géologie, Paleontologie et Hydrologie, 7,* 164–166.

Draghi, J., Parsons T. L., Wagner G. P., & Plotkin J. B. (2010). Mutational robustness can facilitate adaptation. *Nature, 463,* 353–355.

Draghi, J., & Wagner G. P. (2008). Evolution of evolvability in a developmental model. *Evolution, 62,* 301–315.

Drago, L., Fusco, G., & Minelli, A. (2008). Non-systemic metamorphosis in male millipede appendages: Long delayed, reversible effect of an early localized positional marker? *Frontiers in Zoology, 5,* 5.

Drago, L., Fusco, G., Garollo, E., & Minelli, A. (2011). Structural aspects of leg-to-gonopod metamorphosis in male helminthomorph millipedes (Diplopoda). *Frontiers in Zoology*, *8*, 19.

Earl, D. J., & Deem, M. W. (2004). Evolvability is a selectable trait. *Proceedings of the National Academy of Sciences of the United States of America*, *101*, 11531–11536.

Erwin, D. H. (2012). Novelties that change carrying capacity. *Journal of Experimental Zoology (Molecular and Developmental Evolution)*, *318*, 460–465.

Fisher, R. A. (1930). *The genetical theory of natural selection*. Oxford, UK: Clarendon Press.

Flatt, T. (2005). The evolutionary genetics of canalization. *Quarterly Review of Biology*, *80*, 287–316.

Flohr, R. C. E., Blom, C. J., Rainey, P. B., & Beaumont, H. J. E. (2013). Founder niche constrains evolutionary adaptive radiation. *Proceedings of the National Academy of Sciences of the United States of America*, *110*, 20663–20668.

Frazzetta, T. H. (2012). Flatfishes, turtles, and bolyerine snakes: Evolution by small steps or large, or both? *Evolutionary Biology*, *39*, 30–60.

Fusco, G., & Minelli, A. (2010). Phenotypic plasticity in development and evolution. *Philosophical Transactions of the Royal Society B: Biological Sciences*, *365*, 547–556.

Gerhart, J., & Kirschner, M. (1997). *Cells, embryos and evolution: Towards a cellular and developmental understanding of phenotypic variation and evolutionary adaptability*. Oxford, UK: Blackwell.

Gerhart, J. C., & Kirschner M. W. (2007). The theory of facilitated variation. *Proceedings of the National Academy of Sciences of the United States of America*, *104*, 8582–8589.

Gerhart, J. C., & Kirschner, M. W. (2010). Facilitated variation. In M. Pigliucci & G. B. Muller (Eds.) *Evolution: The extended synthesis* (pp. 253–280). Cambridge, MA: MIT Press.

Gibson, G., & Dworkin, I. (2004). Uncovering cryptic genetic variation. *Nature Reviews Genetics*, *5*, 681–691.

Gilbert S. F. (2010). *Developmental biology* (9th ed). Sunderland, MA: Sinauer.

Gilbert, S. F., Loredo, G. A., Brukman, A., & Burke, A. C. (2001). Morphogenesis of the turtle shell: The development of a novel structure in tetrapod evolution. *American Zoologist*, *31*, 616–627.

Goldschmidt, R. (1940). *The material basis of evolution*. New Haven, CT: Yale University Press.

Gompel, N., & Prud'homme, B. (2009). The causes of repeated genetic evolution. *Developmental Biology*, *332*, 36–47.

Greene, E. (1999). Phenotypic variation in larval development and evolution: Polymorphism, polyphenism, and developmental reaction norms. In M. Wake & B. K. Hall (Eds.), *The origin and evolution of larval forms* (pp. 379–410). New York, NY: Academic Press.

Griswold, C. K. (2006). Pleiotropic mutation, modularity and evolvability. *Evolution and Development*, *8*, 81–93.

Hall, B. K. (2005). Consideration of the neural crest and its skeletal derivatives in the context of novelty/innovation. *Journal of Experimental Zoology (Molecular and Developmental Evolution), 304,* 548–557.

Hallgrímsson, B., Jamniczky, H., Young, N. M., Rolian, C., Schmidt-Ott, U., & Marcucio R. S. (2012). The generation of variation and the developmental basis for evolutionary novelty. *Journal of Experimental Zoology (Molecular and Developmental Evolution), 318B,* 501–517.

Hansen, T. F. (2003). Is modularity necessary for evolvability? Remarks on the relationship between pleiotropy and evolvability. *BioSystems, 69,* 83–94.

Hansen, T. F. (2006). The evolution of genetic architecture. *Annual Review of Ecology and Systematics, 37,* 123–157.

Hendrikse, J. L., Parsons, T. E,. & Hallgrímsson, B. (2007). Evolvability as the proper focus of evolutionary developmental biology. *Evolution and Development, 9,* 393–401.

Henriquez, C. L., Arias, T., Pires, J. C., Croat, T. B., & Schaal, B. A. (2014). Phylogenomics of the plant family Araceae. *Molecular Phylogenetics and Evolution, 75,* 91–102.

Houle, D. (1992). Comparing evolvability and variability of quantitative traits. *Genetics, 130,* 195–204.

Hustert, R., Frisch, M., Böhm, A., & Pass, G. (2014). A new kind of auxiliary heart in insects: Functional morphology and neuronal control of the accessory pulsatile organs of the cricket ovipositor. *Frontiers in Zoology, 11,* 43.

Janzen, F. J., & Phillips, P. C. (2006). Exploring the evolution of environmental sex determination, especially in reptiles. *Journal of Evolutionary Biology, 19,* 1775–1784.

Jiménez-Guri, E., Okamura, B., & Holland, P. W. H. (2007). Origin and evolution of a myxozoan worm. *Integrative and Comparative Biology, 47,* 752–758.

Kampourakis, K., & Minelli, A. (2014). Evolution makes more sense in the light of development. *The American Biology Teacher, 76,* 493–498.

Kemp, T. S. (1999). *Fossils and evolution.* Oxford, UK: Oxford University Press.

Kemp, T. S. (2007). The concept of correlated progression as the basis of a model for the evolutionary origin of major new taxa. *Proceedings of the Royal Society B: Biological Sciences, 274,* 1667–1673.

Kirkpatrick, M. (2009). Patterns of quantitative genetic variation in multiple dimensions. *Genetica, 136,* 271–284.

Kirschner, M. (2013). Beyond Darwin: Evolvability and the generation of novelty. *BMC Biology, 11,* 110.

Kirschner, M., & Gerhart, J. (1998). Evolvability. *Proceedings of the National Academy of Sciences of the United States of America, 95,* 8420–8427.

Kirschner, M. W., & Gerhart, J. C. (2005). *The plausibility of life: Resolving Darwin's dilemma.* New Haven, CT: Yale University Press.

Klingenberg, C. P. (2005). Developmental constraints, modules and evolvability. In B. Hallgrímsson & B. K. Hall (Eds.), *Variation: A central concept in biology* (pp. 219–247). Burlington, MA: Elsevier.

Laforsch, C., & Tollrian, R. (2004). Inducible defenses in multipredator environments: Cyclomorphosis in *Daphnia cucullata*. *Ecology, 85*, 2302–2311.

Lande, R. (1986). The dynamics of peak shifts and pattern of morphological evolution. *Paleobiology, 12*, 343–354.

Lee, M. S. Y. (1996). Correlated progression and the origin of turtles. *Nature, 379*, 812–815.

Levy, Y. Y., & Dean, C. (1998). The transition to flowering. *Plant Cell, 10*, 1973–1990.

Lewontin, R. (1983). Gene, organism, and environment. In D. S. Bendall (Ed.), *Evolution from molecules to man* (pp. 273–285). Cambridge, UK: Cambridge University Press.

Leys, S. P., & Riesgo, A. (2012). Epithelia, an evolutionary novelty of metazoans. *Journal of Experimental Zoology (Molecular and Developmental Evolution), 318*, 438–447.

Love, A. C. (2003). Evolutionary morphology, innovation, and the synthesis of evolutionary and developmental biology. *Biology and Philosophy, 18*, 309–345.

Martin, A., & Orgogozo, V. (2013). The loci of repeated evolution: A catalog of genetic hotspots of phenotypic variation. *Evolution, 67*, 1235–1250.

Masel, J., & Trotter, M. V. (2010). Robustness and evolvability. *Trends in Genetics, 26*, 406–414.

Maynard Smith, J., & Szathmáry, E. (1995). *The major transitions in evolution*. Oxford, UK: Oxford University Press.

Maynard Smith, J., Burian, R., Kauffman, S., Alberch, P., Campbell, J., Goodwin, B., Lande, R., Raup, D., & Wolpert, L. (1985). Developmental constraints and evolution. *Quarterly Review of Biology, 60*, 265–287.

Mayr, E. (1960.) The emergence of evolutionary novelties. In S. Tax (Ed.), *The evolution of life* (pp. 157–180). Chicago, IL: University of Chicago Press.

Messina, F. J. (1993). Heritabilty and "evolvability" of fitness components in *Callosobruchus maculatus*. *Heredity, 71*, 623–629.

Minelli, A. (2000). Limbs and tail as evolutionarily diverging duplicates of the main body axis. *Evolution and Development, 2*, 157–165.

Minelli, A. (2003). The origin and evolution of appendages. *International Journal of Developmental Biology, 47*, 573–581.

Minelli, A. (2009). *Perspectives in animal phylogeny and evolution*. Oxford, UK: Oxford University Press.

Minelli, A. (2011a). A principle of developmental inertia. In B. Hallgrímsson & B. K. Hall (Eds.) *Epigenetics: Linking genotype and phenotype in development and evolution* (pp. 116–133). San Francisco: University of California Press.

Minelli, A. (2011b). Development, an open-ended segment of life. *Biological Theory, 6*, 4–15.

Minelli, A. (2014). Developmental disparity. In A. Minelli & T. Pradeu (Eds.), *Towards a theory of development* (pp. 227–245). Oxford, UK: Oxford University Press.

Minelli, A. (2015). Morphological misfits and the architecture of development. In E. Serrelli & N. Gontier (Eds.), *Macroevolution: Explanation, interpretation, evidence* (pp. 329–343). Cham: Springer.

Minelli, A., Chagas, A. J., & Edgecombe, G. D. (2009). Saltational evolution of trunk segment number in centipedes. *Evolution and Development, 11*, 318–322.

Minelli, A., & Fusco, G. (2010). Developmental plasticity and the evolution of animal complex life cycles. *Philosophical Transactions of the Royal Society B: Biological Sciences, 365*, 631–640.

Minelli, A., & Fusco, G. (2012). On the evolutionary developmental biology of speciation. *Evolutionary Biology, 39*, 242–254.

Minelli, A., & Pradeu, T. (Eds.). (2014). *Towards a theory of development.* Oxford, UK: Oxford University Press.

Mitchell, C. H., & Diggle, P. K. (2005). The evolution of unisexual flowers: Morphological and functional convergence results from diverse developmental transitions. *American Journal of Botany, 92*, 1068–1076.

Miura, T. (2005). Developmental regulation of caste-specific characters in social-insect polyphenism. *Evolution and Development, 7*, 122–129.

Moczek, A. P. (2007). Developmental capacitance, genetic accommodation, and adaptive evolution. *Evolution and Development, 9*, 299–305.

Moczek, A. P. (2008). On the origins of novelty in development and evolution. *Bioessays, 30*, 432–447.

Moczek, A.P. (2014). Towards a theory of development through a theory of developmental evolution. In A. Minelli & T. Pradeu (Eds.) *Towards a theory of development* (pp. 218–226). Oxford, UK: Oxford University Press.

Monteiro, A., Brakefield, P. M., & French, V. (1997). Butterfly eyespots: The genetics and development of the color rings. *Evolution, 51*, 1207–1216.

Müller, G. B. (1990). Developmental mechanisms at the origin of morphological novelty: A side-effect hypothesis. In M. H. Nitecki (Ed.), *Evolutionary innovations* (pp. 99–130). Chicago, UK: University of Chicago Press.

Müller, G. B. (2010). Epigenetic innovation. In M. Pigliucci & G. B. Müller (Eds.), *Evolution: The extended synthesis* (pp. 307–332). Cambridge, MA: MIT Press.

Muller, G. B., & Newman, S. A. (2005). The innovation triad: An EvoDevo agenda. *Journal of Experimental Zoology (Molecular and Developmental Evolution), 304*, 487–503.

Müller, G. B., & Wagner, G. P. (1991). Novelty in evolution: Restructuring the concept. *Annual Review of Ecology, Evolution and Systematics, 22*, 229–256.

Müller, G. B., & Wagner, G. P. (2003). Innovation. In B. K. Hall & W. M. Olson (Eds.), *Keywords and concepts in evolutionary developmental biology* (pp. 218–227). Cambridge, MA: Harvard University Press.

Nagy, L. (1998). Changing patterns of gene regulation in the evolution of arthropod morphology. *American Zoology, 38*, 818–828.

Nicoglou, A. (2015). The evolution of phenotypic plasticity: Genealogy of a debate in genetics. *Studies in the History and Philosophy of Biological and Biomedical Sciences, 50*, 67–76.

Nijhout, H. F. (2003). Development and evolution of adaptive polyphenisms. *Evolution and Development, 5*, 9–18.

Nüsslein-Volhard, C., & Wieschaus, E. (1980). Mutations affecting segment number and polarity in *Drosophila*. *Nature, 287*, 795–801.

Obst, M., & Funch, P. (2003). Dwarf male of *Symbion pandora* (Cycliophora). *Journal of Morphology, 255*, 261–278.

Odling-Smee, J. (2010). Niche inheritance. In M. Pigliucci & G. B. Muller (Eds.) *Evolution: The extended synthesis* (pp. 175–207). Cambridge, MA: MIT Press.

Odling-Smee, F. J., Laland, K. N., & Feldman, M. W. (2003). *Niche construction: The neglected process in evolution*. Princeton, NJ: Princeton University Press.

Orr, H. A. (1998). The population genetics of adaptation: The distribution of factors fixed during adaptive evolution. *Evolution, 52*, 935–949.

Pass, G. (1991). Antennal circulatory organs in Onychophora, Myriapoda and Hexapoda: functional morphology and evolutionary implications. *Zoomorphology, 110*, 145–164.

Pass, G. (1998). Accessory pulsatile organs. In F. Harrison & M. Locke (Eds.) *Microscopic Anatomy of Invertebrates, 11. Insecta* (11B, pp. 621–640). New York, NY: Wiley.

Peterson, T., & Müller, G. B. (2013). What is evolutionary novelty? Process versus character based definitions. *Journal of Experimental Zoology (Molecular and Developmental Evolution), 320B*, 345–350.

Pigliucci, M. (2001). *Phenotypic plasticity: Beyond nature and nurture*. Baltimore, MD: John Hopkins University Press.

Pigliucci, M. (2008). Is evolvability evolvable? *Nature Reviews Genetics, 9*, 75–82.

Pigliucci, M., Murren, C. J., & Schlichting, C. D. (2006). Phenotypic plasticity and evolution by genetic assimilation. *Journal of Experimental Biology, 209*, 2362–2367.

Pradeu, T., Laplane, L., Prévot, K., Hoquet, T., Reynaud, V., Fusco, G., . . . Vervoort, M. (2016). Defining "development." *Current Topics in Developmental Biology, 117*, 171–183.

Quayle, A. P., & Bullock S. (2006). Modelling the evolution of genetic regulatory networks. *Journal of Theoretical Biology, 238*, 737–753.

Raff, E. C., & Raff, R. A. (2000). Dissociability, modularity, evolvability. *Evolution and Development, 2*, 235–237.

Raff, R. A. (1996). *The shape of life*. Chicago, IL: University of Chicago Press.

Relyea, R.A. (2004). Fine-tuned phenotypes: Tadpole plasticity under 16 combinations of predators and competitors. *Ecology, 85*, 172–179.

Rice, S. H. (2012). The place of development in mathematical evolutionary theory. *Journal of Experimental Zoology (Molecular and Developmental Evolution), 318,* 480–488.

Riedl, R. J. (1977). A systems-analytical approach to macro-evolutionary phenomena. *Quarterly Review of Biology, 52,* 351–370.

Riedl, R. J. (1978). *Order in living organisms: A systems analysis of evolution.* New York, NY: Wiley.

Rieppel, O. (2001). Turtles as hopeful monsters. *BioEssays, 23,* 987–991.

Rieseberg, L. H., Raymond, O., Rosenthal, D. M., Lai, Z., Livingstone, K., Nakazato, T., . . . Lexer, C. (2003). Major ecological transitions in wild sunflowers facilitated by hybridization. *Science, 301,* 1211–1216.

Rolian, C., Lieberman, D. E., & Hallgrímsson, B. (2010). The coevolution of human hands and feet. *Evolution, 64,* 1558–1568.

Schlichting, C. D. (2008). Hidden reaction norms, cryptic genetic variation, and evolvability. *Annals of the New York Academy of Sciences, 1133,* 187–203.

Schlichting, C. D., & Murren, C. J. (2004). Evolvability and the raw materials for adaptation. In Q. C. B. Cronk, J. Whitton, R. H. Ree, & I. E. P. Taylor (Eds.) *Plant adaptation: molecular genetics and ecology* (pp. 18–29). Ottawa, CA: NRC Research Press.

Schlichting, C. D., & Pigliucci, M. (1998). *Phenotypic evolution: A reaction norm perspective.* Sunderland, MA: Sinauer.

Schlosser, G., & Wagner, G. P. (2003). Introduction: The modularity concept in developmental and evolutionary biology. In G. Schlosser & G. P. Wagner (Eds.), *Modularity in development and evolution* (pp. 1–11). Chicago, IL: University of Chicago Press.

Siddall, M. E., & Whiting, M. F. (1999). Long-branch abstractions. *Cladistics, 15,* 9–24.

Sniegowski, P. D., & Murphy, H. A. (2006). Evolvability. *Current Biology, 16,* R831–R834.

Spencer, C. C., Tyerman, J., Bertrand, M., & Doebeli, M. (2008). Adaptation increases the likelihood of diversification in an experimental bacterial lineage. *Proceedings of the National Academy of Sciences of the United States of America, 105,* 1585–1589.

Stern, D. L. (2010). *Evolution, development, and the predictable genome.* Greenwood Village, CO: Roberts.

Stern, D. L. (2013). The genetic causes of convergent evolution. *Nature Reviews Genetics, 14,* 751–764.

Stern, D. L., & Orgogozo, V. (2008). The loci of evolution: How predictable is genetic evolution? *Evolution, 62,* 2155–2177.

Stern, D. L., & Orgogozo, V. (2009). Is genetic evolution predictable? *Science, 323,* 746–751.

Theißen, G. (2009). Saltational evolution: Hopeful monsters are here to stay. *Theory in Biosciences, 128,* 43–51.

Thompson, J. V. (1830). On the Cirripedes or Barnacles; demonstrating their deceptive character; the extraordinary Metamorphosis they undergo, and the Class of Animals to which they indisputably belong. In J. V. Thompson, *Zoological researches, and illustrations; Or, natural history of nondescript or imperfectly known animals* (Vol. 1, pp. 69–82). Cork: King and Ridings.

Thompson, J. V. (1835). Discovery of the metamorphosis in the second type of the cirripedes, viz. the lepades, completing the natural history of these singular animals, and confirming their affinity with the crustacea. *Philosophical Transactions of the Royal Society, 126,* 355–358.

Thurber, C. S., Jia, M. H., Jia, Y., & Caicedo, A. L. (2013). Similar traits, different genes? Examining convergent evolution in related weedy rice populations. *Molecular Ecology, 22,* 685–698.

Tögel, M., Pass, G., & Paululat, A. (2008). The *Drosophila* wing hearts originate from pericardial cells and are essential for wing maturation. *Developmental Biology, 318,* 29–37.

Turney, P. D. (1999). Increasing evolvability considered as a large-scale trend in evolution. Proceedings of the 1999 Genetic and Evolutionary Computation Conference (GECCO-99) Workshop Program, Orlando, FL. July 13, 1999 (pp. 43–46).

Vavilov, N. I. (1922). The law of homologous series in variation. *Journal of Genetics, 12,* 67–87.

von Dassow, G., & Munro, E. M. (1999). Modularity in animal development and evolution: Elements of a conceptual framework for EvoDevo. *Journal of Experimental Zoology (Molecular and Developmental Evolution), 285,* 307–325.

Waddington, C. H. (1953). Genetic assimilation of an acquired character. *Evolution, 7,* 118–126.

Waddington, C. H. (1957). *The strategy of the genes.* London, UK: Allen & Unwin.

Waddington, C. H., & Robertson, E. (1966). Selection for developmental canalisation. *Genetics Research, 7,* 303–312.

Wagner, A. (2005). *Robustness and evolvability in living systems.* Princeton, NJ: Princeton University Press.

Wagner, A. (2008). Robustness and evolvability: A paradox resolved. *Proceedings of the Royal Society B: Biological Sciences, 275,* 91–100.

Wagner, A. (2011). *The origins of evolutionary innovations: A theory of transformative change in living systems.* Oxford, UK: Oxford University Press.

Wagner, G. P. (1989). The origin of morphological characters and the biological basis of homology. *Evolution, 43,* 1157–1171.

Wagner, G. P. (1996). Homologues, natural kinds and the evolution of modularity. *American Zoologist, 36,* 36–43.

Wagner, G. P., & Altenberg, L. (1996). Complex adaptations and evolution of evolvability. *Evolution, 50,* 967–976.

Wagner, G. P., Booth, G., & Bagheri-Chaichian, H. (1997). A population genetic theory of canalization. *Evolution, 51,* 329–347.

Wagner, G. P., & Laubichler, M. D. (2004). Rupert Riedl and the re-synthesis of evolutionary and developmental biology: Body plans and evolvability. *Journal of Experimental Zoology (Molecular and Developmental Evolution), 302B*, 92–102.

Walsh, B., & Blows, M. W. (2009). Abundant genetic variation + strong selection = multivariate genetic constraints: A geometric view of adaptation. *Annual Review of Ecology, Evolution and Systematics, 40*, 41–59.

West-Eberhard, M. J. (1998). Evolution in the light of developmental and cell biology, and vice versa. *Proceedings of the National Academy of Sciences of the United States of America, 95*, 8417–8419.

West-Eberhard, M. J. (2003). *Developmental plasticity and evolution*. New York, NY: Oxford University Press.

West-Eberhard, M. J. (2005). Developmental plasticity and the origin of species differences. *Proceedings of the National Academy of Sciences of the United States of America, 102*, 6543–6549.

Westheide, W. (1987). Progenesis as a principle in meiofauna evolution. *Journal of Natural History, 21*, 843–854.

Williston, S. W. (1914). *Water reptiles of the past and present*. Chicago, UK: University of Chicago Press.

Wood, T. E., Burke, J. M., & Rieseberg, L. H. (2005). Parallel genotypic adaptation: When evolution repeats itself. *Genetica, 123*, 157–170.

Woods, R. J., Barrick, J. E., Cooper, T. F., Shrestha, U., Kauth, M. R., & Lenski, R. E. (2011). Second-order selection for evolvability in a large *Escherichia coli* population. *Science, 331*, 1433–1436.

Wray, G. A. (2002). Do convergent developmental mechanisms underlie convergent phenotypes? *Brain Behavior and Evolution, 59*, 327–336.

Yang, A. S. (2001). Modularity, evolvability, and adaptive radiations: a comparison of the hemi- and holometabolous insects. *Evolution and Development, 3*, 59–72.

8 "CHANCE CAUGHT ON THE WING"

METAPHYSICAL COMMITMENT OR METHODOLOGICAL ARTIFACT?

Denis M. Walsh

In his landmark book *Chance* and *Necessity: An Essay on the Natural Philosophy of Modern Biology*, Jacques Monod seeks to articulate and then to resolve what he perceives to be a paradox afflicting modern biology.[1] He calls it the "paradox of invariance," and he leaves us in no doubt regarding its significance: "In fact, the central problem of biology lies with this very contradiction" (1971, p. 12).[2] The problem is to be found in a tension that Monod detects between scientific methodology and biological reality. On the one hand, "The cornerstone of the scientific method is the postulate that science is objective." Objectivity entails "the systematic denial that 'true' knowledge can be got at by interpreting phenomena in terms of . . . 'purpose.'" On the other hand, "Objectivity nevertheless obliges us to recognize the teleonomic character of living organisms, to admit that in their structure they act projectively—realize and pursue a purpose" (p. 12).

"Living creatures are strange objects" (p. 17), Monod tells us. They exhibit two extreme—and contradictory—kinds of organizing principles: invariance$_1$ and purpose. They are invariant$_1$ in the sense of possessing an "ability to reproduce and to transmit *ne variateur* the

---

1. Monod's term is a "flagrant epistemological contradiction."

2. Unfortunately, the term "invariance" and its cognates are used in two drastically different ways in the literature drawn on in this paper—awkward. Where the use of these terms cannot be avoided, and the context does not wholly disambiguate, I shall index them: "invariance$_1$," denotes Monod's conception of invariance, and "invariance$_2$," denotes Woodward's.

information corresponding to their own structure: A very rich body of information . . . which . . . is preserved in tact from one generation to the next" (p. 12). Equally, organisms are by their very natures purposive entities. Monod does not elaborate much, but at just about the same time, Ludwig von Bertalanffy made the latter point vividly:

> You cannot even think of an organism . . . without taking into account what variously and rather loosely is called adaptiveness, purposiveness, goal seeking and the like. (Von Bertalanffy, 1969, p. 45)

As Monod sees it, organisms are, and yet could not be, both fundamentally invariant and purposive—and that is the paradox of invariance$_1$.

In order to resolve the tension we must account for one of these evidently antithetical features of organisms in terms of the other. There are two options. We could strive to explain invariance$_1$—in particular the high fidelity of inheritance, the robustness of biological form, and the stability of DNA—as a consequence of organismal purposiveness, in this way: "Invariance is safeguarded, ontogeny guided, and evolution oriented by an initial teleonomic principle, of which all these phenomena are the purported manifestations" (Monod, 1971, p. 24). Alternatively, we could seek to ground the purposiveness of organisms by appeal to invariance$_1$. By this resolution, "all properties of living beings rest on a fundamental mechanism of molecular invariance" (Monod, 1971, p. 116). Scientific methodology—the "objectivity of science"—licenses only one of these strategies: the superordination of invariance$_1$ over purpose. So, in order to render evolutionary biology a science in good standing, we must seek to explain all its proprietary phenomena—the distribution and adaptedness of organismal form, the exquisite functional integration of organisms, their supple robustness—in terms of the preservation and transmission of invariant information.

This too raises a complication. Invariance$_1$ is stasis, but evolution is change. Moreover, *adaptive* evolution is biased change. Through the process of adaptive evolution, populations come to comprise organisms that are increasingly well suited to survival and reproduction in their conditions of existence. There must be a constant source of new variants, and there must be a process that biases evolutionary change. We cannot suppose that the new variants that arise within organisms are biased in favor of their goals or purposes, on pain of violating the "objectivity of science." So, the source of evolutionary novelties, Monod insists, must arise from unbiased—*chance*—alterations to the invariant$_1$ molecular structure that underlies organisms.

The initial elementary events which open the way to evolution in these intensely conservative systems called living beings are microscopic, fortuitous, and utterly without relation to whatever may be their effects upon teleonomic functioning. (p. 118)

Thus, Monod's *Chance and Necessity* elegantly traces modern synthesis biology's commitment to ineluctable chance right back to its methodological source. The choice is between chance or purpose, and our hand is forced by the demands of scientific method, in the form of the "postulate of objectivity." The postulate of objectivity is compulsory for scientific enquiry.

The postulate of objectivity is consubstantial with science; it has guided the whole of its prodigious development for three centuries. There is no way to be rid of it . . . without departing from the domain of science itself. (1971, p. 12)

And it has no truck with purpose. Not that he is reluctant to embrace a commitment to chance:

[Mutations] constitute the only possible source of modifications in the genetic text, itself the sole repository of the organism's hereditary structures, *it necessarily follows that chance alone is at the source of every innovation, of all creation in the biosphere. Pure chance, absolutely free but blind, at the very root of the stupendous edifice of evolution.* (p. 112, emphasis added)

Evolution, in Monod's resonant phrase, is "chance caught on the wing."

My objective here is to question the commitment to ineluctable chance that, as Monod so astutely observes, is a cornerstone of modern synthesis evolutionary biology. Unlike Monod, I contend that the primacy of invariance$_1$ over purpose is not mandated on methodological grounds. Furthermore, I claim that it no longer seems plausible, even on empirical grounds. I wish to explore in outline the prospects for an evolutionary biology that prioritizes purpose over invariance$_1$. I call this "neo-Aristotelian" evolutionary biology and contrast it with the "neo-Democritean" cast of the modern synthesis. One surprising consequence of neo-Aristotelian evolutionary biology is that it relieves biology of the obligation to accord an ineliminable role to chance. From the perspective of neo-Aristotelian biology, evolution is not fundamentally chancy; adaptive evolutionary change is inherent in the adaptiveness of organisms. What from the neo-Democritean biology is a compulsory metaphysical commitment, looks, from the neo-Aristotelian perspective, like an inapposite methodological artifact. Yet, ineluctable chance is so integral to our current neo-Democritean biology that removing that particular cornerstone threatens to undermine much of the edifice of the modern synthesis theory of evolution.

8.1. Neo-Democritean Evolutionary Biology

Monod's project of banishing purpose from biology draws inspiration from the pre-Socratic Atomistic philosophers. The intellectual debt is acknowledged in his title. Monod credits Democritus with the claim that "everything in the world is the fruit of chance and necessity." The Atomists, like Democritus and Empedocles, held that the fundamental units of matter are atoms of different kinds. Each kind of atom has a characteristic kind of motion. The world began in chaos—single atoms moving randomly in the void. Yet our world is clearly not chaotic. It consists of observable regularities among stable, complex, macroscopic entities. The world has become ordered because atoms encounter and combine with one another at random. Some combinations are ephemeral and some enduring. Stable combinations of atoms are differentially preserved as complex entities. These macroscopic entities behave in predictable ways because their characteristic properties are fixed—*necessitated*—by the properties and the arrangements of their constituent atoms. Ultimately, the order in the universe is the result of the chance encounters of randomly moving atoms, and the necessary consequences of those encounters (Hankinson, 1998). These same principles—chance and necessity—suffice to explain order in the biological realm.

> Empedocles, and others of his sort, consistently attempted to understand the features of mature organisms as coincidental outcomes of materially necessitated developmental processes. (Lennox, 2009, p. 357)

Monod's choice of methodological archetype is apt. There are powerful resonances of pre-Socratic Atomism in modern synthesis evolutionary biology (Wicken, 1981). Atomism even has it own theory of the fit and diversity of organic form that, while not strictly *evolutionary*, bears a striking resemblance to our own.

> The generation of whole natured forms able to survive depended on the chance combination of fragments which were well adapted both to the needs of the whole creature and to the environment. The account of

the origin of species by Empedocles is the first recorded account of the theory of natural selection. (Roux, 2005, p. 6)[3]

Modern synthesis biology takes genes or replicators to be evolutionary atoms, whose formation and whose coming together are random events. New replicators, and new combinations of replicators, arise by chance. The production of whole organisms, and their differential survival and reproduction, are causally necessary consequences of the activities of replicators. Modern synthesis biology can rightly be said, in an important sense, to be neo-Democritean.

Monod is to be commended for noticing how closely modern synthesis methodology cleaves to its Atomistic predecessor. The resemblance is so strong that it is tempting to speculate that any deficiencies of the ancestor may have been transmitted *ne variateur* to the descendant. A cogent argument against Atomism might well form the basis of a robust challenge to the modern synthesis.

The most ardent ancient opponent of pre-Socratic Atomism, especially its Democritean variant, is Aristotle. According to Aristotle, the telling deficiency of Atomism can be seen in the role that it accords to chance. The ordered features of the natural world, Aristotle contends, are not ineluctably chancy. Nature only *appears* fundamentally chancy when viewed through the polarizing lens of Atomistic methodology.

Aristotle's argument can be found in part in *Physics* II.5 (Aristotle, 1996). He illustrates it with the story of a man "collecting subscriptions for a feast." Our protagonist happens on a debtor, let us suppose in the marketplace, and extracts from him a payment. This encounter is strictly a matter of chance: "He actually went there for another purpose, and it was only incidentally that he got his money by going there." So, although the encounter fulfills our agent's purpose, it was not done *for the sake of* the agent's purpose. The agent's intentions or purposes form no part of the explanation of how they were fulfilled. On the other hand, if the agent *had* gone to the marketplace *in order to* collect the subscription, then the intention or purpose *would have* entered into the explanation.

This is a complex passage, but part of Aristotle's point is that the same encounter might have occurred as a matter of chance, or for a purpose. Either way it would result in the payment of the debt. Moreover, either way it would get the same Atomistic explanation. The modern analogue would cite the various internal

---

3. Dawkins's (1989) scenario of assembling the best rowing eight by randomly assigning combinations of rowers to a boat, and then selecting the fastest is a nice analogy. It too is meant to illustrate how the random aggregation of parts produces well-functioning ordered, differentially persisting aggregations.

physiological mechanisms that caused our agent's limbs to move, various features of their shared ecological setting, various times and locations and so forth. These mechanisms and their capacity to explain their effects are completely insensitive to whether the outcome is a chance occurrence or a matter of design. If there is an important difference between the occurrence of an event by chance or by purpose, it is invisible to Atomistic explanation.

Yet there *is* a difference between chance events and purposive events. It is most clearly seen in their respective modal profiles. Chance and purposive occurrences are robust across different ranges of counterfactual circumstances. Purposive occurrences are robust across a range of alternate background conditions and mechanisms. For instance, if this were a purposive encounter, we would expect it to be somewhat *insensitive* to initial conditions, like location. Had the debtor been somewhere else in the market, or at the bath, or at the barber, then the encounter might well have occurred anyway, only in a different place, by a different set of mechanisms. Our protagonist would be expected to have done whatever was necessary, under the circumstances, to ensure that the event occurred. In contrast, chance events are highly sensitive to initial conditions. If this were a *chance* encounter, then had the specific spatiotemporal circumstances been different, had the mechanisms been different, then in all likelihood, the event in question—the exchange of money—would not have occurred. Then again, purposive occurrences are highly *sensitive* to goals in a way that chance occurrences are not. If this were a purposive occurrence, then had the agent's goals been different—say to collect money from someone else, to avoid his friend, to walk the dog in the country—the encounter probably would have not occurred at all.

Events that occur because they fulfill a purpose are thus subject to two distinct kinds of explanations, one that identifies the occurrence as resulting from mechanical interactions, the other that identifies it as occurring because it conduces to the fulfillment of a goal (Walsh, 2012, 2013). In contrast, a chance occurrence is subject (at best) only to a mechanistic explanation.[4] Aristotle's claim is that restricting ourselves to only one kind of explanation leads us to miss out on at least one important kind of scientifically explainable regularity. As a result, certain perfectly explainable regularities may be erroneously dismissed as inexplicable chance occurrences.

Neo-Democritean biology—like its Atomistic precursor—is unequipped to recognize the significance of events that might occur because they fulfill purposes.

---

4. Some chance occurrences, like radioactive decay, may have no mechanistic explanation.

Evolution is exclusively to be explained in terms of chance variations and their necessary consequences. This is no mere oversight. It is borne of the conviction that Monod identifies, *viz.* that purpose has no role in scientific explanation. But here neo-Democritean evolutionary biology makes a bold empirical wager, that the pursuit of organismal purposes makes no appreciable contribution to organic evolution. It might look rather more like backing the favorite each way. After all, purpose has been all but expunged from the natural sciences since the advent of the scientific revolution, so why not bet against it. Challenging neo-Democritean biology on these grounds would require showing that natural purpose can earn its keep in biology. That, in turn, requires two things: (1) a compelling case that purposes can figure in genuinely scientific explanations, and (2) demonstrating that such explanations are needed for the project of explaining evolution.

8.2. Natural Purpose

The principal impediment to purposive explanation in the natural sciences seems to be comprehensiveness of mechanism. If every event has a complete mechanistic explanation, mechanism leaves no unexplained residuum. There is nothing left over for purpose to explain.

Immanuel Kant's (2000/1790) discussion of teleology is illuminating in this regard. Kant evinces an uncommon sensitivity to the need for purposive explanation in biology, but still he worries that this form of explanation is unavailable to the natural scientist. Kant reminds us that organisms are unique in the world. They are *self*-building, *self*-nourishing, *self*-reproducing entities (McLaughlin, 1990). Organisms pursue purposes that inhere in their life activities. The parts of organisms are not only appropriate to the production of the organism's characteristic activities, they possess these properties precisely *because* of the organism's pursuit of those activities. So, while the activities of an organism are consequences of the properties and mechanical interactions of its parts, the properties and interactions of the parts are the consequence of the activities of organisms. There is a "reciprocal causation" between organisms as a whole and the activities of the parts. In this organisms are wholly distinctive, and it is this, according to Kant, that makes them *natural purposes*.

> The definition of an organic body is that it is a body, every part of which is there for the sake of the other (reciprocally as an end, and at the same time, means). . . . An organic (articulated) body is one in which each part, with its moving force, necessarily relates to the whole (to each part in its composition). (*Opus postumum*, quoted in Guyer, 2005, p. 104)

I would provisionally say that a thing exists as a natural end if it is cause and effect of itself. (Kant, 2000/1793, p. 371)

Mere nonpurposive mechanism evidently cannot wholly account for the marvelous integration, the supple adaptation, the self-regulating ability that is constitutive of organisms. We can only make sense of these magnificent features of living things by considering them as natural purposes.

The Aristotelian resonances are evident and profound.[5] The reciprocity that Kant sees in organisms as natural purposes is reflected in Aristotle's multiplicity of explanatory modes. For Kant organisms are both causes and effects of the activities of their parts. For Aristotle, the capacities of an organism's parts constitute the efficient cause of organismal purposiveness. And, reciprocally, the purposiveness of organisms is the final cause of the capacities of their parts. Like Kant, Aristotle stresses that the functional unity of organisms calls for special explanation. According to Aristotle "βιος" (way of life) is a teleologically basic capacity that explains the integration of an organism's various parts and activities:

[βιος] accounts for the unity that integrates the many parts of an animal's body and the many different activities those parts perform. That understanding comes from the recognition that an animal's functional parts must make coordinated contributions to a single way of life. (Lennox, 2010, p. 350)

There is a crucial difference between Kant and Aristotle, however, on the implications of this reciprocity for a science of biological form. For Kant the need for teleology puts organisms beyond the ambit of scientific explanation. For him the paradigm of scientific explanation is Newtonian mechanism—to explain a phenomenon is to show it to be a consequence of mechanical law. But the way in which the parts of organisms, and their characteristic activities, arise out of the purposive, self-synthesizing capacities of organisms simply does not conform to the mechanistic mode of explanation.

It is quite certain that we can never adequately come to know the organized beings and their internal possibility in accordance with merely mechanical principles of nature, let alone explain them; and indeed this is so certain that we can boldly say that it would be absurd for humans even to make such an attempt or to hope that there may yet arise a Newton

5. Ginsborg (2004) offers a lovely discussion.

who could make comprehensible even the generation of a blade of grass according to the natural laws that no intention has ordered; rather, we must absolutely deny this insight to human beings. (Kant, 2000, p. 400)

Aristotle, Kant, and Monod are responding to the same problem, the tension between the evident need to explain the features of organisms by appeal to purpose, and the presumed naturalistic stricture against doing so. Yet their respective approaches are radically different. In effect, Kant despairs of any naturalistic solution (Zammito, 2006). Monod's attempted resolution, in contrast, is an orthodox version of modern naturalism. His strategy is simply to deny that the purposiveness of organisms has any ineliminable explanatory role to play. For Aristotle, purposive explanation is in no way antithetical to naturalism. Between them, these authors exemplify three strategies for dealing with the "paradox of invariance$_1$." We can either repudiate naturalism (Kant), repudiate teleology (Monod), or simply deny that there is a paradox of at all (Aristotle).

The Aristotelian strategy requires us to show how purposes could actually figure in scientific explanations. A recent and compelling approach to explanation might help here. In order to explain some event X by appeal to Y, the account goes, there must be an invariance$_2$ relation between X and Y. James Woodward captures the relation between invariance$_2$ and explanation succinctly:

> On my view, the key feature that a generalization must possess if it is to figure in explanations is invariance. Invariance is a kind of robustness or stability property: a generalization is invariant if and only if it would continue to hold under some range of physical changes involving intervention. . . . To explain why an explanandum Y takes some particular value we need to identify some variable X and a generalization G linking X to Y such that, according to G, some range of changes in the value of X that are due to interventions are associated with changes in the value of Y. This requires that the generalization G must be invariant under some interventions on X that change the value of Y. (Woodward, 2001, p. 4)

Invariance$_2$ is just a form of robust counterfactual dependence. An explanation proceeds by demonstrating the way in which the value of the explanandum variable, Y, counterfactually depends on the value of the explanans variable, X, by providing an illuminating description of the relation (Walsh, 2012, 2013).[6]

---

6. Invariance is necessary, but not sufficient, for explanation. In the case of purposive explanations, one must also provide a description of how the means conduces to the end (Walsh, 2012, 2013).

Aristotle's taxonomy of efficient and final cause explanations translates easily into the modern parlance of counterfactual dependence (invariance$_2$) relations. Causes explain their effects because effects counterfactually depend on their (efficient) causes. Similarly, purposive systems manifest a different kind of counterfactual dependence: *Means counterfactually depend on their ends.* If Y is a system's goal and X is a means to the attainment of Y, then intervening on Y (changing the goal) will occasion a change in X (the means). Similarly, where Y is a system's goal and X is a means to Y, then changing the background conditions will occasion a change in X. In either case, the change in X will be predictable, when specified in terms of ends. The change in X will be such as to bring about the end, Y. This purposive invariance$_2$ relation has exactly the same counterfactual form as the mechanistic invariance$_2$ relation.[7]

This modernized version of Aristotelian explanation holds that if invariance$_2$ underwrites mechanistic explanation, it should licence purposive explanations too. Far from succumbing to Kant and Monod's methodological qualms, it urges that as naturalists we should avail ourselves of purposive explanations. It remains to be seen whether this taming of purpose (to coin a phrase) provides any opportunities for modern evolutionary biology.

8.3. Plasticity and Purpose

The purposiveness of organisms is manifested in their adaptive plasticity and robustness. Plasticity is the capacity of an organism to adapt to the vagaries of its conditions of existence by controlling and implementing changes to its own structures and processes. Robustness is the capacity of organisms to maintain constant form and function across a range of circumstances (Kitano, 2004). Plasticity and robustness are two sides of the same coin (de Visser et al., 2003).

> The organism is not robust because it is built in such a manner that it does not buckle under stress. Its robustness stems from a physiology that is adaptive. It stays the same, not because it cannot change but because it compensates for change around it. The secret of the phenotype is dynamic restoration. (Kirschner & Gerhard, 2005, pp. 108–109)

Adaptive phenotypic plasticity is not an optional extra for organisms. *Organism*, as Monod, Kant, and Aristotle all agree, is a purposive category. "Phenotypic

---

7. Except it is time reversed. The earlier event counterfactually depends on the later. Note, this does *not* imply backward causation.

plasticity is a ubiquitous, and probably primal phenomenon of life" (Wagner, 2011, p. 216). These purposive capacities are manifest at practically all levels of organization, from gene networks, to genomes, to cells, to entire organisms (Keller, 2013). And their importance for evolution is increasingly becoming recognized.

8.3.1. Gene Networks

Genes, as we are becoming aware, play their characteristic role in development as parts of complex suites or networks with complex regulatory topologies (Davidson, 2010, Meir, von Dassow, Munro, & Odell, 2002). Ciliberti, Martin, and Wagner (2007) and Wagner (2011, 2012) have revealed some of the startling ways that gene networks contribute to evolution. These studies demonstrate that the robustness of gene networks is crucial both to the origin of evolutionary novelty and to the stability of organismal form.

The gene regulatory networks under study begin with initial input levels of gene products and a particular topology of regulatory relations between elements of the network. These networks settle down into a stable equilibrium output. This is taken to be their "phenotype." "Mutations" are then introduced, as either quantitative or qualitative changes in gene products, or in the regulatory relations among genes in the networks.

Ciliberti et al. (2007) and Wagner (2011) map the outputs of these networks in a vast multidimensional "network space," the axes being the values of an individual gene's function (regulatory activities or products of transcription). Along any given axis, immediate neighbors differ from one another by a single genetic mutation. In general, in these systems there are many more genotypes than phenotypes: "This means any one phenotype typically has many genotypes that form it" (Wagner, 2011, p. 71). The space of viable phenotypes exhibits some remarkable features: it is *clustered* and *connected*. Viable networks are clustered in the sense that the genotypes that produce a particular phenotype all occupy a single small region of the state space. They are connected in the sense that every single network capable of producing the system's characteristic phenotype can be accessed from any other through a series of single mutations without leaving the space of viable networks. Such a connected network of networks is called a "neutral network."

Gene networks are robust. They can compensate for perturbations by producing changes in function that preserve their viable phenotype. Robustness is measured as the number (proportion) of nearest (single-mutation) neighbors in the neutral network that produce the typical phenotype. Gene networks with many viable neighbors are capable of withstanding a considerable degree of perturbation, both mutational and environmental. Wagner (2011) in fact demonstrates the enormous capacity of such systems to maintain their stability under mutational change. For gene regulatory networks of a size typically found in biological systems, two networks producing the same phenotype may share only 20% of their regulatory interactions (Wagner, 2011, p. 721). The constancy of form of organisms, the reliability of development, the transgenerational recurrence of form, are secured in large part by the adaptive robustness of gene regulatory systems.

The robustness of gene networks facilitates adaptive phenotypic evolution in at least three distinct ways. First, one and the same gene network can produce multiple phenotypes. The neutral networks that underwrite different phenotypes overlap. So, a gene network may move from one neutral network (where it produces one phenotype) to another (where it produces a novel phenotype) with no genetic change. Second, typically a gene network will have novel phenotypes as its near neighbors, so phenotypic novelties may be produced as an adaptive response of networks to minor mutations. In these ways, plasticity allows a gene network to innovate.[8] Novel phenotypes may be initiated by mutations or environmental changes. Third, robustness acts as an evolutionary capacitor. A gene network that can maintain its stability across a range perturbations has the capacity to access a large range of novel phenotypes (Wagner, 2012).

One startling feature of the adaptiveness of gene networks is that the production of novelties is biased. To be adaptively plastic *just is* to produce compensatory changes to perturbations precisely because those changes are conducive to the viability of the system. That goes as much for compensations that maintain phenotypic stability as for compensatory changes that produce new phenotypes. The new phenotypes that these systems generate tend to be stable and viable.

8.3.2. Reactive Genomes

Research in developmental genetics over the last 20 years or so has issued in a gradual shift in emphasis from discrete genes as bearers of phenotypic information to genomes, as complex, reactive systems:

> The transition that concerns us has involved genomes rather than genes being treated as real, and systems of interacting macromolecules rather

---

8. Andreas Wagner (2014) calls this capacity of biological systems to innovate "innovability." It is manifest at practically all levels of biological organization—gene networks, genomes, cells, tissues.

than sets of discrete particles becoming the assumed underlying objects of research. (Barnes & Dupré, 2008, p. 8)

Reactive genomes are not repositories of phenotypic information—or "programs"—as such. Instead they are open systems that exploit the resources of their genetic, cellular, extracellular, and environmental circumstances in the production of stable, viable forms. Just as genes influence biological form, biological form affects the functioning of genes. "The passage of information is not simply one-way, from genes to function. There is two-way interaction" (Noble, 2006, p. 50).

Genomes are complex, adaptive systems. They respond to genetic, organismic, and environmental cues. They buffer the organism against perturbations.

At the very least, new perceptions of the genome require us to rework our understanding of the relation between genes, genomes and genetics. . . . It has turned our understanding of the basic role of the genome on its head, transforming it from an executive suite of directorial instructions to an exquisitely sensitive and reactive system that enables cells to regulate gene expression in response to their immediate environment. (Keller, 2013, p. 3)

We cannot understand the way that genes, cells, whole organisms, or environments contribute to the development of an organism unless we understand how genomes react to these influences. The reactive, adaptive dynamics of genomes is the hallmark of a goal-directed, purposive system. Organismal purposes, then are reflected in the dynamics of genomes.

8.3.3. Developmental Plasticity

There are architectural requirements for complex systems to be evolvable, which essentially requires the system to be robust against environmental and genetic perturbations (Kitano, 2004, p. 829). As discussed previously, phenotypic plasticity is "the ability of an organism to react to an environmental input with a change in form, state, movement or activity" (West Eberhard, 2003, p. 37). It might be added that it also consists in the capacity to respond to an *internal* input in the same way. An organism has the capacity to produce a wide range of phenotypes, according to the circumstances.

Through its ancient repertoire of core processes, the current phenotype of the animal determines the kind, amount and viability of phenotypic variation the animal can produce . . . the range of possible anatomical and physiological relations is enormous. (Gerhard & Kirschner, 2007, p. 8588)

As in the case of gene networks and genomes, plasticity allows organisms to maintain stability in the face of genetic and environmental perturbations. But it also permits them to innovate; often enough they produce novel, stable phenotypes when perturbed.

> Adaptive phenotypic adjustments to potentially disruptive effects of the novel input exaggerate and accommodate the phenotypic change. *Without genetic change.* (West-Eberhard, 2005, p. 613, emphasis in original)

The production of a novel phenotype has potentially catastrophically disruptive effects on the organism as a whole. But here again, the plasticity of development buffers the organism against these. This adjustment to phenotypic change is known as "phenotypic accommodation," and it is crucial to development: "Phenotypic accommodation reduces the amount of functional disruption occasioned by developmental novelty" (West-Eberhard, 2003, p. 147).

Accommodation, it turns out, is a prerequisite for adaptive evolution (Schlichting & Moczek 2010). The evolution of complex adaptations requires a significant degree of orchestration between an organism's various systems. For example, the increase in the strength of a striated muscle carries with it demands not just on the muscular system but the associated skeletal, nervous, circulatory, and integumentary systems (Sterelny, 2009). Each system responds adaptively to its changed circumstances, in a way that accommodates the new structure and its altered function. If each system had to wait for a fortuitous mutation in order to produce the appropriate accommodation, complex evolutionary adaptations might never arise.

> In contrast to the rapid response produced by plasticity, if the production of newly favored phenotypes requires new mutations, the waiting time for such mutations can be prohibitively long and the probability of subsequent loss through drift can be high. (Pfennig et al., 2007, pp. 459–460)

Recent developmental biology, then, appears to suggest that the adaptive plasticity of organisms positively contributes to the origin and evolution of organismal form.

> These considerations all point to the importance of adaptive plasticity for the adaptive evolution of organismal form. Plasticity consists in the capacity to respond adaptively to genetic, epigenetic, and environmental

influences, in the production of viable, robust living things. This plasticity is evident at every level of biological organization (Keller, 2013).

Plasticity is merely a manifestation of organismal purposiveness. The purposiveness that makes organisms organisms appears to be a necessary precondition of adaptive evolution.

8.4. Purpose and Genetic Stability

Monod's solution to the "paradox of life," as we saw, was to ground organismal purposiveness in the unchanging (invariant$_1$) structure of the "gene code." "All properties of living beings rest on a fundamental mechanism of molecular invariance" (p. 116). This strategy presupposes an asymmetry; whereas the constancy of genes can explain the purposiveness of organisms, the purposiveness of organisms could not possibly explain the constancy of genes. The explanatory asymmetry is predicated on the conviction that there is no way that the structure of DNA could be changed in response to the needs of an organism.

> There exists no conceivable mechanism whereby any instruction or piece of information could be transferred to DNA. . . . Hence the entire system is totally, intensely conservative, locked into itself, utterly impervious to any "hints" from the outside world. (Monod, 1971, p. 110)

Indeed, the idea that the structure of the genome is stable and impervious to the influences of the organism is a bedrock commitment of the modern synthesis.

Yet, there is mounting empirical evidence that the structural integrity of DNA is not an evolutionary primitive. It is safeguarded by the supple adaptiveness of organisms. The DNA damage repair (DDR) system offers vivid evidence of the ability of organisms to secure the invariance$_1$ of their DNA. An organism's DNA is subject to a battery of endogenous and exogenous insults. These cause lesions—structural alterations—at an estimated rate in humans on the order of 150,000/cell/day (Ciccia & Elledge, 2010). These are by no means neutral or benign.

> These lesions can block genome replication and transcription, and if they are not repaired or are repaired incorrectly, they lead to mutations or wider-scale genome aberrations that threaten cell or organism viability. (Jackson & Bartek, 2009, p. 1071)

Cells (organisms) have evolved elaborate systems for detecting and repairing DNA damage. The systems comprise various mechanisms for determining the nature and severity of the lesion, and marshaling the appropriate response. The response may involve mitigating the consequences to the organism of DNA damage through the disruption of mitosis, or through apoptosis (Branzei & Foiani, 2008), or it may consist in altering the structure of damaged DNA, restoring its former nucleotide sequence.

> The DDR system is a highly responsive, adaptive system. The DNA damage response (DDR) is a signal transduction pathway that senses DNA damage and replication stress and sets in motion a choreographed response to protect the cell and ameliorate the threat to the organism. (Ciccia & Elledge, 2010, p. 180)

The famed stability of the genome, then, is not so much due to its being inherently "intensely conservative," as Monod would have us believe, but to the fact that it is actively monitored and repaired by a highly adaptive goal-directed system that detects errors and implements the appropriate cascade of responses, in ways that safeguard the organism's viability.

There is, then, no particular asymmetry between gene structure and organismal purpose. The structural stability of genes is involved in securing the purposiveness of organisms, but conversely, the purposiveness of organisms is involved in securing the structural stability of genes.

> The stability of gene structure thus appears not as a starting point but as an end-product—as the result of a highly orchestrated dynamic process requiring the participation of a large number of enzymes organized into complex metabolic networks that regulate and ensure both the stability of the DNA molecule and its fidelity in replications. (Keller, 2000, p. 31)

We need to appeal to the purposiveness of organisms to explain the invariant$_1$ structure of DNA. This is an example of the very relation between "purpose" and "invariance" that Monod sought to proscribe. That being so, the envisaged reduction of the purposiveness of organisms to the invariant$_1$ structure of DNA is not viable. Invariance$_1$ is neither conceptually nor causally prior to organismal purposiveness.

8.5. Neo-Aristotelian Evolutionary Biology

Adaptive evolution requires three things: (1) the constancy of form—organismal form must be stable, and reliably generated from generation to generation; (2) a source of novelties; and (3) adaptively biased change. The case for a neo-Aristotelian evolutionary biology can be made by demonstrating that organismal purposiveness is crucially involved in each of these phases.

8.5.1. The Constancy of Form

According to the neo-Democritean modern synthesis theory, form is reliably reproduced and passed on from generation to generation because the processes that produce individual organisms—replication and development—are fundamentally conservative (Lewens, 2009). However, a consideration of the importance of adaptive plasticity in securing the inheritance of phenotypes suggests that the intergenerational stability of phenotypes across a huge range of internal and external perturbations is underwritten by the capacity of organisms to make adaptive adjustments to the vagaries of genome and environment. Moreover, form is reliably inherited, reliably produced, and reliably maintained, not because the processes that occur within organisms are inherently *conservative*, but because they are *adaptive*.

If inheritance of form requires the stability of the genotype, that too is to be explained, to a significant degree, by the adaptive, reactive capacities of organisms. The integrity of the genome is secured by adaptive responses to DNA damage.

8.5.2. Novel Variants

Neo-Democritean biology holds that the only source of evolutionary novelty is genetic mutation. Typically, mutations have little or no effect on phenotype. However, even on those occasions when phenotypic novelties are *initiated* by mutations, the new phenotypes are themselves the consequence of the reactive response of the genome and the organisms' various developmental systems to the perturbations. It appears that phenotypic novelties are principally caused by the adaptive response of organisms to perturbations of any sort, genetic, epigenetic, or environmental.

> Responsive phenotype structure is the primary source of novel phenotypes. And it matters little from a developmental point of view whether the recurrent change we call a phenotypic novelty is induced by a mutation or by a factor in the environment. (West-Eberhard, 2003, p. 503)

Novel, stable phenotypes could not result from perturbations of any sort were it not for the responsiveness of organisms, their genomes, and their gene regulatory networks. The principal cause of novel phenotypes is the plasticity of the organism, whether or not those novelties are *initiated* by mutation or by environmental change.

8.5.3. Adaptively Biased Change

Neo-Democriteanism is committed to the view that the source of phenotypic novelties is adaptively unbiased—mutations are random and "utterly without relation to whatever may be their effects upon teleonomic functioning" (Monod, 1971, p. 118). But, if novelties arise from the reactive response of organisms to perturbations, this claim might require some reconsideration. Phenotypic accommodation enables organisms to mitigate the potentially disruptive consequences of novel phenotypes. Robust gene regulatory networks make the alterations they do to their activities because these alterations preserve the proper functioning (the phenotypic output) of the system. So, while mutations may be random and indifferent to organismal well-functioning, the way in which organisms generate novelties in response to them is not.

It would appear, then, that contra the convictions of Neo-Democritean modern synthesis biology, the purposiveness of organisms—as manifested in their robust adaptive plasticity—is not available to figure in genuine scientific explanations of biological phenomena, it is crucially implicated in each of the factors required for adaptive evolution: the constancy of form, the production of novel heritable variants, and the adaptive bias in form. A neo-Aristotelian evolutionary biology would take seriously the significant role that organismal purposes play in securing the conditions necessary for adaptive evolution.

8.6. Chance on the Wing

The purposiveness of organisms is an empirically observable phenomenon, as Monod himself concedes. Yet, the "postulate of objectivity" prohibits explanations that advert to those purposes. In asserting the postulate of objectivity Monod is reaffirming one of the most fundamental methodological commitments of modern synthesis evolutionary biology. The methodological commitment, in turn, generates what I described earlier as the great empirical wager of the modern synthesis, namely, that organismal purposes make no substantive contribution to the process of evolution. It is becoming increasingly evident that the wager will not pay out, for two reasons. First, the stricture against purposive explanation is not well motivated on methodological grounds. Purposive explanation is available to the natural sciences. Secondly, it is unsustainable on empirical grounds. Organismal purposes make an observable difference to evolution. Advances in biology increasingly suggest that purpose is indispensable to a comprehensive understanding of evolution. The reason is simple, but wholly contrary to modern synthesis orthodoxy: Evolution is adaptive because organisms are purposive.

One of the consequences of the proscription against purpose is that it forces on biology a commitment to the ineluctable role of chance in evolution.

> It necessarily follows that chance alone is at the source of every innovation, of all creation in the biosphere. Pure chance, absolutely free but blind, at the very root of the stupendous edifice of evolution. (Monod, 1971, p. 112)

From the "neo-Aristotelian" perspective, the modern synthesis commitment to chance looks ill-conceived. The "source of every innovation" is not random mutation, but the reactive, adaptive response of an organism's myriad systems to influences from genes, cells, tissues, and environments. The influences may be chancy—mutations really are indifferent to an organism's viability—but the responses are not. They are biased by the capacity of organisms to buffer themselves against perturbations, to adapt, to compensate, to orchestrate, to accommodate, to innovate. These are manifestations of organismal purposes. As a consequence, adaptive evolution is not ineluctably chancy. It is inherent in the purposiveness of organisms. If our scientific methodology fails to countenance purpose, then it renders us incapable of explaining these perfectly real, empirical regularities. The upshot is that what ought to be explicable is dismissed as the whim of capricious chance: "pure chance, absolutely free but blind." This is a mistake. Evolution merely *appears* ineluctably chancy if we are blind to the role of purpose. From the perspective of a neo-Aristotelian evolutionary biology, the modern synthesis conviction that evolution is "chance caught on the wing" does not deserve its status as a bedrock metaphysical commitment. It is an unfortunate methodological artifact.

8.7. Conclusion

Since its inception in the early 20th century, the modern synthesis theory of evolution has been guided by a methodology that explicitly prohibits explanations of phenomena in the natural world that appeal to the fulfillment of goals or purposes. This is hardly surprising. The natural sciences comprehensively eschew purposes. Nevertheless, it sits uneasily with the obvious fact that *organism* is a purposive category. Increasingly, it is becoming apparent that the purposiveness of organisms, as manifest in the robust, reactive, adaptive plasticity of their various systems, from gene networks to entire organisms, is pivotal to the process of evolution. Yet, we still have an evolutionary biology of genes and chance, rather than of organisms and purpose.

The principal impediment to an understanding of the role of organisms in evolution, in my view, is not empirical but methodological, or perhaps more importantly historical. The stricture against natural purpose, Monod's "postulate of objectivity," is one of the great legacies of the scientific revolution. The scientific revolution, of course, was not principally motivated by the needs of biology, much less by an understanding of *evolution*. There is no guarantee that its methods are adequate to the task. Monod's methodological orthodoxy is just one example of a puzzling syndrome evident throughout the growth of the modern synthesis. When 16th-century methodology rubs up against 20th- and 21st-century biology, philosophers and biologists alike have overwhelmingly chosen to side with the former. I think we should feel no compunction about throwing our lot in with the latter.

Acknowledgments

I thank audiences at the University of Missouri-Columbia, University of Calgary, IHPST Paris, and KLI Vienna. The first version of this paper was written in Cannizaro Gardens in Wimbledon and Jardin du Luxembourg, Paris. I would like to thank the groundskeepers in each of these public spaces for creating such propitious surroundings in which to work.

References

Aristotle. (1996). *Physics* (R. Waterfield, Trans.). Oxford, UK: Oxford University Press.
Barnes, B., & Dupré, J. (2008). *Genomes and what to make of them.* Chicago, IL: University of Chicago Press.
Branzei, D., & Foiani, M. (2008). Regulation of DNA repair throughout the cell cycle. *Nature Reviews (Molecular Cell Biology), 9*, 297–308.
Ciccia, A., & Elledge, S. J. (2010). The DNA repair response: Making it safe to play with knives. *Molecular Cell, 40*, 179–204.
Ciliberti, S., Martin, O. C., & Wagner, A. (2007). Robustness can evolve gradually in complex regulatory gene networks with varying topology. *PLOS Computational Biology, 3*, e15. doi:10.1371/journal.pcbi.0030015
Davidson, E. (2010, December 16). Emerging properties of animal gene regulatory networks *Nature, 468*, 911–920. doi:10.1038/nature09645
Dawkins, R. (1989). *The selfish gene.* Oxford, UK: Oxford University Press.

de Visser, J., Hermisson, J., Wagner, G. P., Meyers, L. A., Bagheri-Chaichian, H., Blanchard, J. L., & Chao, L. (2003). Perspective: Evolution and detection of genetic robustness. *Evolution, 57*, 1959–1972.

Gerhard, J. M., & Kirschner, M. (2007). The theory of facilitated variation. *Proceedings of the National Academy of Sciences, 104*, 8582–8589.

Ginsborg, H. (2004). Two kinds of mechanical inexplicability in Kant and Aristotle. *Journal of the History of Philosophy, 42*, 33–65.

Guyer, P. (2005). Organisms and the unity of science. In P. Guyer (Ed.), *Kant's system of nature and freedom: Selected essays* (pp. 86–111). Oxford, UK: Oxford University Press.

Hankinson, J. (1998). *Cause and explanation in ancient Greek thought.* Oxford, UK: Oxford University Press.

Jackson, S. P., & Bartek, J. (2009). The DNA-damage response in human biology and disease. *Nature, 461*, 1071–1078.

Kant, I. (2000/1790). *Critique of the power of judgment* (P. Guyer, & E. Matthew, Trans.). Cambridge, UK: Cambridge University Press.

Keller, E. F. (2000). *The century of the gene.* Cambridge, MA: Harvard University Press.

Keller, E.F. (2013). The post-genomic genome. In S. Richardson & H. Stevens, *The Post-Genomic Age.* Raleigh, NC: Duke University Press.

Kirschner, M., & Gerhart, J. (2005). *The plausibility of life: Resolving Darwin's dilemma.* New Haven, CT: Yale University Press.

Kitano, H. (2004) Biological robustness. *Nature Reviews Genetics, 5*, 826–837.

Lennox, J. (2009) Form, essence and explanation in Aristotle. In G. Anagnostopoulos (Ed.), *A companion to Aristotle* (pp. 348–367). Chichester, UK: Wiley.

Lennox, J. (2010). *Bios* and explanatory unity in Aristotle's biology. In D. Charles (Ed.), *Definition in Greek philosophy* (pp. 329–355). Oxford, UK: Oxford University Press.

Lewens. (2009). What's wrong with typological thinking. *Philosophy of Science, 76*, 355–371.

McLaughlin, P. (1990). *Kant's critique of teleology in biological explanation.* Lewiston, NY: Mellon.

Meir, E. G., von Dassow, G., Munro, E., & Odell, G. (2002). Robustness, flexibility, and the role of lateral inhibition in the neurogenic network. *Current Biology, 12*, 778–786.

Monod, J. (1971). *Chance and necessity* (A. Wainhouse, Trans.) New York, NY: Schopf.

Noble, D. (2006). *The music of life.* Oxford, UK: Oxford University Press.

Pfennig, D. W., Wund, M. A., Snell-Rood, E. C., Cruickshank, T., Ciliberti, S., Martin, O. C., & Wagner, A. (2007). Innovation and robustness in complex regulatory gene networks. *Proceedings of the National Academy of Sciences, 104*, 13591–13596.

Roux, S. (2005). Empedocles to Darwin. In E. Close, M. Tsianikas, & G. Frazis (Eds.), *Greek research in Australia: Proceedings of the Biennial International Conference of Greek Studies, Flinders University April 2003* (pp. 1–16). Adelaide: Flinders University Department of Languages, Modern Greek. http://dspace.flinders.edu. au/jspui/bitstream/2328/8143/1/1_16%20Roux.pdf

Schlichting, C. D., & Moczek, A. P. (2010). Phenotypic plasticity's impacts on diversification and speciation. *Trends in Ecology and Evolution, 25*, 459–467.

Sterelny, K. (2009). Novelty, plasticity and niche construction: The influence of phenotypic variation on evolution. In *Mapping the Future of Biology. Boston Studies in the Philosophy of Science, 266*, 93–110.

Von Bertalanffy, L. (1969). *General systems theory.* New York, NY: Braziller.

Wagner, A. (2011). *The origins of evolutionary innovations.* Oxford, UK: Oxford University Press.

Wagner, A. (2012). The role of robustness in phenotypic adaptation and innovation. *Proceedings of the Royal Society B: Biological Sciences, 279*, 1249–1258.

Wagner, A. (2014). *The arrival of the fittest: Solving evolution's greatest puzzle.* New York, NY: Current Books.

Walsh, D. M. (2012). Mechanism and purpose: A case for natural teleology. *Studies in the History and Philosophy of Biology and the Biomedical Sciences, 43*, 173–181.

Walsh, D. M. (2013). Mechanism, emergence and miscibility: The autonomy of evo-devo. *Synthese.*

West-Eberhard, M. J. (2003). *Developmental plasticity and evolution.* Oxford, UK: Oxford University Press.

West-Eberhard, M. J. (2005). Phenotypic accommodation: Adaptive innovation due to developmental plasticity. *Journal of Experimental Zoology, 304B*, 610–618.

Wicken, J. (1981) Chance, necessity and purpose: Toward a Philosophy of Evolution. *Zygon, 16*, 303–322.

Woodward, J. (2001). Law and explanation in biology: Invariance is the kind of stability that matters. *Philosophy of Science, 68*, 1–20.

Zammito, J. (2006). Teleology then and now: The question of Kant's relevance for contemporary controversies. *Studies in the History and Philosophy of Biology and the Biomedical Sciences.*

INHERITANCE

9 LIMITED EXTENDED INHERITANCE

Francesca Merlin

It has been many years since biologists started investigating the transmission of extranuclear organelles present in the cytoplasm (such as mitochondria and chloroplasts; cf. Sapp, 1987), observing the mitotic stability of methylation marks (Holliday & Pugh, 1975; Riggs, 1975; Singer, Roberts-Ems, & Riggs, 1979), and studying the transmission of behaviors and culture in animals (cf. Heyes & Galef, 1996). Nevertheless, since the early 20th century the concept of inheritance has been restricted to the transmission of genes from parents to offspring,[1] and biological evolution has been defined as a change in gene frequencies over time. It is only during the last 20 years that the idea of organisms inheriting more than just DNA has been seriously considered. Biologists have started using the term "nongenetic inheritance" to refer to a variety of factors and mechanisms (epigenetic, parental, behavioral, ecological, and cultural) that contribute to phenotypic similarity across generations (Bonduriansky & Day, 2009, Danchin et al., 2011; Jablonka & Lamb, 2005). This has opened the door for the extension of the traditional conception of inheritance as well as for an extended evolutionary theory integrating this plurality of forms of inheritance (Pigliucci & Müller, 2010).

This paper aims to show that one of the main assumptions in recent literature is simply taken for granted without proving any evidence or argument to support it. In fact, the authors—mainly biologists and philosophers—presuppose that the new phenomena they are dealing with are new forms of "inheritance", that is, cases of "hereditary" transmission rather than other forms of transmission. Yet they advance no argument in favor of this thesis, but simply stress

---

1. An organism can have one or two parents depending on the way it has been produced, that is, by asexual or sexual reproduction.

that all these phenomena, which they generally call "nongenetic inheritance," have a significant impact on the evolution of natural populations, and as such should be integrated into an extended evolutionary synthesis. In this chapter, I argue against this widespread assumption by showing that the fact that some form of transmission is relevant in evolution does not imply that it is a form of inheritance.

I first analyze several proposals to extend the concept of inheritance, which I divide into two distinct groups (section 9.1). I then consider four distinctions, mainly empirical, that characterize the literature on nongenetic inheritance (section 9.2) and show what we can learn from them. This leads me to question the ways that biologists and philosophers have recently reconceived inheritance, all of which are implicitly based on the assumption that every form of transmission is a form of inheritance. I finally argue for a privileged link between inheritance and reproduction and suggest a tentative redefinition of inheritance in light of recent research (section 9.3). My proposal, I argue, goes well beyond the traditional, gene-centric, account of inheritance and evolution, for it takes into account the fact that organisms inherit (from their parents) more than DNA. At the same time it builds on mistakes made in recent proposals for extending inheritance, and so ends in a narrower concept with respect to these (section 9.4). This is why I characterize it as "limited extended inheritance."

9.1. Recent Proposals for a Concept of Extended Inheritance

One of the fundamental hypotheses of the traditional theory of evolution (i.e., the modern synthesis) comes from classical genetics: It states that only genes are passed on from one generation to the next. Today, this idea is no longer tenable. Several nongenetic factors and mechanisms involved in development have been found to be transmitted between organisms and, as such, are likely to have an impact on evolutionary dynamics (Jablonka & Raz, 2009). Biologists have observed that phenotypic changes resulting from transmissible modifications in gene expression, such as changes in histones and chromatin conformation, in methylation patterns, and in small interfering RNAs, can be passed on between generations of organisms (Allis, Jenuwein, Reiberg, & Caparros, 2007). They have also studied how parental behaviors at the very beginning of the offspring's life can affect gene expression in the offspring, and their behavior, as a consequence (Mousseau & Fox, 1998). The way the environment is modified due to activities of niche construction is also considered to be hereditary, as are changes in the selective pressures that result from it (Odling-Smee, Laland, & Feldman,

2003). Last but not least, phenotypic variation can also be transmitted via social learning, which is considered as necessary for culture (Heyes & Galef, 1996).

To assess the impact of these research advances on the concept of inheritance, we need to focus on how different authors, from the 1990s onward, have proposed we reconceive it. I suggest distinguishing two groups of proposals: (1) Redefinitions of the concept of inheritance, and (2) Classifications of forms of inheritance into distinct channels and systems.

9.1.1. Redefinitions of the Concept of Inheritance

This section considers proposals by Griffiths and Gray and by Sterelny and colleagues.

9.1.1.1. Griffiths and Gray

The most radical of the last 20 years' suggestions is Griffiths and Gray's proposal (1994, 2001). As partisans of the developmental systems theory,[2] they challenge the conventional gene-centric version of evolutionary theory and propose a new definition of inheritance, among other concepts. They suggest a "principled" definition, justified by the metatheoretical claim that "the concept of inheritance is used to explain the stability of biological form from one generation to the next" (2001, p. 196). Thus, according to Griffiths and Gray, inheritance is "the reliable reproduction of developmental resources down the lineage" (2001, p. 214); it refers to "any resource that is reliably present in successive generations, and is part of the explanation of why each generation resembles the last" (2001, p. 196).[3]

Griffiths and Gray's definition entails a radical expansion of the concept of inheritance, which turns out to apply to a large number of genetic and nongenetic factors involved in the ontogeny of an individual organism. Not only are resources from parents (such as genes, endosymbionts, chromatin marks, organelles in the cytoplasm, membrane structures, antibodies, and behaviors) inherited, so are resources collectively generated by members of a population (e.g., behavioral and social traditions, some features of the habitat like nests, dams), resources self-generated by the developing organism itself (e.g., some proteins produced by gene expression), and even environmental-physical resources persisting independently of an organism's activity.

---

2. See Oyama, Griffiths, and Gray (2001).

3. For an elaborated analysis of Griffiths and Gray's conception of inheritance, see Griesemer, Haber, Yamashita, and Gannet, (2005); Merlin (2010).

Another remarkable feature of Griffiths and Gray's concept of inheritance is its holistic character. In fact, they refuse to distinguish multiple systems or channels of inheritance in interaction. Rather, according to them, it is more biologically realistic and productive to conceive inheritance as diffused within the entire developmental matrix.

9.1.1.2. Sterelny and Colleagues

A second proposal, the "extended replicator" view (Sterelny, Smith, & Dickinson, 1996; Sterelny, 2001) suggests a less radical extension of inheritance than the Griffiths and Gray proposal. This view responds to the developmental systems conception of inheritance and evolution. This consists in limiting the extension of inheritance to the transmission of replicators and, at the same time, in extending the notion of replicator[4] (which is at the core of the traditional "gene's eye" view) to hereditary entities other than genes.

Sterelny argues that there are some privileged entities, both from a developmental and evolutionary point of view, known as replicators: "The set of developmental resources that are adapted for the transmission of similarity across the generations" (2001, p. 338). Thus, DNA sequences are replicators, but so are other developmental factors that represent the developmental outcome as "plans represent buildings because that is their function" (p. 387): Centrioles and other nongenetic cellular structures, gut microorganisms and symbionts, nest sites and structures, songs, food preferences, and so forth. These are all involved in copying mechanisms designed by natural selection to ensure intergenerational similarity.

What distinguishes Sterelny's from Griffiths and Gray's view is that it does not put all developmental resources on a par: Some, namely replicators, have causal privilege in development and evolution because they are adapted to play the role they do, that is, their function is to transmit similarity across generations. Moreover, in the "extended replicator" view, nothing prevents us from distinguishing different sorts of replicators and the variety of copying mechanisms they are involved in.

9.1.2. Classifications of Forms of Inheritance into Distinct Channels and Systems

Let us first note that there is a fundamental distinction between channels and systems of inheritance (Lamm, 2014). Channels are paths across generations

---

4. The concept of replicator was introduced by Dawkins in 1976. Some years later Hull suggested a different definition of it (1980). A replicator can be loosely defined as any entity that, making copies of itself, passes on its structure largely intact in successive copying events (i.e., replications), forming lineages. Among other things, Dawkins and Hull differed over which entities are replicators (only genes, according to Dawkins; other entities too, for Hull).

through which hereditary factors pass from parents to offspring. Systems, by contrast, are much more like categories, bringing together hereditary factors and mechanisms with respect to (some of) their common properties. Let us look at a few examples.

1. Odling-Smee (2010) suggests distinguishing two transmission channels. Channel 1 is the internal environment, the route through which genetic and epigenetic factors, as well as some maternal effects, pass from one generation to the next. Channel 1 establishes a direct connection between parents and offspring, transmitting via reproduction. Channel 2 is the external environment, which establishes an indirect connection across generations. Factors like changes in selective pressures due to the activity of niche construction, as well as behaviors and culture, reach the next generation via channel 2.

2. Jablonka and Lamb (2005) and Helanterä and Uller (2010) give two proposals for classifying inheritance in distinct systems or categories. Jablonka and Lamb distinguish four systems: genetic, epigenetic, behavioral, and symbolic inheritance. They classify these according to the way they store and transmit variations, as well as in terms of the hereditary relations they respectively give rise to. Helanterä and Uller, by contrast, classify hereditary factors according to their respective effects on the rate and direction of short-term evolutionary change, and with respect to components of the Price equation.[5] They distinguish the three following categories. Vertical transmission applies to cases where traits are passed on from parents to offspring. Induction happens when the environment is causally involved in a change between parents and offspring. Acquisition, including horizontal and oblique transmission, occurs in situations where traits come from nonparental individuals or other sources.

I could go on presenting proposals for reconceiving inheritance.[6] But every redefinition and classification of the last 20 years has the same distinctive feature which, I argue, remains problematic: The set of hereditary factors is always too large, and so is the extension of the concept of inheritance. The main reason for this lies in the idea, common to all these proposals, that every form of transmission is a form of inheritance. This in turn is partly due to the fact that the distinction between the acquisition of variation and its transmission across generations is seldom considered. Helanterä and Uller represent an exception to

---

5. The Price equation is a mathematical formulation of evolutionary change that makes no assumptions about the nature of the evolving entities and the mechanisms of inheritance (Price, 1970).

6. See, e.g., Danchin et al. (2011).

that; nevertheless, they do not take apart mechanisms through which organisms acquire variation (i.e., the first introduction of some variant in a lineage) and mechanisms of inheritance (i.e., the transmission of the acquired variant from its acquisition onwards, across generations). In other words, for them, these are both mechanisms of inheritance.

In the next section, four relevant distinctions made in the literature about nongenetic inheritance allow me to highlight why the lack of a distinction between inheritance and other forms of transmission is problematic. I argue that treating every form of transmission between organisms as a case of inheritance is not as straightforward as it would seem at first. Actually, this is like putting together disparate things such as oranges, pears, and apples, which are quite different as to their very nature. In particular, this position fails to capture the specific role of inheritance in evolutionary theory.

9.2. Lessons from Four Distinctions in the Literature

The traditional view about biological inheritance is that it is a form of transmission involving generational events. More precisely, it consists in the transfer of some developmentally privileged material—namely DNA, since the mid-1950s— from parents to offspring that occurs at conception. As Mameli (2005) emphasized, what he calls the "conception/donation theory of inheritance of features" is an old view dating back to Hippocrates's and Aristotle's treatises about biological matters and has been used throughout centuries, until now, in order to account for the like-begets-like phenomenon, that is, the fact that biological organisms, through reproduction, *tend*[7] to generate organisms with features similar to their own. Furthermore, it is acknowledged in biology that inheritance is one of the required conditions for evolution by natural selection (Lewontin, 1970).[8] As a guarantee of some degree of continuity across generations, inheritance fulfills its theoretical role in evolutionary biology. More precisely, inheritance accounts for both the maintenance of phenotypic features between genealogically related organisms and the maintenance of their variation, within a population, from one generation to the next. Its role in the Darwinian theory is thus to allow a process

---

7. I stress here that resemblance between parents and offspring is rather a tendency because, as I say later, it is far from being a necessary effect of inheritance mechanisms.

8. The fundamental role of inheritance as a condition of possibility of evolution by natural selection is explicitly stated by Darwin himself in the well-known passage of his letter to Hooker (May 11, 1859): "Inheritance is of fundamental importance to us, for if a variation be not inherited, it is of no signification to us." Why so? Because "From the strong principle of inheritance, any selected variety will tend to propagate its new and modified form" (1859, p. 5), possibly giving rise, in the long term, to complex adaptive traits.

of cumulative selection; more explicitly, it grounds the possibility that (favorable) variation, which is selected at some generation, is passed on to next and can be the target of selection again.

It is worth noting that the concept of inheritance can be used to characterize intergenerational transmission at different levels, at least from cells to multicellular organisms. Genetic mutations are good examples of intergenerational transmission of variation, both at the unicellular and multicellular level. It is more problematic to argue for the transmission across generations of factors other than DNA. For instance, cellular epigenetic changes (such as modifications of DNA methylation patterns) are systematically transmitted from generation to generation in multicellular organisms such as plants. On the contrary, in mammals, they are passed on from one generation of organisms to the next only if they resist the reprogramming of the epigenome, which takes place during meiosis, gametogenesis, and early embryogenesis. Otherwise, the transmission of cellular epigenetic changes is limited to generations of (somatic) cells, which is a form of (cellular) inheritance (Feng, Jacobsen, & Reik, 2010). The identification of the level at which some (genetic or nongenetic) variant is intergenerationally transmitted is crucial in order to advance hypotheses about the impact of mechanisms of inheritance on evolutionary change at some specific level of biological organization.

Now, let us consider four distinctions that characterize the biological and philosophical literature on inheritance.

9.2.1. Intergenerational Transmission, Intergenerational Causal Influences, and Direct Expositions to the Environment

In the literature on nongenetic inheritance, two phenomena are erroneously taken to be cases of intergenerational transmission (i.e., inheritance). First, there are situations where parental development or the interaction of the parents with environmental conditions provokes changes in the characteristics of offspring. For instance, in environments with abundant food supplies, the transmission of extra resources by the mother in her milk can enhance offspring's conditions (Qvarnström & Price, 2001). Another example is provided by reduced sensitivity to drug addiction in offspring, which is apparently due to epigenetic changes in the sperm of fathers who had self-administrated drugs (e.g., cocaine; Vassoler, White, Schmidt, Sadri-Vakili, & Pierce, 2013). These are all cases of intergenerational causal influences, that is, where an older generation (the parents) causally intervenes in its offspring's life and provokes some change that affects it. These are not cases of inheritance, because there's no transmission of some variant from one generation to the next, and so no continuity between parents and offspring in this respect.

There are also cases where parents and offspring (and even grand-offspring)
are directly exposed to the same environment, which provokes similar changes in
their respective features. For instance, male reduced fertility over several generations can be due to the fact that a gestating female (F0 generation) is exposed to
toxicants, along with her male fetus (F1 generation) and the fetal germline at the
origin of the grand-offspring (F2) generation (Skinner, 2011). In this situation,
two or more generations are directly exposed to the same environmental stimuli
at the same time. Thus they change in a similar way.

This first distinction points to a confusion in the literature, which consists in
taking causal influences across generations and direct expositions to the environment as cases of intergenerational transmission (i.e., inheritance). I have shown
why this may not be not the case.

9.2.2. Three Directions of Transmission: Vertical, Horizontal, and Oblique

Vertical transmission is the transfer of resources from parents to offspring. It
includes the transmission of genetic and epigenetic factors as well as cases of
ecological and cultural transmission. Horizontal transmission occurs between
organisms belonging to the same generation. This applies to the transfer of ecological and cultural factors, and sometimes genetic and epigenetic factors too.
Oblique transmission involves organisms belonging to different generations,
which are not linked by direct descent. As for horizontal transmission, it particularly concerns the transfer of ecological and cultural resources.

In the literature, all three directions of transmission are considered "inheritance." This might seem obvious, because it is always a matter of transmitting
something from one organism to another. However, some cases of horizontal and
oblique transmission appear to contradict our commonsense idea of what inheritance is, which is tightly linked to the traditional view of biological inheritance
and to the theoretical role of this concept in evolutionary biology.

Let us take a look at some of these counterintuitive cases of transmission. It
can happen that a bacteria or a virus passes from one organism to another that is
not its offspring. Is this a case of hereditary transmission or simply an example of
infection or contamination? The same question should be asked regarding the
transfer of viruses between organisms belonging to different species (e.g., a virus
could be transmitted by rats [or other animals] to humans, namely through bite
exposure). Organisms belonging to different species can also transfer behaviors to
each other, as in the case of dogs (or other domestic animals) trained by humans.
This can also happen between organisms of different generations, belonging to
the same or different species: from the older to the younger, and vice versa. Are

these cases of hereditary transmission, or just situations where a behavior or a cultural factor is transmitted by a teaching/learning relationship?

Another example of transmission, often cited as a case of inheritance, is lateral gene transfer, particularly in bacteria (Helanterä & Uller, 2010; Ochman, Lawrence, & Groisman, 2000). This consists in the transmission of DNA between individual cells through transformation (the uptake of DNA from the environment and its incorporation through recombination), transduction (the transmission of DNA from one cell to another mediated by bacteriophages), conjugation (the transmission of DNA, in particular plasmids, through cell-to cell contact through a specialized appendage), and many other sorts of mechanisms. Even though these are all situations where DNA is transmitted, as is traditionally accepted in biology, none necessarily involve generational events. Lateral gene transfer rather consists of a variety of mechanisms through which bacteria acquire genetic variants from others or from environmental sources and integrate them into their own genome. In other words, it is a source of new genetic combinations. Mechanisms dedicated to the transmission of the acquired variants, conserving them from that point (the moment of their acquisition) onward (across generations) are distinct: These are mechanisms of inheritance.

If all these examples were cases of hereditary transmission, this would mean that the transmission of something, independently of the direction of the transfer, is a sufficient condition for inheritance. Yet an argument in favor of this position is not immediately obvious, if there ever was one. Moreover, why should we not just think of them as forms of transmission, rather than inheritance (for instances, as cases of contamination or infection, of learning, and of DNA exchange or acquisition)? I suggest that this would not reduce the importance of these phenomena in the study of development and evolution.

Another way to highlight the controversial nature of these cases, in particular lateral gene transfer, as mechanisms of inheritance is to point, again, at the theoretical role of inheritance in evolutionary biology, namely to be a guarantee of some degree of continuity, and so allow episodes of cumulative selection resulting in adaptation. Calling cases of horizontal and oblique transmission "inheritance" implies to assign such a role to a variety of mechanisms that do not guarantee any continuity but rather introduce new combinations of variants in lineages.

9.2.3. Transmission with and Without Material Overlap

Transmission can be material or formal. If there is material overlap, this means that some parts of an organism (the offspring) were once parts of at least another organism (the parent). For instance, DNA, some cellular epigenetic factors (such as proteins reproducing self-sustaining loops, methyl-groups, RNAs) as well as

272 • FRANCESCA MERLIN

ecological factors (e.g., nests, modified environmental conditions) are material entities passed on between organisms. Transmission without material overlap consists in transferring forms or structures. One well-known example is the transfer of molecular and cellular structures through spatial templating, as in the case of prions, cellular membranes, and retroviruses. Behaviors and cultural factors can also be seen as cases of formal or structural transmission, because they are passed on by learning and do not necessarily involve any material overlap (except when some physical support is needed for transmission to occur).

This third distinction suggests that material overlap is not a necessary feature of inheritance. Rather, it may be a guarantee of a more faithful transmission with respect to transmission without material overlap. This is analogous to the transmission of a story, which turns out to be more precise and faithful when the story is conveyed on a piece of paper rather than orally. But material overlap turns out to be an indispensable feature of inheritance via reproduction. Indeed, in line with Virchow's epigram about cellular theory "*omni cellular ex cellula*," the production of a new entity involves some material entity being passed on from parents to offspring. So far, my claim *is not* that material overlap is a necessary condition of inheritance. Rather, this is the case for reproduction (see Griesemer, 2000). And if reproduction has a privileged link with inheritance (in other words, if inheritance only comes with reproduction, as I argue in the next section), inheritance always involves at least some cases of transmission with material overlap.

9.2.4. Inheritance as a Cause and as an Effect

Inheritance can generally be conceived as a set of processes causing the conservation of "something" across generations. It can also be considered as an effect, that is, patterns of phenotypic resemblance between genealogically related organisms and, at the population level, (statistical) correlation in phenotypic variation across generations. This distinction points to a difference between the explanatory and the descriptive (or phenomenological) meaning of the concept of inheritance. Biologists refer to inheritance as an effect when they evaluate the heritability of traits: They statistically measure the phenotypic correlation between relatives due to transmitted variation, and thus can predict the response of different traits to selection; but this does not tell them anything about the causal processes of inheritance.[9] Moreover, resemblance between parents and offspring is not a necessary effect of inheritance processes, as in cases where a specific phenotypic trait skips one generation or more, and so resemblance shows up in grand-offspring

---

9. For a philosophical discussion of the technical concept of heritability, see Keller (2010).

or even later (Godfrey-Smith, 2009, provides several examples); it is just something that is likely to occur when one organism descends from another. This is why I argue that any redefinition of the concept of inheritance should focus on inheritance as a cause.

Many other distinctions could shed light on the issue, but we can already draw some important lessons from the four discussed already:

1. Intergenerational causal influences and direct expositions to the environment are not cases of inheritance (intergenerational transmission).
2. To consider horizontal and oblique transmission as cases of inheritance is sometimes counterintuitive with respect to the theoretical role of the concept of inheritance in evolution. All these cases are better understood as mechanisms to acquire new variants.
3. Material overlap is not a necessary condition of intergenerational transmission, but it is an indispensable feature of inheritance via reproduction.
4. Resemblance is a nonnecessary effect of processes of inheritance. Let us focus on inheritance as a cause!

In the following section, I propose a redefinition of inheritance based on these considerations. This results in a concept of extended inheritance, which is nevertheless limited with respect to the other proposals that I discussed previously.

9.3. Limited Extended Inheritance: A Tentative Definition

My starting point is the idea that the concept of inheritance refers to the set of causal processes connecting generations of organisms, namely, causing the conservation of "something" (something that I wish to keep indefinite for the moment) across generations, and that what is inherited (this "something") ensures and accounts for some degree of continuity (in form and function) along parent–offspring lineages. Remember that the theoretical role of inheritance in evolutionary biology is to guarantee such continuity, which is one of the required conditions for evolution by natural (cumulative) selection to take place (Godfrey-Smith, 2009; Lewontin, 1970).[10] On this basis, I maintain that the notion of transmission is larger than the notion of inheritance. More explicitly, as shown in the previous section, cases of transmission, such as contamination,

---

10. The possibility of natural cumulative selection resulting, in the long term, in adaptation also lies on another condition, namely, the presence and maintenance across generations of some mechanism for the generation of variation. Actually, organisms should be able to generate phenotypic heritable variation, that is, selectable variation, for their population to have the capacity to evolve (i.e., evolvability; cf., for instance, Kirschner & Gerhart, 1998). Otherwise,

infection, social learning, and so forth—where organisms acquire variation getting in contact with another organisms or with environmental sources—are not cases of inheritance because they do not account for any particular continuity across generations, and so cannot fulfill the role of this concept in evolutionary theory.[11] Inheritance is a specific form of transmission and, I argue, it should be extended with respect to the traditional account of it because inheritance is not always and exclusively genetic. But inheritance should also be limited with respect to recent proposals for reconsideration that do not pay enough attention to the four distinctions introduced previously.

I suggest limiting inheritance to intergenerational transmission via reproduction.[12] In other terms, I limit this concept to what accounts for the maintenance of phenotypic features and their variation along parent–offspring lineages. In order to provide such a tentative definition of inheritance, first I should define what reproduction is. To this end, I take inspiration from Griesemer's as well as Godfrey-Smith's accounts of it. Griesemer (2000) places at the core of his definition of reproduction the fundamental recursive character of living systems. From him, I take the idea that reproduction is the multiplication, with material overlap, conferring to the new entities the capacity to acquire (or to have) the capacity to reproduce.[13] Godfrey-Smith (2009) suggests a conception of reproduction that

---

or if the set of possible variations is limited to a few number of variants (as in the case of prions, pathogen proteins that can assume a small number of different three-dimensional conformations), no variation can occur and the population cannot evolve by cumulative selection toward new adaptations. I thank Mathieu Charbonneau (personal communication) for pointing out this aspect.

11. One day, after my talk, a colleague of mine raised the following question: "If I gave you 1 euro to thank you for this talk, could anyone argue that you've inherited from me a 1 euro coin?" It was of course a rhetorical question, whose short answer is: "No one could!" This anecdote sheds light on the crucial difference between the acquisition of something (1 euro) and its inheritance, that is, its transmission in parent–offspring lineages.

12. An argument for reproduction as the dividing line between inheritance and other transmission processes comes from the following ontological difference. Resources passed on via reproduction (Odling-Smee's channel 1) are transmitted through lineages of cells: vertically, from parent to offspring, through a channel consisting of chains of cells produced by cell divisions (mitotically or meiotically). By contrast, resources transmitted outside reproduction (Odling-Smee's channel 2) are diffused in many directions by organisms and acquired by others, and do not involve the same kind of continuity and material overlap. This ontological difference, I argue, is crucial because it accounts for the peculiar way generations are connected: Transmission via reproduction involves a flow of matter creating genealogy (cf. Griesemer, 2000, p. S359), which can be traced along cellular lineages. For more details on this distinction, see Merlin & Riboli-Sasco, 2017.

13. The acquisition of the capacity to reproduce corresponds, for Griesemer, to the minimal evolutionary concept of development. As regards organisms with null (or nearly null) development, we can rather talk of the capacity to *have* such reproductive capacity.

is compatible with our common-sense view of it (i.e., "the production of new individuals which are of the same kind as their parents," p. 69) and, at the same time, accounts for rudimentary and marginal cases where it is difficult to distinguish reproduction from other things such as growth.[14] It corresponds to the idea that "reproduction involves the creation of a new entity, and this will be a countable individual" (p. 86). From him, I retain the "permissive" or "inclusive" character of this definition, which can apply to unicellular and multicellular organisms as well as to other kinds of biological entities (genes, chromosomes, and viruses on the one hand, and collectives of organisms on the other). Putting together Griesemer's and Godfrey-Smith's ideas, I suggest defining reproduction as the production of new (countable) entities that are endowed with the capacity to acquire (to have) the capacity to reproduce.

Now I can define inheritance as follows: It is the intergenerational transmission, via reproduction, of the set of (genetic and nongenetic) factors and mechanisms conferring to the new entity the capacity to acquire (to have) the capacity to reproduce.[15] Note that this is a *minimal* evolutionary concept of inheritance, which focuses on the recursive character of living systems (following Griesemer) and so accounts for the transmission/conservation of what is minimally required for natural populations to go on persisting and evolve (i.e., the capacity to acquire [have], at each generation, the capacity to reproduce). Moreover, as I said earlier, this definition applies to unicellular and multicellular organisms, but is also relevant for other kinds of biological entities (such as genes, chromosomes, and viruses on the one hand, and collective entities like buffalo herds on the other), which are "reproducers" according to a loose conception of reproduction such as Godfrey-Smith's (2009). Last but not least, resemblance between genealogically related organisms is not a necessary effect of inheritance and of reproduction. However, as Griesemer notices, it turns out that "the realization in offspring of

---

14. More precisely, Godfrey-Smith distinguishes three different kinds of reproducers that are all cases of reproduction, in a loose sense, but are not all on a par from an evolutionary point of view. "Simple reproducers," such as cells, are the paradigm cases: They can autonomously produce more things like themselves and are not composed of smaller parts that can reproduce. "Collective reproducers," such as organisms like us, can reproduce using their own machinery, but are composed of things that can also make more things like themselves. "Scaffolded reproducers," such as genes, chromosomes, viruses, and so forth, are things that get reproduced as parts of some larger reproducing unit (a simple reproducer) or, anyway, by using the machinery some other entity external to them.

15. In order for the concept of inheritance to entirely fulfill its theoretical role in evolutionary biology, the set of transmitted factors and mechanisms should also include mechanisms for the generation of variation, conferring to the new entity the capacity to produce selectable variation (see footnote 10). The capacity to acquire the capacity to reproduce thus comes in degrees because it directly depends on the specific set of factors and mechanisms (i.e., of variants) transmitted via reproduction.

the capacity to reproduce will undoubtedly entail many particular trait resemblances" (2000, p. S360). Let me also point out that, as mentioned already, material overlap turns out to be an indispensable feature of inheritance here because reproduction is implicated.

What exactly is inherited, then? Answering this question is not easy, particularly in cases where it is difficult to distinguish reproduction from growth, because what look like new individual entities grow directly from old ones.[16] Despite this problem, which can be minimized by adopting Godfrey-Smith's "permissive" account of reproduction, I want to suggest a possible list of hereditary factors (and mechanisms), which can vary depending on the organism under examination. In the case of mammals, this set could include the following factors and the mechanisms in which they are involved: DNA and protein components of chromatin; cellular epigenetic factors (proteins reproducing self-sustaining loops, microRNAs, methyl-groups, and patterns resistant to the reprogramming of the epigenome); cytoplasmic organelles such as maternal mitochondria; intracellular chemical gradients; nuclear and cellular membranes; some endosymbionts, in particular those coming along with maternal gametes. All these factors are passed on from parents to offspring via reproduction, which involves a certain amount of material overlap. Inheritance thus allows the conservation in parent–offspring lineages of the capacity to acquire (to have) the capacity to reproduce. I underline again that such a capacity results from the particular set of factors and mechanisms that are transmitted via reproduction from one generation to the next; it thus comes in degrees, and so there is variation within a population in this respect. By ensuring and accounting for continuity in form and function across generations, inheritance fulfills its theoretical role in evolutionary biology: It allows evolution by natural selection to take place and, in the long term, to result in adaptations.

9.4. Conclusion

In this chapter, I have shown why we should question current accounts of extended inheritance that aim to integrate nongenetic forms of hereditary transmission. With this same objective in mind, I drew a number of lessons from four distinctions present in the literature. I laid down, on these bases, the foundations for an alternative, minimal, definition of inheritance, taking into account recent research advances. My suggestion is that there are different forms of transmission,

---

16. For instance, what looks like thousands of distinct aspen trees scattered across a large area are in fact connected to each other, under the ground, by a common root system. For further problematic examples, see Godfrey-Smith, 2009.

of which inheritance is one: It takes place during reproduction, involves material overlap, and allows the conservation of the capacity to acquire (to have) the capacity to reproduce in parent–offspring lineages.

Time is ripe for moving beyond the traditional gene-centric conception of inheritance. But this change will be worth the effort if it brings with it some epistemic gain. Expanding inheritance so that it includes the mere persistence of environmental circumstances over time, situations where there is no mechanism of transmission, or cases where our intuition is called into question, does not seem to help us give an enriching answer to the question of what inheritance is. Rather, this position makes the sin of putting together very disparate mechanisms of transmission whose nature and role are quite different, from the evolutionary point of view. As classical American pragmatists argued when protesting against the scholastic turn in the philosophy of their times (cf. Rorty, 1982), the most urgent issue we should struggle to resolve when we address a question is the following: What difference does your answer make to you, or to anyone else?[17] My answer to the question of what inheritance is brings to the fore the theoretical role inheritance has assumed in evolutionary biology, to wit, accounting for the conservation of that which allows populations of organisms to go on persisting across generations, ensures some degree of genealogical continuity, and paves the way for evolution by natural (cumulative) selection to produce adaptations.[18]

References

Allis, D. C., Jenuwein, T., Reiberg, D., & Caparros, M. L. (Eds.). (2007). *Epigenetics.* Cold Spring Harbor, NY: Cold Spring Harbor Laboratory Press.

Bonduriansky, R., & Day, T. (2009). Nongenetic inheritance and its evolutionary implications. *Annual Review of Ecology, Evolution, and Systematics, 40*, 103–125.

Danchin, E., Charmantier, A., Champagne, F. A., Mesoudi, A., Pujol, B., & Blanchet, S. (2011). Beyond DNA: Integrating inclusive inheritance into an extended theory of evolution. *Nature Reviews Genetics, 12*, 475–486.

Dawkins, R. (1976). *The selfish gene.* New York, NY: Oxford University Press.

Feng, S., Jacobsen, S. E., & Reik, W. (2010). Epigenetic reprogramming in plant and animal development. *Science, 330*, 622–627.

Godfrey-Smith, P. (2009). *Darwinian populations and natural selection.* New York, NY: Oxford University Press.

---

17. Here I rephrase Kitcher's words (2012, p. 193).

18. A slightly different version of this chapter is published in French in *Précis de philosophie de la biologie*, edited by T. Hoquet and F. Merlin (Paris: Vuibert, 2014).

Griesemer, J. (2000). Development, culture, and the units of inheritance. *Philosophy of Science, 67*, 2348–2368.

Griesemer, J., Haber, M., Yamashita, G., & Gannet, L. (2005). Critical notice: Cycles of contingency; Developmental systems and evolution. *Biology and Philosophy, 20*, 517–544.

Griffiths, P. E., & Gray, R. D. (1994). Developmental systems and evolutionary explanation. *Journal of Philosophy, 91*, 277–304.

Griffiths, P. E., & Gray, R. D. (2001). Darwinism and developmental systems. In Oyama et al. 2001, 195–218.

Helanterä, H., & Uller, T. (2010). The Price equation and extended inheritance. *Philosophy and Theory in Biology, 2*(201306), 1–17. doi:http://dx.doi.org/10.3998/ptb.6959004.0002.001

Heyes, C. M., & Galef, B. G., Jr. (1996). *Social learning and imitation: The roots of culture*. New York, NY: New York Academic Press.

Holliday, R., & Pugh, J. E. (1975). DNA modification mechanisms and gene activity during development. *Science, 187*, 226–232.

Hull, D. L. (1980). Individuality and selection. *Annual Review of Ecology and Systematics, 11*, 311–332.

Keller, F. E. (2010). *The mirage of a space between nature and nurture*. Durham, NC: Duke University Press.

Kirschner, M., & Gerhart, J. (1998). Evolvability. *Proceeding of the National Academy of Sciences USA, 95*, 8420–8427.

Kitcher, P. (2012). *Preludes to pragmatism: Toward a reconstruction of philosophy*. New York, NY: Oxford University Press.

Jablonka, E., & Lamb, M. (2005). *Evolution in four dimensions: Genetic, epigenetic, behavioral, and symbolic variation in the history of life*. Cambridge MA: MIT Press.

Jablonka, E., & Raz, G. (2009). Transgenerational epigenetic inheritance: Prevalence, mechanisms, and implications for the study of heredity and evolution. *Quarterly Review of Biology, 84*, 131–176.

Lamm, E. (2014). Inheritance systems. In Zalta (Ed.), *Stanford encyclopedia of philosophy* (Spring edition), http://plato.stanford.edu/archives/spr2014/entries/inheritance-systems/

Lewontin, R. C. (1970). The units of selection. *Annual Review of Ecology and Systematics, 1*, 1–18.

Mameli, M. (2005). The inheritance of features. *Biology and Philosophy, 20*, 365–399.

Merlin, F. (2010). On Griffiths and Gray's concept of expanded and diffused inheritance. *Biological Theory, 5*, 206–215.

Merlin, F., & Riboli-Sasco, L. (2017). Mapping biological transmission: An empirical, dynamical, and evolutionary approach. *Acta Biotheoretica*, Online first. doi:10.1007/s10441-017-9305-8

Mousseau, T. A., & Fox, C. W. (1998). *Maternal effects as adaptations*. New York, NY: Oxford University Press.

Ochman, H., Lawrence, J. G., & Groisman, E. A. (2000). Lateral gene transfer and the nature of bacterial innovation. *Nature, 405*, 299–304.

Odling-Smee, J. F. (2010). Niche inheritance. In Pigliucci, M., & Müller, G. B. (2010). *Evolution: The extended synthesis*. Cambridge MA: MIT Press, 175–208.

Odling-Smee, J. F., Laland, K. N., & Feldman, M. W. (2003). *Niche construction*. Princeton, NJ: Princeton University Press.

Oyama, S., Griffiths, P. E., & Gray, R. D., (Eds.). (2001). *Cycles of contingency: Developmental systems and evolution*. Cambridge MA: MIT Press.

Pigliucci, M., & Müller, G. B. (2010). *Evolution: The extended synthesis*. Cambridge MA: MIT Press.

Price, G. R. (1970). Selection and covariance. *Nature, 227*, 520–521.

Qvqrnström, A., & Price, T. D. (2001). Maternal effects, paternal effects and sexual selection. *Trends in Ecology and Evolution, 16*, 95–100.

Riggs, A. D. (1975). X inactivation, differentiation, and DNA methylation. *Cytogenetics and Cell Genetics, 14*, 9–25.

Rorty, R. (1982). *Consequences of pragmatism*. Minneapolis: University of Minnesota Press.

Sapp, Jan. (1987). *Beyond the gene*. New York, NY: Oxford University Press.

Skinner, M. M. (2011). Environmental epigenetic transgenerational inheritance and somatic epigenetic mitotic stability. *Epigenetics, 6*, 838–842.

Singer, J., Roberts-Ems, J., & Riggs, A. D. (1979). Methylation of mouse liver DNA studied by means of the restriction enzymes msp I and hpa II. *Science, 203*, 1019–1021.

Sterelny, K. (2001). Niche construction, developmental systems, and the extended replicator. In Oyama, S., Griffiths, P. E., & Gray, R. D., (Eds.), *Cycles of contingency: Developmental systems and evolution* (pp. 333–350). Cambridge MA: MIT Press.

Sterelny, K., Smith, K. C., & Dickinson, M. (1996). The extended replicator. *Biology and Philosophy, 11*, 377–403.

Vassoler, F. M., White, S. L., Schmidt, H. D., Sadri-Vakili, G., & Pierce, C. R. (2013). Epigenetic inheritance of a cocaine-resistance phenotype. *Nature Neuroscience, 16*, 42–47.

10 HEREDITY AND EVOLUTIONARY THEORY

Tobias Uller and Heikki Helanterä

With an eye only seeing particles, and a speech only symbolizing them, there is no such thing as a study of process possible.

RIDDLE (1909)

Evolution by natural selection is often described as the outcome of three conditions: variation among individuals in their characteristics (phenotypic variation), that different variants leave different number of descendants (differential fitness), and that individuals resemble their parents more than they resemble unrelated individuals (heredity) (Godfrey-Smith, 2009; Lewontin, 1970, 1985). Heredity is therefore fundamental to evolutionary theory. If the characters of offspring bear no relationship to the characters of their parents, fitness differences between individuals will not cause systematic shifts in the distribution of phenotypes in the following generation. Natural selection would be powerless and cumulative adaptive evolution would be impossible. This makes it desirable that we have a firm mechanistic and conceptual understanding of what heredity is, and what are the consequences of variation in the mechanisms of heredity for phenotypic evolution.

Heredity is readily observed in nature and it was accepted by the earliest writers on reproduction (e.g., Aristotle; Lennox, 2000). The mechanisms of heredity remained obscure, however. Darwin observed patterns of shared features among individuals that told him that those features were inherited, but he could not provide a satisfactory explanation for the underlying process by which those patterns were generated. Given that Darwin nevertheless could present a strong case for adaptive evolution by means of natural selection, it may at first seem as if the details of heredity are not that important. As long as traits *are* heritable perhaps it does not matter *how* they are inherited? But in fact disagreement about evolution often stems from disagreement about heredity. Perhaps the most important reason for this is that some mechanisms of heredity can render natural selection a relatively minor contributor to organismal design. In particular, adaptive evolution could be greatly facilitated if organisms were able to acquire new functional traits during their ontogeny via use and disuse and pass on those characters to their offspring. The mechanism of heredity was therefore of major interest to evolutionary biologists following the publication of the *Origin of Species* (e.g., Darwin, 1875; Galton, 1876; Romanes, 1895; Weismann, 1893).

10.1. Heredity as Transmission Genetics

Empirical research at the end of the 19th century and the beginning of the 20th century established that the inheritance of acquired characters through use and disuse was rare at best, and the modern evolutionary synthesis effectively removed it from being part of respectable evolutionary theorizing (Mayr, 1982; Sapp, 1987). In fact, the success of the modern synthesis can partly be explained by a changing concept of heredity. Mendel's work on the inheritance of discrete traits, Weismann's separation of soma and germ line, Johannsen's distinction between genotype and phenotype, and Morgan's breeding experiments with fruit flies all contributed to making heredity synonymous with the passing of trait determinants from parents to offspring (Amundson, 2005; Müller-Wille & Rheinberger, 2012). Under this scheme, parent-offspring similarity is ascribed to the (stable and regular) transmission of genes between generations ("transmission genetics"). Heredity-as-transmission-genetics thus reduces the complex biological process of gamete formation, fertilization, and parent-offspring interactions to a single parameter of importance: the passing of alleles from one generation to the next. As a consequence, heredity can be studied as a pattern without paying attention to developmental process.[1]

Heredity-as-transmission-genetics has been immensely useful and successful in evolutionary biology. There are good reasons for this. Even if transmission genetics is an abstraction that leaves out most of the complexities of reproduction, individuals with similar genotypes tend to have similar phenotypes. The transmission of genes from parents to offspring—leading to similarity of genotypes—is therefore causally important for the phenotypic similarity of parents and offspring.[2] Assuming Mendelian inheritance, it is possible to use crosses between individuals with known phenotypes to establish the number of genes involved in production of a particular phenotype, the location of those

---

1. Morgan and his lab members were particularly important for generating widespread acceptance that heredity could be equated with transmission genetics, for example through the publication of "The Mechanisms of Mendelian Heredity" (Morgan, Sturtevant, Muller, & Bridges, 1915; see Amundson, 2005, pp. 148–152, and Allen, 1978).

2. Transmission genetics can also explain why parent-offspring similarity does not always hold, for example, because of recessive alleles that make some traits occur only under some combinations of paternal and maternal genotypes.

genes relative to the others, sex-linkage, and so on. Stable transmission of genes also enables differences in phenotypes between lineages to be maintained down many generations, which is crucial for building and maintaining complex adaptations.[3] Finally, the transgenerational dynamics of genes within populations can be formalized in mathematical terms (i.e., population genetics; Fisher, 1930). It is difficult to overstate the importance of population genetics for the development of evolutionary theory (Provine, 1971, 1986). It provided a tool that could be used to show that natural selection can drive different genetic variants (and hence phenotypes) to fixation in different environments, maintain polymorphism within populations, that chance events can result in fixation of deleterious traits, that phenotypes that never reproduce still can be favored if they help their relatives to reproduce, that even related individuals can sometimes benefit from harming each other, and that genetic architectures can bias evolutionary outcomes. Predictions from population genetic models could be tested in natural or experimental systems, often with satisfactory results (e.g., summary of Dobzhansky's work in Lewontin, Moore, Provine, & Wallace, 2003).

The success of transmission and population genetics notwithstanding, there has been, and continues to be, dissatisfaction with describing the evolutionary process in purely genetic terms. First, although the fact that allelic similarity predicts phenotypic similarity is strong evidence that DNA is causally important in heredity, this does not mean that a gene-centric view allows a complete description of the inheritance of biological features. Second, even though population genetic theory is often able to predict patterns observed in the wild or in the laboratory, this does not mean that population genetics provides a complete description of the evolutionary process. To show that heredity can satisfactorily be reduced to genetic inheritance in evolutionary theory, it must be shown that interactions between parents and offspring other than transmission of DNA— what we will refer to as nongenetic inheritance[4] (Bonduriansky & Day, 2009)— do not contribute to the rate or direction of phenotypic evolution.

In this chapter we address both of these issues. We argue that we need a concept of heredity in biology that is not gene-centric. First, we review why transmission genetics is causally and explanatory insufficient for both the inheritance of features and the inheritance of differences in features. Having established the

---

3. Vertical genetic inheritance also ensures that relatedness becomes highly correlated across the genome, which facilitates the ability to build cumulative adaptations, in particular with respect to social traits (Grafen, 1985; West & Gardner, 2013).

4. The term "nongenetic" does not of course mean that inheritance mechanisms operate in isolation of DNA; as with all other phenotypes, DNA is involved in their development and function.

need for a nongenetic contribution to heredity, we briefly outline three concepts of heredity in evolutionary biology that allow nongenetic inheritance (i.e., mechanisms in addition to the transmission of DNA): heredity as phenotypic covariance, heredity as intergenerational communication, and heredity as developmental process. These perspectives each emphasize different aspects, and implications, of nongenetic inheritance for phenotypic evolution and we explain what these differences are. The last two sections expand on the role of nongenetic inheritance in evolution, first, using a general description of change within populations (the Price equation) and, second, by discussing the role of nongenetic inheritance in the proposed "extended evolutionary synthesis," which emphasizes evolutionary processes that were marginalized in the modern synthesis.

10.2. The Explanatory Insufficiency of Transmission Genetics

Heredity fundamentally refers to the like-begets-like phenomenon. A theory of heredity therefore needs to explain the reliable recurrence of parental features in offspring. This includes characters that are shared among all members of a lineage, such as human-specific features, but also characters that differ between lineages within populations, such as hair color. Mameli (2005) made this distinction by introducing the concepts of F-inheritance and D-inheritance, where the F stands for *Features* and the D for *Differences* in features (see also Mameli, 2004, 2007). F-inheritance requires reference to the full set of mechanisms that result in recurrence of phenotypes down generations. Although genes obviously contribute to species-typical features, the fact that, for example, a lizard egg differs from the egg of a bird should make it obvious that transfer of DNA from parents to offspring is not causally sufficient to explain why offspring of lizards look like lizards rather than like birds.[5] Experimental transfer of DNA between eggs of closely related species has indeed demonstrated that some species typical features are determined by egg content, not the zygotic DNA (e.g., Sun, Chen, Wang, Hu, & Zhu, 2005). Developmental biologists are increasingly revealing such maternal regulation of early development (e.g., Gilbert, 2010; Li, Zheng, & Dean, 2010; Pelegri, 2003; see East, 1934a, 1934b, and Sapp, 1987, for reviews of the early literature). Similarly, development of species-typical behaviors require parent–offspring interactions that go well beyond the transmission of DNA (e.g., Gottlieb, 1997; Hood, Tucker Halpern, Greenberg, & Lerner, 2010; Moore, 1995; Slagsvold & Wiebe, 2011).

---

5. In fact, development typically fails if DNA is transferred from one species to the other unless the species are closely related.

Adaptive evolution requires transgenerational stability of phenotypic differences, not similarities. Consequently, evolutionary theory has focused on D-inheritance, requiring only reference to those mechanisms that contribute to the recurrence of phenotypic differences down generations. Some causal factors in development that contribute to F-inheritance (such as species-typical environments) can be considered "background conditions" under D-inheritance. It is possible to acknowledge the importance of nongenetic inheritance for heredity of features, while arguing that nongenetic inheritance is of no relevance for adaptive evolution (e.g., Dawkins, 1982). Transgenerationally stable differences between lineages within populations are typically assumed to be due to genetic differences. This is not always the case, however. In a famous example, cross-fostering of rat pups between mothers of different parenting style (assessed by their licking and grooming behaviors) found that these differences are induced and maintained via behavioral interactions between the mother and her pups and not because of genetic differences (Francis, Diorio, Liu, & Meaney, 1999; Weaver et al., 2004). Other examples of nongenetic inheritance of behavioral phenotypes that can be maintained over several generations include differences in migration to breeding or overwintering sites (Brown & Shine, 2007; Harrison et al., 2008), foraging and exploration behaviors (Slagsvold & Wiebe, 2011; Schuett 2013), and preferences for food (Robinson & Méndez-Gallardo, 2010) and sexual partners (Freeberg, 2000). In addition, there is increasing evidence that some epigenetic variants may be transmitted through the germ line and that this contributes to stable differences between lineages within populations (e.g., Roux et al., 2010; reviewed in Jablonka & Lamb, 2014; Jablonka & Raz, 2009; Lim & Brunet, 2013).

A common response to many of these examples is to argue that nongenetic inheritance can be reduced to genetic inheritance of parental traits and hence is best viewed as being genetically determined (e.g., Dawkins, 2004; Dickins & Rahman, 2012; Dobzhansky, 1935; Toyama, 1913). It is of course true that genes are important for development of the parental phenotypes that "transmits" nongenetic factors to offspring, and hence that a full causal explanation for the differences between lineages may also need to refer to the genetics of parents. However, this does not show that genes are causally prior to, or more important than, nongenetic components with respect to the reconstruction of life cycles. This would imply ascribing genetic causes a more important or fundamental role than nongenetic causes not only in evolution, but also in development. As has been pointed out by many authors, this position is not defensible (e.g., Godfrey-Smith, 2000; Griffiths & Gray, 1994; Maynard-Smith, 2000; Nijhout, 1990; Oyama, 1985; Shea, 2007; see Griffiths & Stotz, 2013, for a recent summary).

A second counterargument in favor of gene centrism is that even if nonge‑
netic inheritance in principle could contribute to the recurrence of phenotypic
differences between lineages, those mechanisms do not allow cumulative adapt‑
ive change because they lack certain features that DNA exhibits. Important fea‑
tures of DNA that make it particularly useful as an inheritance system include
stable replication during reproduction, potential for transmission of large
("unlimited") number of messages, and modularity (Bergstrom & Rosvall, 2011;
Maynard Smith & Szathmáry, 1995). These are good reasons for why genetic
inheritance plays a fundamental role in evolutionary models. However, as the
earlier examples show, other mechanisms of inheritance also enable stable inher‑
itance of phenotypes. Thus, the difference between DNA and at least some non‑
genetic inheritance mechanisms is one of degree, not of kind (see e.g., Jablonka
& Raz, 2009, for discussion of tentative examples of epigenetically inherited
phenotypes).

A gene-centric view of heredity therefore fails to hold up to scrutiny. Of
course, given that most populations harbor substantial amounts of genetic var‑
iation, gene transmission is likely to be an important reason for the recurrence
of phenotypic differences between lineages. It remains an open question to what
extent nongenetic inheritance contributes to stable inheritance of differences in
phenotypes. Nevertheless, a complete explanation of both inheritance of features
and inheritance of differences in features from one generation to the next will
have to include all those mechanisms that contribute to parent-offspring similar‑
ity, and not just to the transmission of DNA.

10.3. Alternative Perspectives on Heredity in Evolutionary Theory

That many modern definitions of heredity refer specifically to the transmission
of genes (Table 10.1) reflects that for many biologists transmission genetics is not
just a heuristic that makes modeling the evolutionary process feasible, it is an
accurate and satisfactory description of the mechanism of heredity for the vast
majority of organisms (humans often excluded because of our extensive capac‑
ity for culture). But as we have seen, transmission genetics cannot be taken to
be causally or explanatory sufficient for hereditary phenomena. This suggests
that we need an alternative concept of heredity in evolutionary biology. Here we
briefly discuss three ways to conceptualize heredity that do not force heredity to
be (exclusively) a genetic phenomenon. In the following sections we discuss to
what extent these perspectives are able to capture how nongenetic inheritance
contribute to the evolutionary process.

Table 10.1 Contemporary Dictionary Definitions of Heredity*

Definition	Source
A. The sum of the characteristics and potentialities genetically derived from one's ancestors B. The transmission of such qualities from ancestor to descendant through the genes	Merriam-Webster Dictionary (http://www.merriam-webster.com/dictionary)
The passing on of physical or mental characteristics genetically from one generation to another	Oxford Dictionary (http://oxforddictionaries.com)
A. The transmission from one generation to another of genetic factors that determine individual characteristics: responsible for the resemblances between parents and offspring B. The sum total of the inherited factors or their characteristics in an organism	Collins English Dictionary (http://www.collinsdictionary.com)
A. The genetic transmission of characteristics from parent to offspring B. The sum of characteristics and associated potentialities transmitted genetically to an individual organism.	Free Online Dictionary (http://www.thefreedictionary.com)
A familial phenomenon wherein biological traits appear to be transmitted from one generation to another. [. . .] heredity results from the transmission of genes from parents to offspring [and] offspring therefore tend to resemble their parents [. . .] rather than unrelated individuals	King, R. C., Mulligan, P. K., & Stansfield, W. D. (2013). *A dictionary of genetics* (8th ed.). Oxford University Press.
The transmission of characteristics from parents to offspring via the chromosomes.	(2005). *Oxford Dictionary of Biology* (5th ed.). Oxford University Press.

(continued)

Table 10.1 Continued

Definition	Source
A. The genetic constitution of an individual B. The transmission of genetically based characteristics from parents to offspring	Lawrence, E. (Ed.). (2008). *Henderson's dictionary of biology* (14th ed.). Essex: Pearson Education.

* The *Dictionary of developmental biology and embryology* (2nd ed., Dye, 2012) does not include heredity or inheritance. However, it refers to inheritance of acquired characteristics as "the hypothesis that phenotypic changes in the parental generation can be passed on, intact, to the next generation; may have a mechanism if the inherited morphological alteration can be mediated by epigenetic changes in the DNA methylation of germ cells."

10.3.1. Heredity as Phenotypic Covariance

Lewontin's (1970, 1985) three necessary conditions for evolution by natural selection that opened this chapter imply that adaptive evolution does not rely on particulate inheritance, but that offspring resemble their parents more than they resemble unrelated individuals. Rather than treating the relationship between parents and offspring as transmission of discrete particles, we can treat it statistically in terms of the covariance between the phenotypes of parents and offspring.[6] The covariance between two random variables, X and Y, is defined as

$$\text{cov}(X,Y) = E[(X - E[X])(Y - E[Y])]$$
$$= E[XY] - E[X]E[Y]$$

Where $E[X]$ represents the expected value for variable X.

Using this statistical measure of covariance between the average phenotype of the parents (known as the midparent phenotype) and the phenotype in the offspring gives us the breeder's equation for change in a phenotype, z, from one generation to the next,

$$\Delta z = \frac{\text{cov}(z_o, z_p)}{\text{var}(z_p)} \text{cov}(w, z_p)$$

where subscripts denote phenotypic values in parents, p, and offspring, o, respectively (see, e.g., Falconer & Mackay, 1996; Rice, 2004, for mathematical details).

---

6. The use of a statistical approach to heredity has a long history that goes back to Galton (e.g., Galton, 1876) and the biometricians that clashed with the Mendelians about the nature of heredity following rediscovery of Mendel's work; see Provine (1971) for a historical account.

The covariance between phenotype values in parents and offspring divided by the total phenotypic variance in parents is equal to the slope of the regression of off-spring phenotype on midparent phenotype (i.e., β_{z_o, z_p}), which is also equal to the heritability, usually denoted h^2 (Rice, 2004). This is multiplied by the covariance between phenotype and fitness, which is known as the selection differential. The selection differential measures the change in phenotype due to differential survival or reproduction. The breeder's equation, typically written as $R = h^2 S$, shows that natural selection can be effective, that is, can cause a change in the average phenotype from one generation to the next, as long as the covariance between parents and offspring is nonzero. This equation occurs in virtually all textbooks on evolution.

The concept of heredity as a phenotypic covariance is representative of the field of quantitative genetics (Falconer & Mackay, 1996; Rice, 2004). However, in quantitative genetics, heritability is not always defined as a regression slope, but instead as the ratio of additive (roughly equal to "transmitted") genetic variance over total phenotypic variance, that is, $h^2 = \frac{V_A}{V_P}$. The additive genetic variance cannot be observed directly, but it can be estimated by comparing traits in relatives (e.g., parents and offspring, half- and full-sibs). The focus on additive genetic variance is a consequence of a gene-centric definition of heredity, and it does not imply that the additive genetic variance divided by the total phenotypic variance accurately captures how differences in fitness will translate into phenotypic change from one generation to the next. In fact, the covariance between the phenotype of parents and offspring is only equal to the additive genetic variance when the phenotype is determined additively by genes and environment and there are no correlations between, for example, the genotype of the parent and the environment of the offspring or the environment of parents and offspring (Rice, 2004). Thus, the additive genetic variance of a phenotype is only an estimate of the covariance between the phenotype of parents and offspring.

This insight has some important consequences for understanding the evolutionary implications of nongenetic inheritance. First, any mechanism that allows covariance between phenotypes of parents and offspring can contribute to heritability. This means that it should also be possible to empirically estimate additive epigenetic variance, additive behavioral variance, and so on (Danchin et al., 2011; Furrow, Christiansen, & Feldman, 2011; Tal, Kisdi, & Jablonka, 2010). Together these sum up as the total additive, transmitted, variance (which when divided by the total phenotypic variance represents a quantity termed "inclusive heritability" by Danchin & Wagner, 2010). Second, offspring phenotype is determined not only by the additive components of inheritance and its own environment but also by aspects of the phenotype of its parent that are not "transmitted" additively. This means that some of the variation in offspring phenotype can be statistically attributed to nonadditive genetic and nongenetic variation in parental phenotypes, which can cause the covariance between parents and offspring to be negative despite that a negative heritability is not possible under the additive genetic variance definition of heritability. Empirical studies suggest that a substantial proportion of variance in traits in natural populations can be ascribed to variation in parental, in particular maternal, phenotypes ("parental effects"; e.g., Maestripieri & Mateo, 2009; Mousseau & Fox, 1998).

Quantitative genetic models that incorporate parental effects were first developed in the field of animal breeding (e.g., Dickerson, 1947; Willham, 1963). Over the last decades those models have been put to use for addressing how phenotypes evolve in the presence of nongenetic mechanisms that contribute to the covariance between parental and offspring phenotypes (e.g., Cheverud, 1984; Kirkpatrick & Lande, 1989; Moore et al., 1997; recent reviews in Cheverud & Wolf, 2009; Hadfield, 2012). These models show that parental effects can affect the rate and direction of evolution, which has been confirmed empirically in natural populations of animals (e.g., McAdam & Boutin, 2004; Wilson et al., 2005). The expansion of quantitative genetic models to include epigenetic inheritance, which focuses on the resetting and environmental dependence of epigenetic marks, is ongoing (e.g., Johannes et al., 2008; Tal et al., 2010).

In summary, the heredity-as-phenotypic-covariance perspective treats (at least in principle) all mechanisms that contribute to the covariance between parental and offspring phenotypes as mechanisms of heredity. It is thus conceptually different from heredity as transmission genetics both in that it does not assume particulate heredity and that it does not assume that DNA is causally or explanatorily sufficient for the inheritance of phenotypic differences. Nevertheless, quantitative genetic models often assume genetic inheritance only and reduce the relationship between parents and offspring to a single parameter of evolutionary relevance; heritability, h^2, estimated as the ratio of additive genetic variance over phenotypic variance. More recent models that relax this assumption by allowing parental effects show that nongenetic inheritance can have evolutionary consequences, both for the rate and direction of phenotypic change (Hadfield, 2012).

10.3.2. Heredity as Intergenerational Information Transfer

Heredity is often described as the passing of information between generations. This is true for population genetic models (e.g., Frank, 2009), quantitative genetic models that allow nongenetic inheritance (e.g., Danchin et al., 2011), and cultural evolution models (Boyd & Richerson, 1985; Cavalli-Sforza & Feldman, 1981), and for general discussions about nongenetic inheritance (e.g., Jablonka & Lamb, 2005).[7] The use of information language suggests that heredity could be seen as a form of communication between parents and offspring (Bergstrom & Rosvall, 2011; Shea, 2012). Consequently, inheritance mechanisms could be seen as adaptive features that contribute to the fit between organism and environment by allowing parents to transmit information about the state of the world to their offspring, thereby enabling offspring to match their phenotype accordingly. Effects of the parental phenotype on offspring phenotype are often interpreted in this way in behavioral ecology, where they are referred to as maternal (or parental) effects (adopting the terminology from quantitative genetics[8]) (Uller, Nakagawa, & English, 2013). A number of recent studies have taken this information perspective on the evolution of nongenetic inheritance (English et al., 2015; Leimar & McNamara, 2015; Rivoire & Leibler, 2014; Uller et al., 2015).

The rationale for treating inheritance as parent–offspring communication is that mechanisms of inheritance can carry correlational information about the state of the world (Bergstrom & Rosvall, 2011; Jablonka, 2002; Shea, 2011; Shea, Pen, & Uller, 2011). Correlational information is found whenever some entity's being in a particular state changes the probability that some other entity is in another particular state. For example, the presence of smoke increases the probability that there is a fire nearby. As Figure 10.1 shows, different mechanisms of inheritance can also carry information about the state of the world. For DNA, correlational information can arise because DNA is transmitted down generations unchanged (with the exception of mutations), which enables natural selection to build up statistical correlations between genotypes and environments (Figure 10.1). Thus, the passing of DNA from parents to offspring also passes information about the historical state of the environment, which makes selection an information-generating process (Frank, 2009; Kimura, 1961). DNA is not the only information-carrying entity in heredity, however. Parental phenotypes

---

7. For example, two of the strongest proponents of nongenetic inheritance, Eva Jablonka and Marion Lamb, tend to define inheritance in terms of information transmission, see, for example, the Prologue in *Evolution in Four Dimensions* (Jablonka & Lamb, 2005).

8. It is a potential source of confusion that maternal effects in quantitative genetics refer to a proportion of variance in offspring phenotype attributed to variation in maternal phenotype (which can further be divided into variation due to genetic and environmental differences between mothers), whereas in behavioral ecology it tends to refer to a causal, potentially adaptive, effect of the maternal phenotype on offspring phenotype. Recently, quantitative geneticists have suggested that a causal rather than statistical definition should be adopted (Wolf & Wade, 2009), which largely avoids this problem. (For more on the relationship between parental effects and "nongenetic inheritance" see Bondurisansky & Day, 2009; Uller, 2012.)

can also carry information about the state of the environment that the offspring are likely to experience (Figure 10.1). For example, maternal hormone levels during breeding can carry information about the quality of the local habitat (e.g., Tschirren, Fitze, & Richner, 2007). Offspring could capitalize on this information if there are mechanisms that enable development of alternative phenotypes, for example, dispersive versus nondispersive behavior, in response to hormone exposure in utero or in the egg yolk. Parents (the signalers) may also evolve to increase the information content of the hormonal signal, for example, by modification of the timing, strength, or duration of their endocrine response to appropriate cues (Badyaev & Oh, 2008).[9] Other mechanisms of nongenetic inheritance, including epigenetic and behavioral mechanisms, can carry information in similar ways (English et al., 2015; Shea et al., 2011).

The information perspective on heredity thus establishes a difference between mechanisms of inheritance that is based on the underlying cause(s) for mechanism X to carry correlational information about future environmental state Y. Shea et al. (2011) named these two ways by which inheritance mechanisms carry information selection-based—when the information is generated through selection on stably transmitted variants—and detection-based—when the parent responds to an adaptively relevant feature of the environment in ways that communicate the state of the world to the offspring (Figure 10.1). The distinction between selection-based and detection-based information transmission helps to evaluate some claims regarding the evolutionary function of nongenetic inheritance. Specifically, it shows that even if different mechanisms of heredity can be on a par in terms of their causal effects on development, they need not be on a par with respect to their role in evolution (Shea, 2011). Several authors have pointed out that DNA is very good at storing and transmitting an arbitrary sequence and hence that it may have been under selection for its ability to generate long-run heredity of selected phenotypes (Bergstrom & Rosvall, 2011; Maynard-Smith & Szathmary, 1995). This would make DNA an inheritance system in a more strict sense than, say, maternal hormones that are less able to sustain consistent lineage differences in phenotypes down generations (partly because they are so sensitive to context). However, not only DNA is an inheritance system in this strict sense. Any mechanism that enables variants to be faithfully passed on can result in selection-based information (Shea, 2011). For example, some epigenetic variants are replicated with sufficiently high fidelity to suggest that they acquire information through a selective process (Jablonka & Raz, 2009). More complex

---

9. Parents could also change the state of the world to fit the offspring phenotype (Odling-Smee, Laland, & Feldman, 2003; see section 10.3.3).

(a)

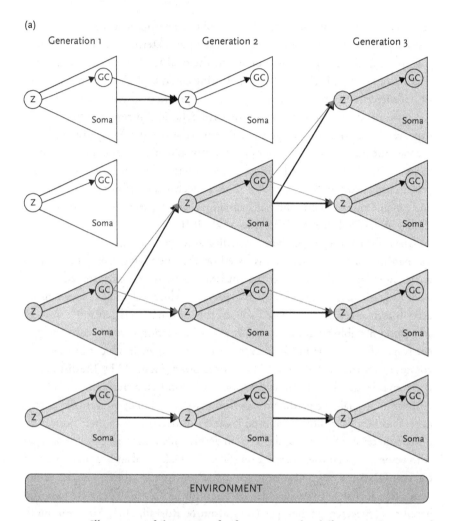

FIGURE 10.1 Illustration of the origin of information under different mechanisms of inheritance. Development begins with a zygote (z), which gives rise to germ cells (GC) and soma. Both the germ cells and the soma of the parent contribute to the zygote of the next generation. White and gray colors denote different phenotypes that are caused by corresponding differences in germ cells or soma and that affect the development of the subsequent generation (black vs. gray arrows). a) *Selection-based information transmission.* Stably inherited differences, here in germ cells (GC; e.g., DNA or epigenetic variation), generate differences in phenotypes (gray vs. white). Gray phenotypes have higher fitness, which causes gray germ cells to increase in frequency and hence establish a correlation between germ cell type and environment. b) *Detection-based information transmission.* Individual responses to the environment result in correlations between the phenotype and the selective context, independently of the color of the germ cells. These differences in parental phenotypes can be exploited by development through nongenetic mechanisms of inheritance, resulting in transgenerationally stable phenotypes within environments.

(b)

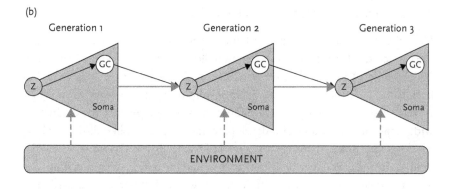

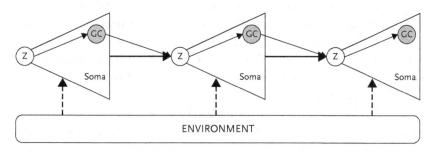

FIGURE 10.1 Continued

interactions between parents and offspring also enable information to be generated in the same way, as in instances of behavioral imitation where complex parental behaviors are faithfully replicated in offspring (Jablonka & Lamb, 2005; Shea, 2009; Weaver et al., 2004). On this account, DNA plays a special, but not unique, informational role in heredity and evolution (Shea, 2011).

In summary, considering heredity as transmission of information between generations emphasizes an important role for both genetic and nongenetic inheritance. In contrast to heredity as phenotypic covariance, which focuses on the evolution of phenotypes subject to different forms of inheritance, heredity as intergenerational communication emphasizes the adaptive evolution of inheritance mechanisms themselves.[10] However, both perspectives single out statistical properties of parent–offspring relations as the key to understanding evolutionary phenomena. In contrast, our last perspective on heredity attempts to explain heredity in a causal-mechanistic way.

---

10. Note that selection is not always concordant across generations and hence that there is potential for "parent-offspring conflict" (Trivers, 1974), which can influence the evolution of nongenetic inheritance (e.g., Uller & Pen, 2011).

10.3.3. Heredity as Developmental Process

Until the 18th century, heredity and development were not seen as the two distinct processes that are so entrenched in biological thinking today (Amundsen, 2005; Müller-Wille & Rheinberger, 2007, 2012). In fact, their separation has been hailed as a crucial step in advancing evolutionary theory (e.g., Mayr, 1982). Nevertheless, some biologists and philosophers of biology view the distinction with suspicion. Advocates of a developmental perspective, like those ascribing to "developmental systems theory" (DST; Oyama, 2000; Oyama et al., 2001), tend to view all causal mechanisms that contribute to parent–offspring similarity as inheritance in a broad sense. They thus view heredity not as transmission of adaptively tuned information through discrete channels, but as the entire process of reconstruction of life cycles to which the parents contribute (Badyaev, 2011; Griffiths & Gray, 2001; Oyama, 2000, 2001, ch. 4). This of course involves the replication and transmission of DNA to the gametes, but it also includes all other interactions that contribute to the reliable recurrence of phenotypic development down generations.

This developmental perspective on heredity is associated with several major conceptual differences compared to the standard narratives of development and evolution. Perhaps the most conspicuous is a rejection of the notion that some aspects of the organism can be considered to be due to nature and some due to nurture (Moore, 2013; Oyama, 2000). If life cycles are reconstructed, one cannot point to any single causal mechanism and say that it is prior to, or more fundamental than, the rest of the developmental system (Griffiths & Gray, 1994; Laland et al., 2013; Oyama, 1985). Developmental system theorists thus reject that genes have a privileged role as causes in development, which implies they are also not privileged as causes of heredity. Heredity cannot be reduced to the transmission of traits; what is transmitted are the developmental means that enable the reconstruction of life cycles, and this includes more than DNA (Oyama, 2000). Selection is suggested to be redefined as "changes in the distribution of developmental systems that occur when traits are differentially associated with different lineages and the variants interact with their environment in ways that confer on them different probabilities of being perpetuated" (Oyama, 2000, p. 81). Natural selection can generate adaptations because the organism itself contributes to the reconstruction of life cycles, which enables cumulative evolution of organism-environment complexes. Evolution is not defined genetically but instead as any changes in the composition of the developmental systems that enable perpetuation of life cycles (Oyama, 2000, ch. 4).

Considering heredity a developmental process extends the role of non-genetic inheritance in evolution beyond heritability and information transmission (Laland et al., 2015). Specifically, it raises the possibility that nongenetic mechanisms of heredity also can contribute to the origin of novelties (Badyaev, 2008, 2009; Rice, 2012). This can be due to changes in genes that contribute to parental transfer of developmental resources ("maternal effect genes"; Gilbert, 2010, ch. 6) but, as West-Eberhard (2003) and others have argued, environmentally induced phenotypes may have even greater evolutionary potential. Responses to novel environmental input are often accommodated in functional ways ("phenotypic accommodation") and, in contrast to DNA mutations, can affect many individuals within a single population. If the ability to respond is heritable, selection can result in the spread and further modification of environmentally induced phenotypic accommodations. Such heritable variation will often be due to genetic differences between individuals,[11] but the retention and spread of new variation can also be due to nongenetic inheritance, including epigenetic and behavioral mechanisms (e.g., by offspring imitation of parental behavior). This raises the possibility that heredity itself evolves as a result of increasing stabilization of life cycles under natural selection (Badyaev, 2009), but to our knowledge this process has not yet been formally modeled.

In summary, it is possible to reject a fundamental distinction between heredity and development and consider heredity as the developmental process by which offspring come to resemble their parents. This may have some potentially radical consequences for evolutionary theory. In contrast to heredity as phenotypic covariance and heredity as information transmission, heredity as developmental process emphasizes the origin of adaptations in addition to their spread within populations, and downplays the adaptive function of inheritance in terms of information transmission. Indeed, developmental systems theorists are suspicious of the notion that any mechanism of inheritance can usefully be considered to transmit information (e.g., Oyama, 1985).

10.4. Heredity in Evolutionary Theory Revisited

The four perspectives on heredity that we have discussed look quite different (Table 10.2), and it is not immediately obvious how they are related. A comprehensive analysis is beyond the scope of this chapter. Our aim is instead to

---

11. This is often described as variation among individuals in the slope of reaction norms (Schichtling & Pigliucci, 1998). Lande (2009) presents a quantitative genetic model of evolution following environmental change that explicitly considers genetic variation in norms of reaction.

Table 10.2 Summary of the Four Different Perspectives on Heredity in Evolutionary Biology Discussed in This Chapter. NGI = Nongenetic inheritance

	Heredity as transmission genetics	Heredity as parent-offspring covariance	Heredity as intergenerational information transfer	Heredity as developmental process
Representative research community/ approach:	Evolutionary population genetics	Evolutionary quantitative genetics	Behavioral ecology	Developmental systems theory
NGI is considered as:	Parentally transferred instructions for development under genetic control	Source of variation in offspring phenotype ("parental effects")	Cues or resources that enable adaptive transfer of information across generations	Components of the reconstruction of life cycles that contribute to transgenerational stability and variation of phenotypes
Research emphasis concerning NGI:	Adaptive evolution of nongenetic inheritance	Evolution of traits subject to nongenetic inheritance	Adaptive evolution of nongenetic inheritance	Evolutionary transitions between variation—retention-stabilization of life cycles
Primary role of NGI in adaptive evolution:	None or as a genetic adaptation	Affect the response to selection by affecting parent-offspring covariance	Mechanism that facilitates adaptation to heterogeneous environments	Contribute to the development, selection, and heredity of phenotypes

clarify the evolutionary implications of nongenetic inheritance and the explanatory sufficiency of different perspectives on heredity in two ways. First, we show that even mechanisms that generate transient or partial inheritance are of evolutionary significance and that those mechanisms can contribute to phenotypic evolution even if they do not affect parent–offspring covariance or adaptively transmit information. Importantly, this analysis also helps to clarify differences and similarities between different mechanisms of inheritance, which we exemplify by discussing epigenetic and ecological inheritance. Second, we explain what is needed of a concept of heredity to encompass changes in the structure of evolutionary theory that are associated with calls for an extended evolutionary synthesis. Both of these exercises point toward considering heredity a developmental process and we end with a brief discussion of the consequences of this for evolutionary theory.

10.4.1. Heredity and the Price Equation

We have seen that the breeder's equation shows that an evolutionary response to selection requires a covariance between parental and offspring phenotypes. Nongenetic inheritance can contribute to this covariance in both the short (e.g., parental effects) and long term (e.g., "epialleles"), and could therefore be important for predicting how populations evolve. This formulation of evolutionary change is based on some simplifying assumptions, however. Here, therefore, we start with a general description of change in average phenotype between two populations (e.g., ancestors and descendants) (see also Day & Bonduriansky, 2011; Helanterä & Uller, 2010; Uller & Helanterä, 2013). This is known as the Price equation, which can be written as

$$\Delta \bar{z} = \frac{1}{\bar{w}}[\text{cov}(w,z) + E(w\Delta z)] \tag{1}$$

where $\Delta \bar{z}$ is the change in the average phenotype in the population, $\bar{w}$ is the mean number of descendants per individual, $\text{cov}(w,z)$ is the covariance between fitness and trait value, and $E(w\Delta z)$ is the expected value of the product of fitness and the average phenotypic difference between parent and offspring in the absence of selection (Price, 1970, 1972). The last two terms can be interpreted as the change due to differential reproduction and survival and the change that occurs as a result of reproduction and the mechanisms of inheritance, respectively (Rice, 2004; it is important to note that the expected value is also affected by external changes between generations that cause shifts in phenotypes, such as environmental change). Division by the mean number of descendants means that fitness is relative and not absolute.

Because the Price equation does not make any assumptions of the underlying mechanisms of parent–offspring similarity it can be used to derive the standard single-locus and quantitative genetic formulations of evolutionary change (Rice, 2004). However, none of the terms in Eq. (1) specifically refer to the covariance between parental and offspring phenotypes, which makes this difficult to see. We therefore rewrite the equation as

$$\Delta z = \frac{1}{\bar{w}}(\beta_{z_o,z}\mathrm{Cov}(w,z) + E(\Delta z) + \mathrm{Cov}(w,z_o|z)) \qquad (2)$$

This decomposes change in population mean phenotype into three components (Figure 10.2; see Heywood, 2005; Lynch & Walsh, 2013, for mathematical details on how to get from Eq. 1 to Eq. 2). The first term in the parenthesis on the right hand side consists of $\beta_{z_o,z}$, the best linear slope of the parent–offspring regression or the heritability of the phenotype (using selected parents; Heywood, 2005), times the covariance between fitness and trait value, that is, the selection differential. If the remaining terms are zero, this equation corresponds to the breeder's equation discussed earlier (Falconer & Mackay, 1996). The second term represents the expected phenotypic change in the absence of fitness differences, which is often referred to as transmission bias (e.g., Heywood, 2005). Mathematically, this is represented by the intercept of the parent–offspring regression and could be, for example, changes that occur because of intergenerational environmental change that affects phenotypic development (Rice, 2004). The third term is the covariance between the residuals for the regression of fitness on parental phenotype and for the residuals for parent–offspring regression (named "spurious response to selection" by Heywood, 2005; Figure 10.2). Why would the residuals covary? As shown in Figure 10.2, this can happen because when we describe evolutionary change in this form of the Price equation we are forcing the slope of the regression to be linear (Lynch & Walsh, 2013, ch. 13). But this is not always true (e.g., Gimelfarb, 1986; see Rice, 2012). If the relationship is nonlinear, the residuals are biased across parental phenotypic values (Figure 10.2). Because fitness may also show nonlinear relationships with phenotype, the residuals may be correlated and hence $\mathrm{Cov}(w,z_o|z)$ may be nonzero. An alternative reason for why $\mathrm{Cov}(w,z_o|z)$ could be nonzero is that, even if one or both regressions are linear, the residuals are correlated via a third variable (Heywood, 2005). Heywood (2005) discusses a case with breeding date in birds, where there is a spurious response to selection even when both regressions are linear. This is because a third variable, nutritional status, covaries with both the residuals of breeding date on fitness and the residuals of parental breeding date on offspring breeding date.

In summary, Equation 2 describes the change in the population mean phenotype from one generation to the next in terms of the product of the parent–offspring regression ("heritability") and the covariance between phenotype and fitness ("selection differential"), and two terms that can be affected by

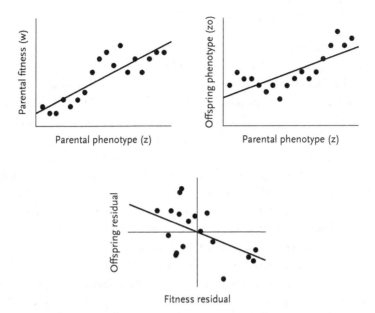

FIGURE 10.2 Illustration of a situation that gives rise to a "spurious response to selection." The relationship between parental phenotype and fitness (top left) and the relationship between parental and offspring phenotype (top right) are nonlinear. As a result, the residuals for the best-fitting linear regressions are nonrandom. This results in a negative covariance between the residuals (bottom graph), which is the spurious response to selection, $\mathrm{Cov}(w, z^0 | z)$, in Equation 3. Figure from Uller and Helanterä (2013), originally adopted with modifications from Heywood (2005) and Lynch and Walsh (2013).

mechanisms of inheritance and environmental effects ("transmission bias" and "spurious response to selection"). Quantitative genetics typically assume that the last two terms are zero (hence the breeder's equation), but they may be nonzero even under pure genetic inheritance (Heywood, 2005; Lynch & Walsh, 2013). Explicitly addressing how the mechanisms of heredity and development influence each of the components of the Price equation helps to establish the consequences of different forms of inheritance for phenotypic evolution. For example, by separating evolutionary change into that produced by genetic and nongenetic inheritance, Day and Bonduriansky (2011) developed a series of models that exemplify how different nongenetic inheritance mechanisms can affect evolution via their effects on phenotypic covariance and transmission bias. Furthermore, Rice (2008, 2012) has shown how a stochastic version of the Price equation makes explicit the importance of nongenetic inheritance for evolution because of its effect on the shape of the parent–offspring phenotype distribution. Here we exemplify these points by discussing two "inheritance systems," epigenetic inheritance and ecological inheritance.

10.4.2. Epigenetic Inheritance

Epigenetic inheritance, such as DNA methylation, differs from genetic inheritance in several ways (Jablonka & Lamb, 1995, 2005). Although epigenetic variants can be reliably inherited through meiosis in some multicellular organisms, their stability appears to be relatively short-lived compared to transmission of DNA sequence variation. Epigenetic variation can be environmentally induced, but, unlike DNA mutation (which can also vary across environments), a broader range of environments are apparently able to modify epigenetic states, perhaps in nonrandom directions (Rando & Verstepen, 2007; Jablonka, 2012a). The degree to which offspring pass on the same "epiallele" as they received from their parents can therefore depend on the similarity of environments across generations, the parental phenotype, and perhaps the epigenetic state itself (Jablonka, 2012b; Jablonka & Lamb, 2005; Richards, 2006; Uller, 2012).

From Equation 2 we can see that epigenetic mechanisms can contribute to phenotypic change in several ways. First, direct transmission of epigenetic variants means that epigenetic mechanisms can cause offspring to resemble their parents, that is, that the (linear) slope of regression of offspring phenotype on parental phenotype is nonzero. Thus, as mentioned previously, epigenetic mechanisms contribute to the overall heritability of a character (e.g., Danchin et al., 2011; Furrow et al., 2011; Tal et al., 2010). The long-term consequences of epigenetic inheritance will depend on the stability of these variants and their pattern of inheritance (e.g., non-Mendelian), which itself can be a function of phenotypic and environmental change. Day and Bonduriansky (2011) have shown that differences in the stability of epialleles can affect both evolutionary trajectories and equilibria of genotype and phenotype values within populations (see also Geoghegan & Spencer, 2013).

Second, the environment-dependence of epigenetic inheritance is likely to cause significant transmission bias, which makes $E(\Delta z)$ nonzero as well (Day & Bonduriansky, 2011; Helanterä & Uller, 2010). This affects the predicted evolutionary change from one generation to the next because epigenetic inheritance, or environmental epigenetic effects, means that phenotypes can change more or less predictably from one generation to the next even in the absence of parent–offspring covariance (e.g., due to a common plastic response in the population). This shows that mechanisms that are shared among all lineages of the population, but do not contribute to inheritance of differences in phenotypes, nevertheless have consequences for how populations evolve.

Finally, epigenetic inheritance may be more likely than genetic inheritance to generate a spurious response to selection. The stochastic nature of epigenetic inheritance and its dependence on the phenotypic character state of the parent and the environment may make it more likely that there will be nonlinear relationships between parent and offspring phenotype or biased distribution of residuals of the regressions in Figure 10.2. Nonlinearity could actually be common whenever there are parental effects, as they tend to skew the distribution of phenotypes from that expected under additive genetic variance (Rice, 2012). For example, biased transmission stability of DNA methylation may result from passive loss of methylation with age. This can result in a spurious response for both reasons mentioned above (Figure 10.2). First, it could lead to nonlinear relationships between phenotypes in parents and offspring and hence residuals may become correlated even in the absence of a causal link. Second, age may covary both with the residuals of fitness for a focal trait (e.g., older individuals may be more experienced and thus have higher breeding performance for a given trait value) and the residuals of the parent–offspring regression (e.g., older parents may be less likely to transmit the same epigenetic mark as they themselves inherited because of stochastic loss of DNA methylation with age). This line of reasoning suggests that establishing the pattern of parent–offspring similarity (e.g., if it is linear), and its underlying mechanism (e.g., if there is environment-specific transmission of epigenetic states), is an important task if we are to understand and predict the extent to which epigenetic mechanisms contribute to short- and long-term evolution (Day & Bonduriansky, 2011; Rice, 2012).

10.4.3. Ecological Inheritance

Parents (or more generally ancestors) do not only "transmit" resources to their offspring (descendants) but also modify the environment that the offspring encounter by choosing nest sites, constructing burrows, and so on ("niche construction"; Odling-Smee et al., 2003). As a result, the environmental context of the offspring is partly determined by the phenotypes of the parents, a phenomenon that has been termed "ecological inheritance" (Odling-Smee et al., 2003). Just like epigenetic inheritance, the evolutionary consequences of ecological inheritance for a particular phenotypic trait (not necessarily the niche constructing trait itself) can appear through its effect on the linear slope of the parent-offspring regression ("heritability"), its intercept ("transmission bias"), or the covariance between the residuals of the two regressions in Figure 10.2 ("spurious response"). First, when ecological inheritance causes environmental similarity within lineages, but maintains environmental differences between lineages, it increases the parent–offspring covariance of phenotypes that show environmental-dependence in their expression, and hence heritability (e.g., Furrow et al., 2011). Such effects occur in organisms where resources are unevenly distributed and passed on to offspring, as occurs in species where offspring take over the territory of their parents. Second, even without differences among lineages within populations, the collective actions of organisms can make the environment change in ways that influence offspring development. For example, as population densities increase, more frequent encounter rates with other individuals can stimulate development of more aggressive behaviors, which could result in directional shifts in aggression even in the absence of parent–offspring covariance. Human niche construction is also a good example of transmission bias, since collective cultural inheritance may underlie directional changes in many traits independently of a direct response to selection (e.g., changes in body size, sexual behaviors, language use; Boyd & Richerson, 1985; Laland & O'Brien, 2011). Third, ecological inheritance could cause $\mathrm{Cov}(w, z_o | z)$ to be nonzero. For example, the deviation from the expected value of offspring size from parental size may be a function of the available resources as determined by territory quality. If larger individuals benefit disproportionally from having good territories, the residuals from a size–fitness relationship could also be biased across territory qualities. Ecological inheritance of territories would thus cause a spurious response to selection. Of course, ecological inheritance may also cause nonlinear relationships between parents and offspring that, as outlined above, can cause residuals to correlate even in the absence of a causal link.

An added complexity with ecological inheritance is that its definition[12] emphasizes another mechanism by which it can have evolutionary consequences. Specifically, the niche-constructing activities of organisms do not only change the "transmission terms" of the Price equation but also potentially cause $\mathrm{Cov}(w, z)$ to be a function of z, or other phenotypic traits, in immediate or more distant ancestors. For example, assume that we follow the evolutionary change in a specific phenotype, P, that originally has a mean value of z . Further assume that the trait is heritable and that phenotypic covariance is only due to genetic variation and that there is no transmission bias. The trait

12. Odling-Smee et al. (2003, p. 42) define ecological inheritance as "any case in which organisms encounter a modified feature-factor relationship between themselves and their environment where the change in selective pressures is a consequence of the prior niche construction by parents or other ancestral organisms." Odling-Smee (2010, pp. 180–181) defines ecological inheritance as occurring "whenever the environmental consequences of prior niche constructing activities of organisms (e.g., the presence of burrows, mounds, and dams or, on a larger scale, changed atmospheric states, soil states, substrate states, or sea states) persist or accumulate in environments as modified natural selection pressures, relative to successive generations of organisms."

would predictably evolve in response to natural selection according to equation 2 with $E(\Delta z) = 0$ and $\text{Cov}(w, z^0|z) = 0$. Now assume that the average trait value in the population affects the external environment in ways that does not affect how P is inherited, but that change what value of P has the highest fitness. For example, if z is large the environment in the next generation could deteriorate, which favors a lower z in the next generation. But when z is low the environment may improve, which favors a larger z in the next generation. That the average trait value in a population would have strong effects on selection across two generations is probably unlikely, but it illustrates the importance of ecological inheritance for the dynamics of covariance between phenotype and fitness (Laland et al., 1996, 1999; Odling-Smee et al., 2003). Importantly, traits do not evolve in isolation. This means that evolution of one particular character can contribute to the dynamics of $\text{Cov}(w, z)$ for other characters, which implies that predicting evolutionary dynamics will often be difficult and require explicit consideration of trait interactions and the spatial structure of populations (Silver & Di Paolo, 2006).

As emphasized by Odling-Smee et al. (2003), "ecological inheritance" does not only contribute to parent–offspring resemblance, that is, how most biologists would understand the terms "inheritance" or "heredity," but also to patterns of fitness, that is, selection. Thus, the term "ecological inheritance" can be somewhat misleading since it contributes to phenotypic evolution in so many ways: (1) it could affect heritability (regression slope) of a particular phenotype (D-inheritance), (2) it could affect the expected change in the trait in the population in addition to the response to selection via transmission bias or spurious responses caused by niche-constructing activities (which can include F-inheritance *sensu* Mameli, 2005), and (3) it could affect the covariance between phenotype and fitness of future generations. All of these are potentially important, but not easily captured by the term "inheritance." The diversity of effects on the evolutionary process that stems from ecological inheritance may be representative also for the evolutionary implications of other nongenetic systems of inheritance. Indeed, although the effect on selection is emphasized for the term "ecological inheritance," it is important to note that other forms of genetic and nongenetic inheritance also modify covariance between phenotype and fitness of the subsequent generation (Badyaev & Uller, 2009). More generally, because what matters in evolution is how fitness of parents relates to the distribution of phenotypes in offspring, any developmental process that influences either fitness or parent–offspring distributions can be evolutionarily consequential (Rice, 2004, 2012). Nongenetic mechanisms of inheritance clearly have the capacity to do both.

10.5. Heredity and the Extended Evolutionary Synthesis

The development of the modern synthesis did not only adopt a genetic definition of heredity and evolution. It also came with a shift in what was considered to be sufficient as an *evolutionary explanation*. Whereas previous debates often centered on how novel adaptive characters can arise from existing characters (e.g., "the arrival of the fittest"; de Vries, 1904), the reduction of heredity to the transmission of genes implied that one can study evolution without reference to development (e.g., Mayr, 1961). In fact, phenotypes take on a limited role in this modern synthesis view of evolution, which is well exemplified by the redefinition of evolution as cross-generational change in gene frequencies, and not in phenotypes (Dobzhansky, 1937). The notion of the genome as a blueprint and Mayr's introduction of the distinction between proximate and ultimate causes (Mayr, 1961) further cemented the view that nongenetic mechanisms could not do any explanatory job in evolutionary biology, but rather should be seen as execution of functions encoded in the genome (see Dawkins, 2004; Dickins & Dickins, 2008; Dickins & Rahman, 2012; Haig, 2007; Scott-Phillips et al., 2011, for recent versions of this argument; see Laland et al., 2011, 2013, and Mesoudi et al., 2013, for criticism; see also Bonduriansky, 2012; Uller, 2013).

These views are at odds with contemporary evolutionary biology that emphasizes the importance of the developmental origin of novel, potentially adaptive, variants, the many processes that may promote their recurrence and spread, and what maintains the ability for further evolution. This includes discussions of the role of plasticity as an initiator of evolutionary change in novel environments (e.g., West-Eberhard, 2003), the importance of developmental bias promoting directional and perhaps lineage-specific evolutionary trajectories (e.g., Arthur, 2004), how organism–environment interactions contribute to selective dynamics in time and space (e.g., Odling-Smee et al., 2003), and how developmental mechanisms influence evolvability (e.g., Stern, 2009). Together these advances in evolutionary thinking have suggested to some authors that there is a new "extended evolutionary synthesis" emerging (see Laland et al., 2015, for an overview).

The conceptual structure of such an extended evolutionary theory—if it ever is realized—is debated (Laland et al., 2015; Pigliucci & Muller, 2010). However, a fundamental feature is that it is a theory of phenotypes rather than of genotypes. Treating heredity as a process makes it take center place in evolutionary scenarios that involve developmental plasticity or developmental bias. In fact, we suggest that an extended synthesis necessitates a shifting concept of heredity, away from transmission genetics and toward the reconstruction of life cycles. This shift in emphasis makes it explicit that nongenetic inheritance—the transference from parents to offspring of developmental resources that contribute to the reconstruction of life cycles—enters evolutionary theory through all three of Lewontin's conditions. Nongenetic inheritance contributes to the origin of variation (condition one) and those variants are inherited because the parents reconstruct the developmental niche for the offspring in recurrent "cycles of contingency" (Oyama et al., 2001) (condition three), and not only because they transfer genes. This parental transference of developmental resources does not only affect offspring phenotype and its recurrence within populations, it also contributes to the relationship between phenotype and survival or reproductive success (Lewontin's second condition). As ecological inheritance in particular makes explicit, selection, or an absence of selection, partly arises from the actions of ancestors (Odling-Smee et al., 2003). It is a major focus of the extended evolutionary synthesis to establish how selection arises through the interactions between organism and environment. Both genetic and nongenetic inheritance will affect what, and in what form, phenotypic accommodations to novel genetic and environmental input are recurrent down generations. Because induction of phenotypic variation in offspring through nongenetic inheritance has been channeled through a responsive phenotype (i.e., the parent), this may further enhance the functionality of such variation (Badyaev, 2009). Once expressed, parental transference of developmental resources can facilitate or increase transgenerational persistence of induced phenotypes, for example via behavioral mechanisms of inheritance (Badyaev & Uller, 2009; 2012). For example, cross-fostering of great and blue tit chicks suggests that a suite of species differences in sexual preferences and foraging behavior may have originated, spread, and been maintained as culturally inherited phenotypes (Slagsvold & Wiebe, 2011). In the extended evolutionary synthesis view on evolution, therefore, some adaptive phenotypes may initially be inherited through mechanisms with low fidelity, and hence exhibit a low parent–offspring covariance, and only gradually become stably inherited as combinations of genes and nongenetic components of parent–offspring interactions that increase this covariance are accumulated under stabilizing selection (Badyaev, 2009, 2011; Laland et al., 2015; Schmalhausen, 1949; Waddington, 1957; West-Eberhard, 2003).

10.6. Summary and Outlook

Heredity is fundamental to evolution. We have argued that it cannot be reduced to genetic inheritance and that the causal-mechanistic perspective offered by heredity as developmental process is the only of the four concepts of heredity that we have discussed that also is causally and explanatorily sufficient in evolutionary biology. This perspective recognizes that recurrence of phenotypes within lineages, and differences among lineages, require reference to the recurrence of both genetic and nongenetic causes of *development*. Heredity occurs precisely because parents transfer a variety of developmental resources that enable reconstruction of life cycles and hence phenotypes. Nongenetic inheritance refers to this transference, which is mediated through a variety of epigenetic, physiological, and behavioral mechanisms. These concepts of heredity and inheritance avoid the problems inherent in a gene-centric perspective and allows for a richer understanding of the reasons for why offspring resemble their parents. Importantly, it shows that mechanisms of nongenetic inheritance contributes to all three of Lewontin's (1970, 1985) conditions for evolution by natural selection.

Treating heredity as a developmental process makes nongenetic inheritance fundamental to a phenotype-oriented evolutionary framework. What ultimately matters for phenotypic evolution is the relationship between the fitness of parents and the phenotypes of their offspring (Rice, 2012). All the mechanisms by which parents contribute to the reconstruction of life cycles can potentially affect the origin, fitness, and inheritance of phenotypes. The Price equation helps to establish how genetic and nongenetic inheritance contributes to intergenerational phenotypic change (Day & Bonduriansky, 2011; Uller & Helanterä, 2013). Nongenetic mechanisms of inheritance can affect the regression of offspring phenotype on parental phenotype (i.e., heritability). But we have seen that this is not all there is to transgenerational phenotypic change. Nongenetic inheritance can cause transmission bias and spurious responses to selection, partly by causing nonlinear parent–offspring relationships. Thus, "parental effects" should not be treated as confounding environmental noise, but instead as a real cause for parent–offspring resemblance that can have evolutionary consequences at both short and long time scales. Nongenetic inheritance can also link the phenotypes in one generation with selection in future generations (Odling-Smee et al., 2003), which makes the covariance between phenotype and fitness dynamic and evolving rather than a static property. This is the fundamental point of niche construction theory and shows that niche construction is essentially a developmental process occurring in an ecological context (Odling-Smee, 2010). Although these complexities can make the mathematics complicated, nongenetic inheritance (and development more generally) can be incorporated into formal models of phenotypic evolution (e.g., Day & Bonduriansky, 2011; Cavalli-Sforza & Feldman 1981; Rice, 2004, ch. 8, 2008).

Does this mean that we should abandon transmission genetics in our evolutionary models? We believe not. It remains entirely valid to use abstraction in theoretical model building. Reducing the complexity of heredity to transmission of genes will remain a useful way to model the evolutionary process. This is not surprising, considering that DNA has features that make it particularly well suited for long-run heredity. But it is an abstraction that leaves out some important features that can affect evolutionary dynamics. One therefore needs to be aware that by reducing inheritance to transmission genetics, one leaves out potentially important evolutionary processes. This is not unusual in evolutionary biology. For example, it is acknowledged that phenotypic optimality models do not account for the underlying genetics (Grafen, 1984), which makes it necessary to treat them with caution. Confusing the biological mechanisms of heredity with a heuristic abstraction, namely, transmission genetics, may have delayed recognition of the importance of development (including nongenetic inheritance) in evolutionary theory (Rice, 2012). Nongenetic inheritance is not a just a proximate mechanism of no evolutionary significance on its own, it is an essential part of the reconstruction of life cycles on which evolution relies (Badyaev & Uller, 2009; Griffiths & Stotz, 2013).

Our discussion also sheds doubt on the value of interpreting all forms of inheritance as transmission of information. Information emerges through the contingencies of development and evolution. That inheritance mechanisms carry information is therefore a derived state, a consequence of the adaptive evolution of life cycles. This can make it useful to explicitly link inheritance and information because it provides a condition (or maximand) for evaluating the adaptive value of different mechanisms of inheritance.[13] However, it may also detract from the importance of nongenetic inheritance in the origin and recurrence of novel characters through developmental plasticity.

These final points suggest to us that treating heredity as a developmental process invites a pluralistic stance with respect to how heredity is treated in formal models. But it also implies that nongenetic inheritance cannot be fully integrated into evolutionary theory without an integration of development and evolution. In fact, we suggest that a wider concept of inheritance is a necessary and fundamental component of the extended evolutionary synthesis. At the very least, as our understanding of the role of developmental processes in evolution becomes more sophisticated, the parts of those processes that underlie heredity should take on a more central role in evolutionary theory (Badyaev & Uller, 2009; Griffiths & Stotz, 2013; Odling-Smee, 2010; Rice, 2012).

---

13. English et al. (2015) show how adaptive evolution of developmental switches tends to maximize mutual information between phenotype and environment. This maximization can occur through differential response to inherited genes, nongenetic inheritance, or direct environmental input, which are all on a par in terms of their effect on development even if the processes that cause them to carry information differ. See also Shea et al. (2011).

Acknowledgments

TU was supported by the Royal Society of London, the Knut and Alice Wallenberg Foundation, and the European Union's Seventh Framework Programme (FP7/2007-2011) under grant agreement n° 259679. HH was supported by the Academy of Finland (grant number 135970 to HH and 252411 to the Centre of Excellence in Biological Interactions) and the Kone Foundation. We are grateful to the editors and Kevin Laland for comments on a draft chapter.

References

Allen, G. E. (1978). *Thomas Hunt Morgan: The man and his science.* Princeton, NJ: Princeton University Press.

Amundsen, R. (2005). *The changing role of the embryo in evolutionary thought: Structure and synthesis.* New York, NY: Cambridge University Press.

Arthur, W. (2004). The effect of development on the direction of evolution: Toward a twenty-first century consensus. *Evolution and Development, 6,* 282–288.

Badyaev, A. V. (2008). Maternal effects as generators of evolutionary change a reassessment. *Year in Evolutionary Biology, 1133,* 151–161.

Badyaev, A. V. (2009). Evolutionary significance of phenotypic accommodation in novel environments: An empirical test of the Baldwin effect. *Philosophical Transactions of the Royal Society B: Biological Sciences, 364,* 1125–1141.

Badyaev, A. V. (2011). Origin of the fittest: Link between emergent variation and evolutionary change as a critical question in evolutionary biology. *Proceedings of the Royal Society B: Biological Sciences, 278,* 1921–1929.

Badyaev, A. V., & Oh, K. P. (2008). Environmental induction and phenotypic retention of adaptive maternal effects. *BMC Evolutionary Biology, 8,* 3.

Badyaev, A. V., & Uller, T. (2009). Parental effects in ecology and evolution: Mechanisms, processes and implications. *Philosophical Transactions of the Royal Society B: Biological Sciences, 364,* 1169–1177.

Bergstrom, C. T., & Rosvall, M. (2011). The transmission sense of information. *Biology and Philosophy, 26,* 159–176.

Bonduriansky, R. (2012). Rethinking heredity, again. *Trends in Ecology and Evolution, 27,* 330–336.

Bonduriansky, R., & Day, T. (2009). Nongenetic inheritance and its evolutionary implications. *Annual Review of Ecology, Evolution, and Systematics, 40,* 103–125.

Boyd, R., & Richerson, P. J. (1985). *Culture and the evolutionary process.* Chicago: University of Chicago Press.

Brown, G.P., & Shine, R. (2007). Like mother, like daughter: Inheritance of nest-site location in snakes. *Biology Letters, 3,* 131–133.

Cavalli-Sforza, L. L., & Feldman, M. W. (1981). *Cultural transmission and evolution: A quantitative approach.* Princeton, NJ: Princeton University Press.

Cheverud, J. M. (1984). Evolution by kin selection: A quantitative genetic model illustrated by maternal performance in mice. *Evolution, 38,* 766–777.

Cheverud, J. M., & Wolf, J. B. 2009. The genetics and evolutionary consequences of maternal effects. In D. Maestripieri & J. M. Mateo (Eds.), *Maternal effects in mammals* (pp. 11–37). Chicago, IL: Chicago University Press.

Danchin, É., Charmantier, A., Champagne, F. A., Mesoudi, A., Pujol, B., & Blanchet, S. (2011). Beyond DNA: Integrating inclusive inheritance into an extended theory of evolution. *Nature Reviews Genetics, 12,* 475–486.

Danchin, E., & Wagner, R. H. (2010). Inclusive heritability: Combining genetic and non-genetic information to study animal behavior and culture. *Oikos, 119,* 210–218.

Darwin, C. (1875). *Variation of animals and plants under domestication* (2 Vols.). London, UK: John Murray.

Dawkins, R. (1982). *The extended phenotype: The long reach of the gene.* Oxford, UK: Oxford University Press.

Dawkins, R. (2004). Extended phenotype—but not too extended. A reply to Laland, Turner, and Jablonka. *Biology and Philosophy, 19,* 377–396.

Day, T., & Bonduriansky, R. (2011). A unified approach to the evolutionary consequences of genetic and nongenetic inheritance. *American Naturalist, 178,* E18–E36.

De Vries, H. (1904). *Species and varieties: Their origin by mutation.* Chicago, IL: Open Court.

Dickerson, G. E. (1947). Composition of hog carcasses as influenced by heritable differences in rate and economy of gain. *Iowa Agricultural Experiment Station Research Bulletin, 354,* 492–524.

Dickins, T. E., & Dickins, B. J. A. (2008). Mother nature's tolerant ways: Why non-genetic inheritance has nothing to do with evolution. *New Ideas in Psychology, 26,* 41–54.

Dickins, T. E., & Rahman. (2012). The extended evolutionary synthesis and the role of soft inheritance in evolution. *Proceedings of the Royal Society B: Biological Sciences, 279,* 2913–2921.

Dobzhansky, T. (1935). Maternal effect as a cause of the difference between the reciprocal crosses in Drosophila psuedoobscura. *Proceedings of the National Academy of Sciences USA, 21,* 443–446.

Dobzhansky, T. (1937). *Genetics and the origin of species.* New York, NY: Columbia University Press.

Dye, F. J. (2012). *Dictionary of developmental biology and embryology* (2nd ed.). Hoboken, NJ: Wiley-Blackwell.

East, E. M. (1934a). The nucleus-plasma problem. *American Naturalist, 68,* 289–303.

East, E. M. (1934b). The nucleus-plasma problem. II. *American Naturalist, 68,* 402–439.

English, S., Pen, I. R., Shea, N., & Uller, T. (2015).The adaptive value of non-genetic inheritance in plants and insects. *PLoS One, 10*, e0116996.

Falconer, D. S., & Mackay, T. F. C. (1996). *Introduction to quantitative genetics.* Harlow, UK: Pearson Education.

Fisher, R. A. (1930). *The genetical theory of natural selection.* Oxford, UK: Clarendon Press.

Francis, D., Diorio, J., Liu, D., & Meaney, M. J. (1999). Nongenomic transmission across generations of maternal behaviour and stress responses in the rat. *Science, 286*, 1155–1158.

Frank, S. A. (2009). Natural selection maximizes Fisher information. *Journal of Evolutionary Biology, 22*, 231–244.

Freeberg T. M. (2000). Culture and courtship in vertebrates: A review of social learning and transmission of courtship systems and mating patterns. *Behavioural Processes, 51*, 177–192.

Furrow, R. E., Christiansen, F. B., & Feldman, M. W. (2011). Environment-sensitive epigenetics and the heritability of complex diseases. *Genetics, 189*, 1377–1387.

Galton, F. (1876). A theory of heredity. *Journal of the Anthropological Institute, 5*, 329–348.

Geoghegan, J. L., & Spencer, H. G. (2013). Exploring epiallele stability in a population-epigenetic model. *Theoretical Population Biology, 83*, 136–144

Gilbert, S. F. (2010). *Developmental biology* (9th ed.). Sunderland, MA: Sinauer Associates.

Gimelfarb, A. (1986). Offspring-parent genotypic regression: How linear is it? *Biometrics, 42*, 67–71.

Godfrey-Smith, P. (2000). Explanatory symmetries, preformation, and developmental systems theory. *Philosophy of Science, 67*, S322–S331.

Godfrey-Smith, P. (2009). *Darwinian populations and natural selection.* New York, NY: Oxford University Press.

Gottlieb, G. (1992). *Individual development and evolution: The genesis of novel behavior.* New York, NY: Oxford University Press.

Gottlieb, G. (1997). *Synthesizing nature-nurture: Prenatal roots of instinctive behaviour.* Mahwah, NJ: Erlbaum.

Griffiths, P. E., & Gray, R. D. (1994). Developmental systems and evolutionary explanation. *Journal of Philosophy, 91*, 277–304.

Griffiths, P. E., & Gray, R. D. (2001). Darwinism and developmental systems. In S. Oyama, P. E. Griffiths, & R. D. Gray (Eds.), *Cycles of contingency: Developmental systems and evolution* (pp. 195–218). Cambridge, MA: MIT Press.

Griffiths, P., & Stotz, K. (2013). *Genetics and philosophy. An introduction.* New York: Cambridge University Press.

Grafen, A. (1984). Natural selection, kin selection and group selection. In J. R. Krebs & N. B. Davies (Eds.), *Behavioural ecology: An evolutionary approach* (pp. 62–84). Oxford, UK: Oxford University Press.

Grafen, A. (1985). A geometric view of relatedness. *Oxford Surveys in Evolutionary Biology, 2,* 28–89.

Hadfield, J. (2012). The quantitative genetic theory of parental effects. In N. Royle, P. Smiseth, & M. Kölliker (Eds.), *Evolution of parental care* (pp. 268–284). Oxford, UK: Oxford University Press.

Haig, D. (2007). Weismann rules! OK? Epigenetics and the Lamarckian temptation. *Biology and Philosophy, 22,* 415–428.

Harrison, X. A., Tregenza, T., Inger, R., Colhoun, K., Dawson, D. A., Gudmundsson, G. A., & Hodgson, D. J. (2008). Cultural inheritance drives site fidelity and migratory connectivity in a long-distance migrant. *Molecular Ecology, 19,* 5484–5496.

Helanterä, H., & Uller, T. (2010). The Price equation and extended inheritance. *Philosophy and Theory in Biology, 2,* 1–17.

Heywood J. S. (2005). An exact form of the breeder's equation for the evolution of a quantitative trait under natural selection. *Evolution, 59,* 2287–2298.

Hood, K. E., Tucker Halpern, C., Greenberg, G., & Lerner, R. M. (2010). *Handbook of developmental science, behaviour and genetics.* Chicester, UK: Wiley-Blackwell.

Jablonka, E., & Lamb, M. (1995). *Epigenetic inheritance and evolution: The Lamarckian dimension.* New York, NY: Oxford University Press.

Jablonka, E., & Lamb, M. (2005). *Evolution in four dimensions: Genetic, epigenetic, behavioral and symbolic variation in the history of life.* Cambridge, MA: MIT Press.

Jablonka, E., & Lamb, M. (2014). *Evolution in four dimensions: Genetic, epigenetic, behavioral and symbolic variation in the history of life* (2nd ed.). Cambridge, MA: MIT Press.

Jablonka, E., & Raz, G. (2009). Transgenerational epigenetic inheritance: Prevalence, mechanisms, and implications for the study of heredity and evolution. *Quarterly Review of Biology, 84,* 131–176.

Jablonka, E. (2002). Information: Its interpretation, its inheritance, and its sharing. *Philosophy of Science, 69,* 578–605.

Jablonka E. (2012a). Epigenetic variations in heredity and evolution. *Clinical Pharmacology & Therapeutics, 92,* 683–688.

Jablonka E. (2012b). Epigenetic inheritance and plasticity: The responsive germline. *Progress in Biophysics and Molecular Biology.* Online early.

Johannes, F., Porcher, E., Teixeira, F. K., Saliba-Colombani, V., Simon, M., Agier, N., . . . Colot, V. (2008). Assessing the impact of transgenerational epigenetic variation on complex traits. *PLos Genetics 5.*

Kimura, M. (1961). Natural selection as the process of accumulating genetic information in adaptive evolution. *Genetics Research*, 2, 127–140.

Kirkpatrick, M., & Lande, R. (1989). The evolution of maternal characters. *Evolution*, 43, 485–503.

Laland, K. N., Odling-Smee, F. J., & Feldman, M. W. (1996). The evolutionary consequences of niche construction: a theoretical investigation using two-locus theory. *Journal of Evolutionary Biology*, 9, 293–316.

Laland, K. N., Odling-Smee, F. J., & Feldman, M. W. (1999). Evolutionary consequences of niche construction and their implications for ecology. *Proceedings of the National Academy of Sciences USA*, 96, 10242–10247.

Laland, K. N., & O'Brien, M. J. (2011). Cultural niche construction: An introduction. *Biology Theory*, 6, 191–202.

Laland, K. N., Sterelny, K., Odling-Smee, J., Hoppitt, W., & Uller, T. (2011). Cause and effect in biology revisited: Is Mayr's proximate-ultimate dichotomy still useful? *Science*, 334, 1512–1516.

Laland, K. N., Odling-Smee, J., Hoppitt, W., & Uller, T. (2013). More on how and why: Cause and effect in biology revisited. *Biology and Philosophy*, 28, 719–745.

Laland, K. L., Uller, T., Feldman, M., Sterelny, K., Müller, G. B., Moczek, A., . . . Odling-Smee, J. (2015). The extended evolutionary synthesis: its structure, core assumptions, and predictions. *Proceedings of the Royal Society of London*, 282, 20151019

Lande, R. (2009). Adaptation to an extraordinary environment by evolution of phenotypic plasticity and genetic assimilation. *Journal of Evolutionary Biology*, 22, 1435–1446.

Lennox, J. G. (2000). *Aristotle's philosophy of biology: Studies in the origins of life science.* New York, NY: Cambridge University Press.

Leimar, O., & McNamara, J. M. (2015). The evolution of transgenerational integration of information in heterogeneous environments. *American Naturalist*, 185, E55–E69.

Levins, R., & Lewontin R. C. (1985). *The dialectical biologist.* Cambridge, MA: Harvard University Press.

Lewontin R. C. (1970). The units of selection. *Annual Review of Ecology, Evolution, and Systematics*, 1, 1–18.

Lewontin R. C. (1985). Adaptation. In R. Levins, & R. C. Lewontin (Eds.), *The Dialectical Biologist* (pp. 65–84). Cambridge, MA: Harvard University Press.

Lewontin, R. C., Moore, J. A., Provine, W. B., & Wallace, B. (2003). *Dobzhansky's genetics of natural populations. I-XLIII.* New York, NY: Columbia University Press.

Li, L., Zheng, P., & Dean, J. (2010). Maternal control of early mouse development. *Development*, 137, 859–870.

Lim J. P., & Brunet, A. (2013). Bridging the transgenerational gap with epigenetic memory. *Trends in Genetics*, 29, 176–186.

Lynch M., & Walsh J. B. (2013). *Evolution and selection of quantitative traits: I. Foundations.* http://nitro.biosci.arizona.edu/zbook/NewVolume_2/newvol2.html

Maestripieri, D., & Mateo, J. M. (2009). *Maternal effects in mammals.* Chicago, UK: Chicago University Press.

Mameli, M. (2004). Nongenetic selection and nongenetic inheritance. *British Journal of Philosophy of Science, 55,* 35–71.

Mameli M. (2005). The inheritance of features. *Biology & Philosophy, 20,* 365–399.

Mameli, M. (2007). Genes, environments and concepts of inheritance. In P. Carruthers, S. Laurence, & S. Stich (Eds.), *Innateness and the mind: Foundations and the future* (pp. 37–54). Oxford: Oxford University Press.

Maynard Smith, J. (2000). The concept of information in biology. *Philosophy of Science, 67,* 177–194.

Maynard Smith, J., & Szathmáry, E. (1995). *The major transitions on evolution.* New York: Oxford University Press.

Mayr, E. (1961). Cause and effect in biology. *Science, 134,* 1501–1506.

Mayr, E. (1982). *The growth of biological thought: Diversity, evolution and inheritance.* Cambridge, MA: Belknap Press of Harvard University Press.

McAdam, A. G., & Boutin, S. (2004). Maternal effects and the response to selection in red squirrels. *Proceedings of the Royal Society of London, Series B., 271,* 75–79.

Mesoudi, A., Blanchet, S., Charmantier, A., Danchin, E., Fogarty, L., Jablonka, E., . . . Pujol, B. (2013). Is non-genetic inheritance just a proximate mechanism? A corroboration of the extended evolutionary synthesis. *Biology Theory, 7,* 189–195.

Moore, A. J., Brodie E. D., III, & Wolf, J. B. (1997). Interacting phenotypes and the evolutionary process. I. Direct and indirect genetic effects of social interactions. *Evolution, 51,* 1352–1362.

Moore, C. L. (1995). Maternal contributions to mammalian reproductive development and the divergence of males and females. *Advances in the Study of Behavior, 24.*

Moore, D. S. (2013). Current thinking about nature and nurture. In K. Kampourakis (ed) *The philosophy of biology: A companion for educators* (pp. 629–652). Dordrecht, Netherlands: Springer Verlag.

Morgan, T. H., Sturtevant, A. H., Muller, H. J., & Bridges, C. B. (1915). *The mechanisms of Mendelian heredity.* New York, NY: Holt.

Mousseau, T. A., & Fox, C. W. (1998). *Maternal effects as adaptations.* New York, NY: Oxford University Press.

Müller-Wille, S., & Rheinberger, H.-J. (2007). *Heredity produced: At the crossroads of biology, politics, and culture, 1500–1870.* Cambridge, MA: MIT Press.

Müller-Wille, S., & Rheinberger, H.-J. (2012). *A cultural history of heredity.* Chicago, UK: Chicago University Press.

Nijhout, H. F. (1990). Metaphors and the role of genes in development. *BioEssays, 12,* 441–446.

Odling-Smee, F. J., Laland, K. N., & Feldman, M. W. (2003). *Niche construction: The neglected process in evolution*. Princeton, NJ: Princeton University Press.

Odling-Smee, F. J. (2010). Niche inheritance. In M. Pigliucci & G. B. Müller (Eds.), *Evolution: The extended synthesis* (pp. 175–208). Cambridge, MA: MIT Press.

Oyama, S. (1985). *The ontogeny of information: Developmental systems and evolution* (2nd ed., 2000). Durham, NC: Duke University Press.

Oyama, S. (2000). *A systems view of the biology-culture divide*. Durham, NC: Duke University Press.

Oyama, S., Griffiths, P. E., & Gray, R. D. (2001). *Cycles of contingency: Developmental systems and evolution*. Cambridge, MA: MIT Press.

Pelegri, F. (2003). Maternal factors in zebrafish development. *Developmental Dynamics, 228*, 535–554

Pigliucci, M., & Müller, G. B. (2010). *Evolution: The extended synthesis*. Cambridge, MA: MIT Press.

Price G. R. (1970). Selection and covariance. *Nature, 227*, 520–521.

Price G. R. (1972). Extension of covariance selection mathematics. *Annals of Human Genetics, 35*, 485–490.

Provine, W. B. (1971). *The origins of population genetics*. Chicago, IL: University of Chicago Press.

Provine, W. B. (1986). *Sewall Wright and evolutionary biology*. Chicago, IL: University of Chicago Press.

Rando, O. J., & Verstrepen, K. J. (2007). Timescales of genetic and epigenetic inheritance. *Cell, 128*, 655–668.

Rice, S. H. (2004). *Evolutionary theory: Mathematical and conceptual foundations*. Sunderland, MA: Sinauer Associates.

Rice, S. H. (2008). A stochastic version of the Price equation reveals the interplay of deterministic and stochastic processes in evolution. *BMC Evolutionary Biology, 8*, 262.

Rice, S. H. (2012). The place of development in mathematical evolutionary theory. *Journal of Experimental Zoology Part B: Molecular and Developmental Evolution, 318*, 480–488.

Richards, E. J. (2006). Inherited epigenetic variation: Revisiting soft inheritance. *Nature Reviews Genetics, 7*, 395–401.

Riddle, O. (1909). Our knowledge of melanin color formation and its bearing on the Mendelian description of heredity. *Biological Bulletin Wood's Hole, 16*, 316–351.

Rivoire, O., & Leibler, S. (2014). A model for the generation and transmission of variations in evolution. *Proceedings of the National Academy of Sciences of the United States of America, 111*, E1940–E1949.

Rivoire, O., & Leibler, S. (2014). A model for the generation and transmission of variations in evolution. *Proceedings of the National Academy of Sciences of the United States of America, 111*, E1940–E1949.

Robinson, S. R., & Méndez-Gallardo, V. (2010). Amniotic fluid as an extended milieu intérieur. In K. E. Hood, C. T. Halpern, G. Greenberg, & R. M. Lerner (Eds.), *Developmental science, behaviour, and genetics* (pp. 234–284). Cambridge, MA: Wiley-Blackwell.

Romanes, G. J. (1895). *Darwin, and after Darwin: Volume 2. Post-Darwinian questions—heredity and utility*. London, UK: Longmans, Green & Co.

Roux, F., Colome-Tatche, M., Edelist, C., Wardenaar, R., Guerche, P., Hospital, F., . . . Johannes, F. (2010). Genome-wide epigenetic perturbation jump-starts patterns of heritable variation found in nature. *Genetics, 188*, 1015–U1402.

Sapp, J. (1987). *Beyond the gene: Cytoplasmic inheritance and the struggle for authority in genetics*. New York, NY: Oxford University Press.

Schlichting, C. D., & Pigliucci, M. (1998). *Phenotypic evolution: A reaction norm perspective*. Cambridge, MA: Sinauer Press.

Schmalhausen, I. I. (1949). *Factors of evolution: The theory of stabilizing selection*. Philadelphia, PA: Blakeston.

Schuett, W., Dall, S. R. X., Wilson, A. J., & Royle, N. J. (2013). Environmental transmission of a personality trait: Foster parent exploration behaviour predicts offspring exploration behaviour in zebra finches. *Biology Letters, 9*, 20130120.

Scott-Phillips, T. C., Dickins, T. E., & West, S. A. (2011). Evolutionary theory and the ultimate-proximate distinction in the human behavioral sciences. *Perspectives on Psychological Science, 6*, 38–47.

Shea, N. (2007). Representation in the genome and in other inheritance systems. *Biology and Philosophy, 22*, 313–331.

Shea, N. (2009). Imitation as an inheritance system. *Philosophical Transactions of the Royal Society of London B, 3364*, 2429–2443.

Shea, N. (2012). Two modes of transgenerational information transmission. In K. Sterelny, R. Joyce, B. Calcott, & B. Fraser (Eds.), *Cooperation and its evolution* (pp. 289–312). Cambridge, MA: MIT Press.

Shea, N. (2011). Developmental systems theory formulated as a claim about inherited representations. *Philosophy of Science, 78*, 60–82.

Shea, N., Pen, I., & Uller, T. (2011). Three epigenetic information channels and their different roles in evolution. *Journal of Evolutionary Biology, 24*, 1178–1187.

Silver, M., & Di Paolo, E. (2006). Spatial effects favour the evolution of niche construction. *Theoretical Population Biology, 70*, 387–400.

Slagsvold, T., & Wiebe, K. L. (2011). Social learning in birds and its role in shaping a foraging niche. *Philosophical Transactions of the Royal Society B, 366*, 969–977.

Stern, D. L. (2009). *Evolution, development and the predictable genome*. Greenwood Village, CO: Roberts & Company.

Sun, Y. H., Chen, S. P., Wang, Y. P., Hu, W., & Zhu, Z. Y. (2005). Cytoplasmic impact on cross-genus cloned fish derived from transgenic common carp (*Cyprinus carpio*) nuclei and goldfish (*Carassius auratus*) enucleated eggs. *Biology* of *Reproduction*, *72*, 510–515.

Tal, O., Kisdi, E., & Jablonka, E. (2010). Epigenetic contribution to covariance between relatives. *Genetics*, *184*, 1037–1050.

Toyama, K. (1913). Maternal inheritance and mendelism. *Journal of Genetics*, *2*, 351–404.

Tschirren, B., Fitze, P. S., & Richner, H. (2007). Maternal modulation of natal dispersal in a passerine bird: An adaptive strategy to cope with parasitism? *American Naturalist*, *169*, 87–93.

Trivers, R. L. (1974). Parent-offspring conflict. *American Zoologist*, *14*, 249–264.

Uller, T. (2013). Non-genetic inheritance and evolution. In K. Kampourakis (Ed.), *Philosophy of biology: A companion for educators*. Dordrecht: Springer Verlag.

Uller, T. (2012). Parental effects in development and evolution. In N. Royle, P. Smiseth, & M. Kölliker (Eds.), *Evolution of parental care* (pp. 247–266). New York, NY: Oxford University Press.

Uller, T., & Pen, I. (2011). A theoretical model of the evolution of maternal effects under parent-offspring conflict. *Evolution*, *65*, 2075–2084.

Uller, T., & Helanterä, H. (2013). Integrating non-genetic inheritance in evolutionary theory: A primer. *Non-Genetic Inheritance*, *1*, 27–32.

Uller, T., Nakagawa, S., & English, S. (2013). Weak evidence for anticipatory parental effects in plants and animals. *Journal of Evolutionary Biology*, *26*, 2161–2170.

Uller, T., English, S., & Pen, I. (2015). When does natural selection favour incomplete epigenetic resetting in germ cells? *Proceedings of the Royal Society of London B*, *282*, 20150682.

Waddington, C. H. (1957). *The strategy of the genes: A discussion of some aspects of theoretical biology*. London, UK: George Allen & Unwin.

Weaver, I., Cervoni, N., Champagne, F. A., D'Alessio, A., Sharma, S., Seckl, J. R., ... Meaney, M. J. (2004). Epigenetic programming by maternal behaviour, *Nature Neuroscience*, *7*, 847–854.

Weismann, A. (1893). *The germ-plasm: A theory of heredity*. London: Walter Scott.

West, S. A., & Gardner, A. (2013). Adaptation and inclusive fitness. *Current Biology*, *23*, R577–R584.

West-Eberhard, M. J. (2003). *Developmental plasticity and evolution*. New York, NY: Oxford University Press.

Willham, R. L. (1963). The covariance between relatives for characters composed of components contributed by related individuals. *Biometrics*, *19*, 18–27.

Wilson, A. J., Coltman, D. W., Pemberton, J. M., Overall, A. D. J., Byrne, K. A., & Kruuk, L. E. B. (2005). Maternal genetic effects set the potential for evolution in a free-living vertebrate population. *Journal of Evolutionary Biology*, *18*, 405–414.

Wolf, J. B., & Wade, M. J. (2009). What are maternal effects (and what are they not)? *Philosophical Transactions of the Royal Society B: Biological Sciences*, *364*, 1107–1115.

11 SERIAL HOMOLOGY AS A CHALLENGE TO EVOLUTIONARY THEORY

THE REPEATED PARTS OF ORGANISMS FROM IDEALISTIC MORPHOLOGY TO EVO-DEVO

Stéphane Schmitt

The question of interspecific similarities has been at the center of transformist theories since their emergence in the 18th century. Indeed, the existence of similar structures in organisms—independent of their function—belonging to different species was one of the principal elements of proof in favor of a common phylogenetic origin for these species; in turn, phylogeny offered a coherent explanatory framework for these similarities. This is why the concept of homology, which describes this genre of morphological relationships, is for the most part associated with the notion of evolution in current biology—in spite of the numerous problems there are in defining it, and the disagreements that surround it (see, e.g., Hall, 1994; Minelli & Fusco, 2013; and the vast bibliography given by these authors). However, its history is much older than that of the evolutionary sciences; and it was within a still largely nontransformist context in the middle of the 19th century that the English naturalist Richard Owen, bringing order to a fluctuating terminology, tried to define a "homologue" as "the same organ in different animals under every variety of form and function," in opposition to an "analogue," which was "a part or organ in one animal which has the same function as another part or organ in a different animal" (Owen, 1843, pp. 379, 374; see also Owen, 1848, p. 7).

This definition by Owen was the culmination of a long tradition of anatomical and embryological research concerning the structural similarities between species, whose roots went back to at least Aristotle, but which had rapidly developed since the end of the 18th century—particularly in Germany, in the context of *Naturphilosophie*;

and France, through the work of Étienne Geoffroy Saint-Hilaire (see e.g., Appel, 1987; Balan, 1979; Cole, 1944; Nyhart, 1995; Russell, 1916; Schmitt, 2004, 2006). Owen took on the old idea of "unity of plan": According to him, all of the real animal species (at least within the large groups, such as vertebrates) were variations on a fundamental and ideal form—"the archetype" (Rupke, 2009; Sloan, 1992). And thus he defined two types of homologies: "special homology," which is the relationship between two parts or organs corresponding to two real species; and "general homology," which is the link between the organ of a real species and the corresponding organ in the archetype (Owen, 1848, p. 7). A little later, when the theory of evolution became widespread, these concepts were very easily reinterpreted within this new context: The archetype became the common ancestor of the current species; and special and general homologies, mixed together from this time onward, were explained by the genealogical relationships between these species (see Coleman, 1976; Desmond, 1982; Di Gregorio, 1995; Richards, 1992).

But Owen also defined a third type of homology, "serial homology," which—contrary to the preceding two—was not concerned with corresponding parts between different organisms, but similar parts *within* the same organism. This third type of homology was based on the idea that the archetype was composed of repeated parts, or segments, that were essentially identical but possibly modified. In Owen's mind, serial homology was not fundamentally different from the other types of homology insofar as, for all purposes, it was concerned with ideal relationships that were nongenealogical. But from the moment that the notion of homology became intimately associated with that of phylogeny, the case of serial homology created problems that were altogether specific.

In the present chapter, we present some of the elements of the history of this problem of similar parts within one same organism, before and after Owen. Notably, we show how the notion of serial homology, after having emerged at the end of the 18th century within an essentialist context, managed over the course of the following centuries, and leading up to the present, to retain a central position in the life sciences—in spite of the major transformations that science underwent (such as the spread of transformist theories; and the emergence of new fields and new disciplines such as experimental embryology, genetics, and molecular biology).

11.1. Repeated Parts of Organisms in an Essentialist Context

If the existence of similar and repeated structures within one same organism, with some modifications (for example, the vertebra and the ribs of a vertebrate; or the segments of an annelid or an arthropod), is a clear and immediate given, it was only toward the end of the 18th century that the idea emerged that this particularity could be the object of scientific investigation. From the outset, this question was narrowly correlated to the examination of similarities between the parts of different organisms.

One of the first scientists to have proposed these types of studies was the French physician and anatomist Félix Vicq d'Azyr. This author was one of the principal promoters of comparative anatomy starting from the 1770s. He not only developed the concept of unity of plan between different animal species but also formed a veritable program of research about the anatomical correspondences between these species (Schmitt, 2009). Furthermore, he extended this notion of a structural uniformity to the comparison of parts within a single organism. He wrote:

> If comparative Anatomy has done such important services, could one not institute another one, which would only deal with the connections of the parts of the same individual? Would these new considerations not throw more light on the uses and the mechanism of the parts composing it? Would it be not possible that they would make visible amazing analogies? And if the parts that differ the most in appearance resemble each other in essence, could one not conclude with more certainty that there is only a single set, a single essential form, and that one recognizes everywhere this fecundity of Nature which seems to have given to all beings two characteristics not at all contradictory, namely, the constancy of the type and the variety in the modifications. (Vicq d'Azyr, 1778, p. 254)

He surely considered this kind of study as "a pure curiosity" (Vicq d'Azyr, 1778, p. 270), but his work in comparing the anatomy of arms and legs showed that very concrete research could be done on this subject.

A dozen years later, the German poet, writer, and scientist Johann Wolfgang von Goethe developed similar viewpoints; and thus he himself contributed to the study of similarity between parts—between different organisms as well as within one same organism. In the morphology of living beings as in literature, he attempted to find a unique "type" behind the diversity of forms, and to place this concept at the center of research in the life sciences (see, e.g., Amrine et al., 1987; Brauning-Oktavio, 1956, 1959; Peyer, 1950; Richards, 2002, pp. 325–502; Schmitt, 2004, pp. 91–130; Wells, 1967). For example, this principle led him to the discovery of the human intermaxillary bone: This bone was known to exist in most mammals, but not in man—in whose case it is fused to its neighboring bones and thus difficult to recognize. Goethe showed that this bone did indeed exist in human beings; and in this way, he considered himself to have contributed in proving the existence of an osteological type that was common to all animals (Goethe, 1987 [1795–1796], pp. 225–282).

But like Vicq d'Azyr, he also sought to show that the parts of one same organism could be brought back to a common ideal type. He applied this principle in two cases, concerning plants and animals respectively. In the first case, in his *Versuch die Metamorphose der Pflanzen zu erklaren* (1790), he made great efforts to demonstrate that all of the vegetable organs, notably the floral parts (sepals, petals, stamens, and pistils), were fundamentally similar to one organ-type: the leaf, from which they were derived by an (ideal) process of "normal or progressive metamorphosis" (Goethe 1987 [1790], pp. 109–151; see, e.g., Guédès, 1969). According to him, many aberrations in plants, like the replacement of stamens by petals in double flowers (whether natural or cultivated), could be explained by defects in this metamorphosis which became "abnormal" or "regressive."

The second case in which Goethe was particularly interested was that of the organization of the skull in vertebrates. While walking on the dunes of the Lido, in 1790, he found a sheep skull whose bones were slightly dislocated in such a way that he thought he recognized a number of vertebrae that were hypertrophied, modified, and fused together (Goethe, 1987, p. 928). A few years later, this revelation led him to the formulation of the vertebral theory of the skull (see, e.g., Goethe, 1987, p. 504): According to this theory, the heads of vertebrates, as well as their trunk, were essentially composed of six vertebrae that were adjoined (with their appendices, namely, the cephalic equivalent of the ribs) and transformed in such a way as to be able to welcome the skull and the sense organs. Consequently, not only were vertebrates built according to a unique type but also this type was itself entirely constituted, from tail to head, of standard vertebral unities that were repeated and arranged serially.

This vertebral theory found considerable success in Germany, notably in the context of *Naturphilosophie*. This intellectual current, initiated among others by the philosopher Schelling in the final years of the 18th century, inspired many scientists in the course of the following decades. Indeed, it promoted a certain number of themes that enabled the rise in the sciences of theoretical developments— as well as concrete research programs. Lorenz Oken, a disciple of Schelling, based his entire conception of nature on general Schellingian principles. One of his preferred themes was the fact that all natural objects and phenomena resulted from constructive tensions between opposed elements (Oken, 1805, 1809–1811; see Breidbach, Fliedner, & Ries, 2001; Breidbach & Ghiselin, 2002; Schmitt, 2004, pp. 35–86). This process was not only always the same, but was also iterative— since each phenomenon, once created, gave rise to a new tension between these opposed components, a tension that resolved itself in a new process or a new object that was superior to the preceding one. It resulted in the existence of a complex network of similarities and correspondences within and between all natural objects and processes—while also creating the idea of a hierarchy between these objects and processes.

Concretely, Oken, like other *Naturphilosophen*, saw in embryonic development a repetition of the "great chain of nature" (that is to say, the notion of an ascending succession of living species, from the lowest to the highest, so that every natural production could be defined both by its hierarchical level and its relationship to the others). This idea of repetition did not have a truly transformist connotation because the movement of elevation within the chain of beings had an ideal character; but several years later, this idea naturally led to the law of recapitulation ("ontogeny recapitulates phylogeny") which would be very successful from the 1860s onward.

Correlated to the notion of type, the notion of repetition permitted the establishment of morphological correspondences between different living species. But Oken also applied this notion to the parts of a single organism. Independent of Goethe, it seems, he proposed his own version of the vertebral theory of the skull (Oken, 1807). According to him, the head of vertebrates was fundamentally segmented, like the rest of their bodies, and formed by three (or four: see Oken, 1829) "vertebrae" (in the broad sense of the word), which were composed of the bones and associated organs (notably, the nervous and sensory ones). More generally, Oken established an entire network of correspondence between cephalic parts and other parts of the organism: The mouth, with its sense of taste, corresponded to the abdomen and its digestive functions; the region of the nose and its sense of smell corresponded to the thorax and respiration; as for the upper part of the head, it was specifically cephalic and possessed the sharpest of the senses (hearing and sight); and thus the head appeared as a refined recapitulation of the body as a whole. At the same time, the different parts of the body of a higher animal, each characterized by a particular function, also corresponded with inferior animal groups—which were themselves each characterized by the predominance of the same function. Thus, the head of a mammal was conceived of as the ennobled recapitulation of the body, while the body itself was a recapitulation of the animal world—not only in its adult morphology but also in its embryonic development, which repeated (and carried further) the development of inferior species.

This version of the vertebral theory of the skull was taken up again and modified over the course of the following years by two other zoologists influenced by *Naturphilosophie*, Johann Baptist von Spix (1815) and Ludwig Heinrich Bojanus (1818, 1819), and then by a great number of authors during the first half of the 19th century (see Peyer, 1950). For example, Carl Gustav Carus—who also adhered to

Schelling's ideas and was very influential (from a scientific as well as institutional standpoint) in Germany until his death in 1869—argued that the morphology of all of the animal species derived from one and the same fundamental form: the vertebra (Carus, 1828). The entire body of each animal could thus be divided into a succession of more or less modified vertebrae; and in this manner, it was possible to reconstitute a segmented archetype.

This school of thought was extremely active in the decades before Darwin. It played a considerable role in the history of the morphological sciences. Indeed, beyond the Schellingian system itself, this research stimulated theoretical reflection on similarities—not only between species but also between the parts of a single organism. Furthermore, these scientists did not confine themselves just to theoretical considerations: Contrary to widespread opinion, the *Naturphilosophen* rejected pure speculation and greatly insisted on the necessity of founding their ideas on observations. Numerous programs of research concerning animal anatomy reflected each other in their efforts to concretely determine the structural correspondences between organs—in regard to different animals as well as within one same animal. For example, to demonstrate the vertebral theory of the skull, the anatomists mentioned earlier looked for signs of segmentation in the heads of vertebrates not only in regard to bones, but also within the nervous system (the posterior brain and the cranial nerves) and other organs.

In a different context, many French scientists, following the work of Vicq d'Azyr, also became interested in anatomical similarities after 1800. The most engaged among them was Étienne Geoffroy Saint-Hilaire, who attempted to show the existence of a "unity of composition" between all animals (see in particular Geoffroy Saint-Hilaire, 1818–1822, 1830; on his works and thought, see, e.g., Cahn, 1962; Geoffroy Saint-Hilaire, 1847; Le Guyader, 2004). For example, according to him, the structural organization of insects could be identified as that of vertebrates: The only difference was that insects were "reversed" dorsoventrally (with a ventral nervous chain as opposed to the dorsal one in true vertebrates) and that their "vertebrae" (that is, their segmented exoskeleton) were on the exterior—rather than the interior—of their organs. Likewise, he tried to show that the anatomy of mollusks was fundamentally identical to that of vertebrates. This put him in opposition to Georges Cuvier, who considered that the four great structural plans of the animal kingdom (those "embranchements" being "vertébrés," "mollusques," "articulés," and "radiés") were absolutely isolated from each other (see especially Cuvier, 1812). There was a live public debate between the two scientists in 1830; it ended without a winner or loser but contributed in spreading the ideas of Geoffroy even further (Appel, 1987).

Beyond these general ideas, Geoffroy played a large role in the history of research concerning anatomical similarities. In particular, he tried to define precise criteria by which one could recognize them. And thus he proposed the "principle of connections," according to which organs can undergo all sorts of transformations, alterations, or atrophies—without changing their arrangement in relation to each other. Consequently, it is possible to establish correspondences between two organs from different species, no matter how dissimilar they are, by comparing the organs that surround them. Another important principle of Geoffroy's was the "law of balance of organs," according to which the organism would have at its disposal a sort of fixed "budget" that was linked to its nutritional intake during its course of growth—meaning that if an organ were more developed, it would be at the expense of one or a number of others. These two principles had already been sketched by scientists in the past, such as Vicq d'Azyr, but Geoffroy Saint-Hilaire formalized them with precision.

Terminologically speaking, Geoffroy defined "analogues" as organs from two different species that were fundamentally identical (this is what Owen would call "special homologues" in 1848). On the other hand, like his contemporary German anatomists, he reserved the adjective "homologue" for similar parts within one same organism (Owen's "serial homologues"). In fact, while not adhering to the entire system of *Naturphilosophie*, Geoffroy allowed for the existence of similarities between the parts of an animal. He notably recognized the vertebral theory of the skull, of which he proposed his own version (Geoffroy Saint-Hilaire, 1824), as well as other correspondences (between the various sensory organs, for example). He wrote on the subject:

> The sensory organs are homologues, as German Philosophy would call it, meaning that they are analogous in their mode of development—if there actually does exist within them one same principle of formation, a uniform tendency to repeat themselves, to reproduce themselves in the same fashion: and this I cannot doubt; for example, it reveals itself in the construction of the eye and the nostril of a fish. Indeed, simply by looking at them, one sees that a carp's nostril is a hollow globe whose bottom is wrinkled, just like the retina at the bottom of an eyeball. (Geoffroy Saint-Hilaire, 1825)

On the other hand, work on segmentation and repeated parts multiplied in France in the first decades of the 19th century. Henri-Marie Ducrotay de Blainville (1819, 1820), Marie-Jules-César Lelorgne de Savigny (1816, vol. 1, pp. 1–38), and Pierre-André Latreille (1820) investigated the similarities both between different groups of arthropods and between different segments of one same arthropod—demonstrating, for example, that the different appendices of a crustacean or of an insect (their legs and mouthparts) were built according to one same model. Jean-Victor Audouin, applying the principles of Geoffroy Saint-Hilaire, even proposed the idea of an arthropodian archetype composed of a repetition of fundamentally identical segments each provided with a pair of appendices (Audouin, 1824).

The ideas we have just brought up have been principally concerned with the morphology of adult animals, but others specifically dealt with the development of the repeated parts of organisms. In the first decades of the 19th century, these studies were undertaken chiefly in Germany—where comparative embryology witnessed a rapid and unprecedented development (Churchill, 1991). The idea spread that embryology could reveal undetectable connections in the adult state, and that homologous parts (in Owen's sense, meaning both serial homologues and special homologues) had to have a common developmental origin. In this context, many discoveries were made that reinforced the idea of both a similarity between species and a fundamentally repetitive structural organization in animals. And thus Martin Heinrich Rathke brought to light—in regard to the embryos of mammals and birds—the existence of a succession of gill slits identical to those of adult selachians (Rathke, 1825a, 1825b). This discovery completely went with the theory of parallelism between embryonic development and the scale of nature, since it suggested that the embryos of higher vertebrates possessed characteristics of inferior species. On the other hand, it cleared the way for a whole series of embryological studies about the anterior part of the body of vertebrates; these studies not only permitted the establishment of connections between different groups but also showed that the heads of these animals were segmented like the rest of their bodies—which appeared to be a confirmation of the vertebral theory of the skull. In particular, the embryologist Emil Huschke, a disciple of Oken and a fervent *Naturphilosoph*, showed that the embryos of higher vertebrates possessed aortic arches—which represented both a developmental repetition of equivalent structures within inferior vertebrates, and a cephalic repetition in the segmentation of the trunk (Huschke, 1826, 1828, 1832).

Many similar embryological studies were undertaken in the following decades, which gave rise to important discoveries. For example, Karl Bogislaus Reichert (1837, 1838), when studying the visceral arches in vertebrates, demonstrated that the ossicles of the middle ear in mammals corresponded to some of the bones in reptile jaws. Not all of these works were as influenced by *Naturphilosophie* as that of Huschke: Thus, when Carl Ernst von Baer[1] studied the formation of aortic arches, he did not absolutely accept the notion of recapitulation (which was, on

---

1. Von Baer's role in the history of embryology is considerable. In particular, he discovered the mammal egg and many embryonic structures such as the notochord; he described the development of all the classes of vertebrates and showed the universality of the germ layer theory—which was strong evidence for epigenesis and against preformation. He also fought the law of

the contrary, very much in favor with the disciples of Schelling); but he recognized all the same the existence of a connection between the different vertebral groups and the fundamental segmentation of the head in these animals that was revealed through embryology (Baer, 1827).

In a general manner, in spite of rare criticism (Cuvier, 1835, vol. 2, pp. 711–712; Vogt, 1842), the theory of the vertebral skull was largely accepted until the end of the 1850s, and numerous arguments—anatomical as well as embryological—were gathered in its favor. Yet, within its narrowest sense, it was shaken when Thomas Huxley, after a reexamination of these arguments, showed that cephalic structures (cranial ones, in particular) could not be interpreted as modified vertebrae (Huxley, 1858). However, the idea of a fundamental segmentation in the heads of vertebrates continued to be widely accepted and even gained new vigor with the spread of the theory of evolution after 1859 (see later).

But alongside these ideas, another approach to repeated parts, founded on the colonial conception of organisms, appeared around 1830. The Montpellier physician and naturalist Alfred Moquin-Tandon, struck by the repetitive organization of segments in leeches—not only in their external morphology, but also in their internal anatomy—looked into the individuality of these segments. He ended up suggesting that certain segmented animals (such as annelids) might be essentially made up of a coalescence of simple small animals, endowed with a certain autonomy, that he called "zoonites" (Moquin-Tandon, 1827, pp. 89–91; see Perru, 2000; Schmitt, 2004, pp. 195–235). He was no doubt influenced in this by the work of contemporaries such as Adelbert von Chamisso, Savigny, Lesueur, and others, about diverse colonial organisms (such as salps or certain compound ascidians) which showed that some apparently unitary organisms were in fact made up of many individuals that had first been combined together and then had developed functional specializations (Nyhart, 2011; Winsor, 1976). In the same fashion, Moquin-Tandon thus considered that annelids possessed "two types of life": "individual life, that of each *zoonite*, and general life, that of the collective."

Antoine Dugès pursued still further these reflections on the notion of the zoonite (Dugès, 1832). He tried to demonstrate that all animals were fundamentally composed of zoonites[2]—not only the species showing an apparent external segmentation (such as annelids and arthropods), but also the vertebrates

parallelism and proposed alternative embryological laws, according to which general characteristics developed before special ones.

2. Contrary to Moquin-Tandon, Dugès did not think that a zoonite corresponded to an entire segment, but rather to the right or left half of a segment: He therefore considered, for example, that annelids and vertebrates were repetitions of pairs of zoonites (not of single ones) following an anteroposterior axis.

and animals with radial symmetry (echinoderms, jellyfish, etc.). In this respect, according to him, the only specificity of vertebrates was the large degree of both coalescence and zoonite specialization, which made them sometimes unrecognizable (in the head, for example). Dugès clearly accepted the vertebral theory of the skull: In spite of the "nearly complete fusion" of the zoonites, he noted the existence of structures, such as the cranial nerves, that obviously revealed a primitive segmentation in the cephalic region. In addition, thanks to his theory, he thought it was possible to reconcile Cuvier's position with that of Geoffroy Saint-Hilaire: According to him, all animals shared a certain structural unity (an "organic conformity"), as Geoffroy described it, since they were all made up of elementary unities that were essentially identical. But Cuvier was also correct in claiming that each large "embranchement" was characterized by a particular arrangement of these zoonites (either an arrangement according to one sole axis, or a radial one, or an irregular one), an arrangement that was entirely independent of those of the other "embranchements." Thus, only the zoonites themselves were comparable from one group to another—and not the general organization plan of animals. In addition, Dugès considered that the degree of integration of zoonites, and the division of labor between them, was a sign of the position of a group within the great chain of beings. In a very interesting manner, he established a parallel between this ladder of nature and the progress of human societies—which happened, according to him, through an ever increasing specialization of individuals that was directed toward a common good.

This theory of zoonites is thus of considerable interest in the history of conceptions of repeated parts. First, it established a link between this question and that of individuality—since each module was essentially an animal in itself, and the individual (in the ordinary sense of the word) was in reality a colony. Second, according to this theory, there was no difference between the anatomical connections between different organisms (Owen's special homology) and those within one same organism (serial homology) since—in these two cases—one was comparing autonomous "zoonites." Third, these reflections on the division of labor and on the comparison between animal structure and human societies were going to spread considerably after 1830, notably with the work of Henri Milne-Edwards (see D'Hombres, 2012; Limoges, 1994).

Thus, toward 1850, the question of repeated parts within organisms and of serial homology had taken an important position within the life sciences, and an entire conceptual arsenal was set up around the subject—as well as a methodology founded on anatomical (the principle of connections) or embryological criteria. By a large majority, although the theory of evolution of species had been discussed since the end of the 18th century, all of these ideas were part of a nontransformist context: The homologies of Owen or of *Naturphilosophen* were not conceived of as the consequence of genealogical relationships between organisms, but rather as ideal connections. Likewise, the colonial theory of Dugès had no transformist connotation: The processes of coalescence and zoonite specialization were not supposed to be *actually* produced over time. But the appearance of *On the Origin of Species* in 1859, and the widespread interest in the theory of evolution that followed, cleared the way for a reinterpretation of all of these notions (see Amundson, 1998).

11.2. The Reinterpretation of Serial Homology in Evolutionary Biology After 1859

Darwin himself, as early as 1859, insisted on the importance of all of the work accomplished within the morphological sciences in the preceding decades for the demonstration of his theory (Darwin, 1859, pp. 434–458). In fact, as we have already said, all of the anatomical and embryological similarities which had been highlighted between the animal species took on a new sense in the light of evolution and could be considered from now on as the result of a common phylogeny. Thus, the morphological sciences had no difficulty in assimilating the transformist hypothesis—which spread very rapidly in these disciplines, and brought about a redefinition of most of the concepts of typological biology. We give here some examples of the manner in which serial homology was reinterpreted within this framework.

Edwin Ray Lankester, a disciple of Darwin, gave a new definition of the concept of homology: "Without doubt the majority of evolutionists would agree that by asserting an organ A in an animal α to be homologous with an organ B in an animal β, they mean that in some common ancestor κ the organs A and B were represented by an organ C, and that α and β have inherited their organs A and B from κ" (Lankester, 1870, p. 36). However, Lankester himself was rather reluctant to use the term "homology," which he thought was too intimately associated with the "Platonic" (that is to say, idealist) viewpoints of scientists such as Owen; he thus preferred using the term "homogeny" to designate the resemblance between different animal parts due to a common phylogenetic origin. In a very subtle manner, he discussed the problems created by this notion of "homogeny," indicating its relative character: For example, the anterior limbs of all vertebrates could be qualified as "homogenous" only at a very general level (simply as "anterior limbs"); but if one considered tetrapods only, whose ancestors possess a characteristic digited limb, this homogeny could be extended into anatomical details.

But it was the case of serial homology that posed a particular problem for Lankester. If one compares, for example, the anterior and posterior limbs of a mammal, one clearly notices a conformity of structure. However, when one imagines the organism as being composed of two individuals—one represented by the trunk and legs, the other by the head and arms—"no genetic identity can be established between the fore and hind limbs" (Lankester, 1870, p. 38). Thus, with this in mind, how does one interpret the formal similarities that one notices between organs without a genealogical link—whether they are within one organism, or between different species? Lankester suggested environmental influence: "When identical or nearly similar forces, or environments, act on two or more parts of an organism which are exactly or nearly alike, the resulting modifications of the various parts will be exactly or nearly alike" (Lankester, 1870, p. 39). Otherwise said, if one imagines a common ancestor for all vertebrates that is without limbs, one can think that the action of "forces" and of identical conditions on totally undifferentiated lateral mesodermic masses led to the formation of limbs (which were very rudimentary in the beginning) of an approximately identical structure. Thus, in the end, the posterior and anterior limbs do not have direct genealogical links, but show comparable anatomies all the same. This explanation can be easily transposed to the case of homologous (but not "homogenous") organs from different species—like, for example, the four chambers of the heart in birds and mammals. Indeed, with Lankester, these chambers are not homogenous between the two classes, even with the heart described as such: There certainly was a common cardiac organ in the ancestor of birds and mammals; but it did not possess these four chambers, which appeared independently along the two lineages in response to identical conditions of life. All of these cases were baptized by Lankester under the name "homoplasy."[3] One can subsequently ask what the difference is between homoplasy and analogy (as it was defined by Owen). Lankester anticipates this and explains that analogy has a larger meaning than homoplasy: It can be applied to any organs that show a similarity in function (like the wing of a bird and that of an insect), while homoplasy additionally implies a certain morphological conformity—even if it is not part of a common heritage.[4]

Lankester thus considered the majority of cases of serial homology as being relevant to homoplasy; and he essentially attributes mechanical causes to them, or at least physiological ones: For example, the influence of forces or similar conditions—but also "correlations of growth," which he calls "homotrophies"—leading to the repetition of similar structures. He therefore strictly dissociates

---

3. This term has been borrowed by modern cladistics to designate any trait that is shared by two or many species as a result of convergent or parallel evolution.

4. In that respect, Lankester was probably inspired by the notion of laws of correlation as defined by Darwin in the fifth chapter of the *Origin of Species*.

interspecific homologies, which can possibly (when they are homogenies) have a phylogenetic meaning, from serial homologies, which do not have one.

Other biologists did not establish such a clear distinction. Several years after *On the Origin of Species*, Ernst Haeckel—a fervent partisan of Darwin in Germany—published a work in which he founded his conception of evolution on relative reflections of individuality (Haeckel, 1866; on Haeckel and his thought, see especially Rinard, 1981, and the two recent biographies: Di Gregorio, 2005, and Richards, 2008). From a morphological point of view, he recognized six levels of individuals within the composition of organisms. According to him, the morphological individual of the first order, the "plastid," is the smallest extant living unity. It can present itself in two forms: the cell (which has a nucleus), and the "cytode" (which does not have one). All organisms, whatever their degree of perfection, are composed of plastids. These plastids join with each other to form the morphological individuals of the second order, the "organs," in conformity with two essential laws: the law of aggregation and the law of differentiation (sometimes called the law of division of labor). In turn, the organs assemble and differentiate, thus making up the morphological individuals of the third order, the "antimeres," or "homotypical parts." Animals with bilateral symmetry are either composed of two antimeres, or of numerous pairs of antimeres, while radiated animals and plants have a larger number that are arranged in a circle: for example, four in most jellyfish, five in echinoderms. The constituent antimeres of a living being can be perfectly identical or can show certain differences (notably of an adaptive nature) between them.

Contrary to antimeres, which are placed alongside each other, the entities of the fourth order, "metameres" (or "homodynamic parts"), are arranged one after the other. For example, they can be found in the rings of segmented animals and the protovertebrates of vertebrates and their associated organs—as well as in the internodes of flowering plants, the segments of the arms of echinoderms, and so forth. As a result, these metameres (which are primitively independent) can arrange themselves in chains within the most perfect organisms to form an entity of the fifth order, a "person" (or a "prosope"). This last order often corresponds to individuals in the ordinary sense of the word. But in certain species, "persons" can join together to form the morphological individuals of the sixth order—that of "colonies" or "corms." This is notably the case with higher plants, siphonophores, and certain echinoderms such as ophiurids (each of their arms being the equivalent of a segmented worm; or rather, a "person").

Each of these six levels of morphological individuality, from plastids to colonies, can exist within a physiological independent organism—which Haeckel calls a "bion" (plural "biontes"). For example, among the biontes, protists correspond to the morphological individuals of level 1 ("plastids"); mollusks to the morphological individuals of level 4 ("metameres"); vertebrates or arthropods to the morphological individuals of level 5 ("people"). The degree of perfection of these biontes is as large as their morphological degree is high and their constituent parts are different, specialized, and dependent on each other—which is to say, as polymorphism, the division of labor, and the centralization of the organism are important.

All of these reflections on individuality anchored themselves in the idealist biology of the first half of the 19th century. But Haeckel added a temporal dimension. In fact, in addition to these morphological and physiological individualities (which are defined within space), Haeckel introduced a new notion—that of a temporal or genealogical individuality. This innovation responds to a double characteristic of the living world: the development of organisms (biontes), and the development of species—which are not immutable but rather are born, blossom, and die.

The first level of genealogical individuality corresponds *grosso modo* to the life of the bion from its initial state as a fertilized egg to the period of reproduction and the production of new similar biontes.[5] In the same manner as morphological individuals, these elementary genealogical individuals organize themselves into unities of higher orders. The first among these is the species—the genealogical individual of the second order (Haeckel, 1866, vol. 2, p. 30). But the species itself is subjected to variations over the course of geological time; thus, there exists a higher genealogical unit—the lineage or "phylum"—which is consequently the genealogical individual of the third order. And so there exists a hierarchy of genealogical individualities that is exactly in the same manner as the hierarchy of morphological individualities. Haeckel can therefore introduce the idea of a triple parallelism between the three levels of individuality, which is the heart of the "fundamental biogenetical law"—that is, the law of recapitulation. Ontogenesis is the repetition of phylogeny because these phenomena both correspond to the development of a genealogical individual: An individual of the first order in the case of ontogenesis, an individual of the second or third order in that of phylogeny. Thus the old idea of a parallelism between embryonic development and the scale of nature (see above) is adapted to the new transformist context.

Furthermore, there is a relationship between—on the one hand—the succession of the stages of development, whether phylogenetic or ontogenetic; and on the other hand, the hierarchy of morphological individuals. Every bion must successively pass through all the lower levels of individuality—starting with the

5. The animals and plants that do not reproduce sexually, or that show generational alterations between being sexual or asexual (such as aphids), pose a problem: Haeckel thus considers the genealogical individual as a set of accomplished stages in the course of the cycle of life.

plastid (which generally corresponds to the fertilized egg)—in order to reach its own level of completion:

> Man, for example, like all vertebrates, originates from an egg, a morphological individual of the first order. He reaches the second degree when, through the segmentation of the egg, a cell mass is produced which has the morphological value of an organ. With the formation of the embryonic anlage and the appearance of the primitive streak (the axial plan), it divides itself into two individuals of the third order (antimeres). With the budding of the primitive vertebrae begins the segmentation of the trunk, the division into metameres; and with this differentiation, it arrives at the formation of a person, the morphological individual of the fifth order—which remains in the state of a physiological individual. (Haeckel, 1866, vol. 1, p. 267)

Phylogenetic progress consequently corresponds to the acquisition of an additional level of individuality—which happens in an aggregation of identical parts that is possibly followed by a differentiation of these parts, and then a distribution of tasks among them.

In this fashion, Haeckel argues that the entire theory of evolution is dependent on the morphological sciences. He establishes, in a way, a quadruple parallelism: The ontogeny of a bion (from the moment the egg is fertilized) is a recapitulation not only of its phylogeny (from the ancestral "monera" onward) but also of the hierarchy of the actual species (starting with the protists); and in addition, this bion in its adult state shows in its morphology a hierarchy of its constituent parts (from the plastids upward) which is like a fixed image—in its own anatomy—of this entire progression. Consequently, if one considers a complex organism—such as an adult mammal—each of the morphological levels by which it is composed (for example, a metamere) possesses a counterpart (1) in the developmental history of this organism, (2) in its evolutionary history, and (3) in the series of the actual species.

As Haeckel indicates himself, these ideas are for the most part inspired by *Naturphilosophie*, notably by Oken. However, contrary to Oken, he places them within a radically materialist context, and thus the processes he describes are real and historical (and not idealist). In his system, serial homology takes on a strong meaning at a simultaneous ontogenetic, phylogenetic, and systematic level.

In the same way, the colonial theory of Moquin-Tandon and Dugès is reformulated many times within the evolutionist paradigm. From this point on, the idea that the composition of higher animals emerges from the coalescence of elementary animals is no longer conceived of as an ideal vision but rather as a real

evolutionary scenario. Many authors propose different versions of this new evolutionist colonial theory (see, e.g., Bernard, 1900; Brode, 1898); here, we focus on
one proposed by the French biologist Edmond Perrier, whose efforts are among
the most in-depth within this particular field.

Perrier was one of the leaders of French zoology and the neo-Lamarckian
school of thought at the end of the 19th century (see Blanckaert, 1979; Loison,
2010). Like Haeckel, he developed a very complete conception of the hierarchical organization of life (Perrier, 1881). According to him, the "protoplasm"
(that is, the living substance) "can only exist as masses of feeble size that are
distinct from each other." It thus organizes itself into plastids (i.e., into cells),
which cannot exceed a certain size. This is why all growing plastids are forced—
at a given moment—to divide into two new individuals. And from the moment
these individuals remain joined to each other, they acquire an additional property: organization. Since each plastid is given a certain variability that permits it
to adapt to the possible modifications of its surrounding environment, it establishes within the "societies of plastids" (which are the animals and the plants)
a diversification of structures and functions. Indeed, as long as plastids occupy
a similar position within the colony—as, for example, with the colonial algae
Volvox—they remain completely identical to each other. But if the size of the
colony grows, and it for example takes on the form of a gastrula, there is a rup-
ture of the initial symmetry; the different plastids no longer have the same relationships between them and with the exterior environment, and they differentiate as a result. And so those that are now in an external position—and that are
thus directly in contact with the exterior—acquire a function of protection and
exchange with the environment, forming a sort of ectoderm. At the same time,
the most internal plastids, which enjoy relative tranquility, are able to focus on
digestion and forming the endoderm. In correlation, the plastids lose their original autonomy, becoming more and more interdependent and having to manifest
a "solidarity" that is tighter and tighter. Perrier compares this phenomenon to
the case of human populations that are forced to split up to form colonies when
they reach a certain size.

According to him, the same process happens at a higher anatomical scale—at
the level of gathered plastids, which he names merids. Like plastids, they have
a limited size and are forced to divide when they exceed this limit, thus giving
rise to asexual reproduction through budding (or "metagenesis"). Just as plastids
that are produced through division can remain stuck together side by side within
merids, these same merids can form colonial associations ("zoids") in which they
demonstrate the laws of association and of the independence of anatomical elements, followed by those of specialization, of solidarity, and of the division of
physiological work. In the end, the same process replicates itself at a final level,

and many zoids can form a "deme"—which represents a colonial animal in the habitual sense.

In broad terms, the scenario imagined by Perrier hardly differs from the one offered by Haeckel, the primary contrast being that the number of degrees of organization is lower: four with Perrier (plastid, merid, zoid, and deme) and six with Haeckel (plastid, organ, antimere, metamere, person, corm). However, two original ideas emerge from his system. The first concerns how the elements of the colony are arranged according to the initial element's mode of life. One finds two clear types of arrangements: arborescent and linear. According to Perrier, in regard to a colony of merids, the choice between these two types is directly determined by the locomotive abilities of the initial individual (the "protomerid"). If at a given moment, this protomerid loses the faculty of moving and attaches itself to the ground or a foreign body, the colony resulting from its budding will be irregular (as we see with sponges, bryozoans, or composed ascidians). On the other hand, if the protomerid continues to move freely, it is "mechanically forced, so to speak, to form linear colonies whose transformations have produced the most highly elevated groups within the Animal Kingdom." If the size of fixed irregular colonies seems to be without limit (as we see in coral reefs), that of linear colonies is rapidly limited by the constraints of both locomotion, and the polymorphism reveals itself in an increasingly marked manner. Thus, these free linear colonies have a tendency to produce more coherent individuals, their merids are associated with each other more closely—to the point that one can sometimes nearly no longer distinguish the metameric nature of the whole (as, for example, in the case of mollusks; or, to a lesser degree, that of arachnids or of vertebrates). The nature of the protomerid is therefore a determining factor in the general organization of the colony. Furthermore, in the linear colonies, this initial individual will continue to rule over all of the merids that come from it—since it represents the head, or at least its most anterior part.

Another important aspect of Perrier's thought is his reflection on the meaning of the association of organisms, and of what he calls the "law of reciprocal adaptation." He compares the formation of zoids from primitively independent merids with the establishment of strong physiological links between certain organisms belonging to different species (as, for example, with lichen symbiosis). The analogy with human societies is a constant presence within the discourse of Perrier, which introduces an entire metaphoric terminology on this theme. It addresses the existence within the colony of "functionaries," "parasites in the sense that they do not look for or prepare their own subsistence, parasites in the sense that they end up by losing the organs that would be necessary for them to accomplish these acts; but whose usefulness for the colony can be of the first order, because they become the instruments of its power and definitively insure its victory in the struggle for life."

This idea of animal composition implies a redefinition of homologies. Flouting Owen's terminology, Perrier returns to the vocabulary of Geoffroy Saint-Hilaire. Like him, he reserves the notion of "homology" for merids within the same zoid, or for zoids within the same deme (that is, for repeated parts); while he deems as "analogues" only similar parts of different organisms that derive from one same lineage—provided that these parts are "of the same nature; that is, if they are formed by similar elements, and placed similarly." In fact, for him, "homology" (at least as he understands it) takes on a meaning that is both phylogenetically and ontogenetically broader than what he calls "analogy."

Undeniably, Perrier's ideas were greatly inspired by Haeckel; but they were inspired further still by French predecessors such as Dugès—from whom he borrowed the most important part of his colonial theory. Indeed, he readily admits to this (Perrier, 1884). His conceptions allow us to illustrate the manner by which the colonial theory—such as it had been formulated within an idealist context— could be reused without much change within a transformist framework.

Evolutionary morphology—such as it was established in the 1860s—thus found considerable success, which would last until the beginning of the 20th century. In this context, all anatomical and embryological studies (which, in a concrete sense, hardly differed from those that were performed within idealist biology) now aimed to reconstitute the phylogenetic history of living beings. For example, the Haeckelian law of recapitulation immediately suggested an embryological program of research, since the development of an animal was supposed to represent the entire evolution of its phylum.

The three examples of Lankester, Haeckel, and Perrier show that the notion of serial homology (whatever the name one gives it) was able—along with the ideas of special homology and the archetype—to go beyond the paradigm shift from idealist biology to transformist biology. Thus, in addition to being an element of continuity between the two halves of the 19th century, it also acquired great significance during this transformation in the life sciences—since it was now being interpreted as the consequence of a phylogenetic reality, which was either the colonial origin of organisms according to Haeckel and Perrier, or the existence of evolutionary constraints exercised in a similar manner at different places within one same organism according to Lankester. In the same fashion, all of the other problems concerning the repeated parts of organisms could be reformulated in the new context. The question of the existence of a primitive segmentation in vertebrates, and that of the relationship between this segmentation with the metameres of other animals (arthropods, annelids), now mixed in with the attempt to reconstitute the genealogical tree of animals and the common ancestor of each lineage. For example, the old vertebral theory of the skull reappeared across an entire series of anatomical and morphological work on the cephalic metameres of vertebrates and their origins—which gave rise to very lively debates (see Schmitt, 2004, pp. 361–411).

However, even more profound transformations were produced in the life sciences at the end of the 19th century—particularly with the wide diffusion of the experimental method in areas such as embryology, and the emergence of new fields and disciplines such as genetics. These changes initially threatened the status of homologies in biology; but in the end, homologies managed to maintain their relevance.

11.3. New Trends in Biology as a Challenge to Serial Homology After 1880

Attacks against the methods of evolutionist morphology began as early as the 1870s. Among other things, they were founded on the internal difficulties of the discipline (in particular on the tensions between embryology and anatomy), and on its inability to free itself from consensual phylogenetic scenarios. For example, if we look at the question of the origin of vertebrates, their ancestors were sometimes theorized to be ascidians, or annelids, or arachnids, or nemerteans; in fact, practically all of the invertebrate groups were proposed at one time or another as candidates. These divergences weakened the Haeckelian method—which was strongly contested by a growing number of biologists. For example, Wilhelm His denied the explanatory value of the law of recapitulation, scoffing at its extremely speculative character and instead advocating research into the proximal (mechanical) causes of development (His, 1874; on the crisis of evolutionary morphology, see Nyhart, 1995, pp. 243–305). Several years later, on a similar basis, Wilhelm Roux proposed a new approach to embryology—the "mechanics of development" (*Entwicklungsmechanik*)—which went on to find rapid success in Europe and in the United States (Maienschein, 1991a, 1991b). At the same time, the growing interest in the mechanisms of heredity culminated around 1900 with the founding of a new scientific discipline: genetics.

Initially, experimental embryology and genetics were sciences that were a priori not conducive to work concerning homologies: The former focused on mechanical causes, paying little attention to phylogeny, and advocating the study of determined experimental models rather than comparative studies concerning a large number of species; the latter was interested above all in the transmission of characteristics, and was hardly interested in the characteristics themselves and the structure of organisms in general. The rapid development of these two disciplines thus raised a challenge for the concept of homology in general—and for serial homology in particular. Hans Spemann, who was one of the principal representatives of *Entwicklungsmechanik* and was particularly known for his work on embryonic induction, thus made the following critique of the concept of homology in light of the recent advances in biology (Spemann, 1915). He demonstrated certain difficulties that it encountered when confronted with certain experimental results. For example, the embryologist Gustav Wolff had shown in 1895 that if one removed the lens from the eye of a salamander, an identical new one would be regenerated from the cells of the iris; however, they would have a totally different embryological origin—since the normal lens is produced by the thickening and invagination of the epidermis induced by the optic vesicle, while the iris comes from this vesicle itself. The two types of lenses were fundamentally identical once they reached completion, both in their structure and in their position in regard to the surrounding organs; but as they came from different germlayers, the notion of homology here made no sense for Spemann: According to him, one could consider only that the initial lens and the regenerated lens came from "anlagen endowed with identical potentialities" and were under the same "influence" from the optic vesicle.

Spemann's attitude clearly illustrates the difficulties that *Entwicklungsmechanik* posed for the concept of homology—such as it was defined by an evolutionist morphology that supposed a fundamental similarity (that is, a genealogical continuity between organs that was totally impervious to causal embryology). And thus his attempt to get rid of this notion and to replace it with the concept of homoplasy as defined by Lankester (see above): According to Spemann, homoplasy was more acceptable since it focused on a mechanistic rather than phylogenetic explanation of embryonic development.

Conversely, many authors associated the methods of experimental embryology with the phylogenetic approach and reintegrated homologies into the field of experimental embryology. In this regard, the case of homeosis is of particular note (see Schmitt, 2003). This concept had been introduced by William Bateson (1894) to characterize spontaneous variations that result in the replacement of one organ by another (for example, antennas that are replaced by legs in insects). Bateson, who was unsatisfied with the phylogenetic approach, was in search of a new method and wanted to catalog all of the natural variations in order to better understand the mechanics of evolution (see Peterson, 2008). But very rapidly, other authors applied this concept of homeosis to certain cases of abnormal regeneration where organs that had been removed for the sake of experimentation were replaced by others. These cases were well suited for mechanistic analysis in the sense that it was possible to investigate the physical (proximal) causes that led to the formation of one organ rather than another. But at the same time, homeosis, both spontaneous and regenerative, was interpreted as atavism—that is, a return to an ancestral state of the organ in question; it was therefore considered as evidence of a homology between the replaced organ and the replacing organ[6] (see, e.g., Bateson, 1894, pp. 29–30; Przibram, 1910; Schultz, 1905). Thus, thanks to these studies on homeosis, experimental embryology not only reembraced the concepts of serial and special homology but also enriched them with new criteria.

Likewise, when genetics emerged as an autonomous field after 1900, it first tried to reject the idea of homology, which seemed obsolete and did not make sense to it. A good example of this is given by the Scottish geneticist Francis Crew, who wrote:

> The feather of the bird is not necessarily the modified scale of a reptile but may be a distinctly different characterization based upon an entirely different genotype. A certain genotype results in a certain characterization—scales—; mutation—alteration—in this genotype results in a new genotype and thus leads to another characterization. The old genotype is transformed into the new but the old characterization is not transformed, it disappears and is replaced. Scales and feathers are not homologous structures—homology attempts to establish a similarity in origin and nature of structures seemingly different and is based on the assumption that during the course of evolution structures have undergone transformation yet remain fundamentally the same. In fact, this conception of homologous structures cannot be accommodated by the chromosome hypothesis until it can be experimentally demonstrated that the genes themselves can pass through a process of gradual modification. (Crew, 1925, p. 152; see also Duerden, 1923–1924)

For Crew, as for other geneticists of the first decades of the 20th century, the fundamentally discontinuous character of the genetic mutations—and above all, the absence of any simple relationship between the modifications of the genotype and those of the phenotype—made it impossible to establish homologies between macroscopic structures.

Thus, as it was conceived by evolutionary comparative anatomy, homology seemed to be very far from the scope and approaches of the first geneticists. But some of them soon found different ways to reuse this concept by adapting it to

---

6. This generally concerned serial homology, when an organ was replaced by another organ from the same animal; but also sometimes special homology, if the replacing organ was the same as the replaced one but with the characteristics of a different species.

the new field. As early as 1920, Alexander Weinstein introduced the notion of "homologous genes" to designate genes that belonged to different species but had similar positions on the chromosomes (in relation to other genes), and whose mutations had similar phenotypic effects. This definition was similar to the definition of homology by anatomists—but at the molecular level. The homology of genes, as with the homology of organs, expressed their essential similarity (resulting from a common evolutionary origin). Similar criteria could be used in order to identify homologies between organs or between genes—since Weinstein considered the relation of each gene with its neighbors, much like Geoffroy Saint-Hilaire with his introduction of the principle of connections. Those criteria could be used for genes of different species or for genes in one same organism, which was emphasized by Walter Fitch (1970) in the context of molecular evolution and the reconstruction of phylogenetic trees from the comparison of protein and DNA sequences: for this, Fitch introduced the notions of paralagous and orthologous genes, respectively corresponding to serial and special homology at the chromosomic scale.

But genetics was also able to acknowledge homology at the scale of organs very early on, using genes as a new kind of criteria. In the 1930s, some authors began to try to find signs of homologies by analyzing the genes that work during the development of the different organs. American zoologist Alan Boyden, professor at Rutgers University (and a rather heterodox evolutionist), suggested that "knowledge of gene action should therefore be able to illuminate and explain homology, and might even provide more precise criteria by which it might be recognized. [...] The facts clearly indicate that the structure of any particular organ is determined by the interaction of many if not by all of the genes present in the particular individual. [...] However discontinuous the changes in individual genes may be the fact is that through the interaction of many genes the organ is still essentially and fundamentally similar," so that "the determination of taxonomic relationships and of homology could then be based on methods of testing the identity of genes" (Boyden, 1935). Homology at the microscopic (genetic) level was thus connected to homology at the macroscopic (anatomical) level. This link was not only a theoretical one but also suggested new experimental approaches to find homologies and new programs of research.

But Boyden (1943) establishes a strict difference between serial homology and special homology. His argument is based on viewing development as a dialogue between a "nucleoplasm" (which is made up of all the genes) and the polarized and differentiated cytoplasm of the egg. For him, it is necessary to consider these two elements if one wants to define homology in a satisfying manner, and the error of geneticists such as Crew was to only focus on genes. In effect, homology (whether serial or special) deals with both cytoplasm and genes—but not in the same proportion. Thus, serial homologues (which Boyden prefers to call "homotypes") are the results of the action of identical genes in different zones of a polarized cytoplasm: The resemblance of these organs expresses the identity of the genes involved in their formation—while their differences are the consequence of a prior differentiation of the cytoplasm. On the other hand, special homology essentially explains itself through changes that have taken place at the genetic level: Homologous organs between two different species resemble each other because the properties of the zonation of the egg's cytoplasm vary only slightly from one species to another; while the differences correspond to genetic mutations that took place during the divergence of the two species in the course of evolution. Given that each organ is governed by a large number of genes, differences in genetic order express themselves in a much more discreet manner than cytoplasmic differences. Thus, according to Boyden, homotypes differ between themselves much more than homologues—as is shown when one compares the series of appendices in crustaceans: The difference between the mouth parts and the legs within one same individual is greater than the difference between the legs of different species.

On the contrary, the American ichthyologist Carl Hubbs rejects this separation of serial and special homology. He contests the fact that homotypes are more dissimilar than homologues; and he argues that since the similarities between the parts of one same individual "are presumably due to common genetic basis," they may be used "as evidence in phylogenetic research; they surely have the same sort of evolutionary significance that other types of homology furnish. [. . .] If we admit the homology between any scale x of an individual trout and any scale, say y, of a salmon, and between this scale y in the salmon and scale z in the trout, then how can we logically deny that a homology exists between scales x and z on the body of the same trout" (Hubbs, 1944, pp. 293–294). He additionally suggests to extend the notion of homology to functional aspects and processes which, in the same way as structures, can reveal similarities of ontogenetic and phylogenetic origin—whether within one same organism or between individuals or different species.

In spite of their differences, Hubbs and Boyden both testify to how the concept of homology was integrated into genetics within the first decades of the discipline's existence. In 1938, this process was already so advanced that the English biologist Gavin de Beer judged it necessary to critique what he considered to be an abusive use of the concept of homology within embryology and genetics. From a genetic standpoint, he emphasized that "*characters controlled by identical genes are not necessarily homologous*" since, for example, "in poultry [. . .], a gene controlling the formation of a crest of feathers also produces cerebral hernia. In the wild type, the gene behaves as a dominant in respect of crest, but as a recessive for hernia. The action of the gene on hernia can be entirely suppressed in certain gene-complexes (such as the Japanese silky fowl) while the production of crest is unaffected. There is no homology between crest and hernia." On the other hand, "*homologous characters need not be controlled by identical genes*," so that "*the homology of phenotypes does not imply the similarity of genotypes.* [. . .]" The interesting paradox remains that, while continuity of homologous structures implies affinity between organisms in phylogeny, it does not necessarily imply similarity of genetic factor or of ontogenetic processes in the production of homologous structures" (De Beer, 1938, pp. 66–67, 71).

But in spite of these reservations, embryological and genetic arguments continued to be used widely in establishing phylogenetic relationships between macroscopic structures. More generally, homologies—both special and serial—have remained at the center of every attempt to combine genetics, development, and evolution from the 1930s onward. The concept of homeotic genes played a major role in this regard: First, in the work of such pioneers as the Russian School of Genetics, Conrad Waddington, and Richard Goldschmidt (Schmitt, 1999, 2000, 2004, pp. 437–476); then with the emergence of developmental genetics after World War II; and last, with the rise of evolutionary developmental biology ("evo-devo") in the last decades of the 20th century.

Most of the conceptual foundations established at the beginning of the 19th century to describe the similarities within and between organisms therefore ended up being found (in an adapted format) in modern biology. And thus serial homology has a role to play at many levels—for example, throughout the studies concerning the developmental genes that control the formation and diversification of body parts, like the *Hox* genes. Even if most of these authors agree that "the issue at stake when we establish a comparison between different parts of the same organism is different from a comparison of structures belonging to individuals of different species" (Minelli & Fusco, 2013), the existence of striking morphological similarities between the parts of one same individual continues to be a subject of discussion. For example, this is the case with anterior and posterior limbs in tetrapods—which was the precise starting point for the work of Vicq d'Azyr in 1774, and which continues to be both a developmental and evolutionary enigma. To resolve it, Alessandro Minelli suggested introducing the concept of paramorphism "according to which the appendages are a sort of duplicates of the main body axis. If so, the serial homology between the anterior and posterior limbs of a vertebrate would derive, by transitivity, from the fact that both are partial homologues of the main body axis" (Minelli & Fusco, 2013). New versions of the colonial theory have recently even found defenders, like Thurston C. Lacalli (1997, 1999) and Ruth Ann Dewel (2000). The first proposed a theory of the evolution of deuterostomes according to which the group's ancestor had a two-generation cycle—where free and monomeric asexual phases alternated with fixed phases that were colonial and sexual. A free individual would have thus begun to bud before its fixation, thereby forming a sort of swimming dimer, which resembled certain actual echinodermal larvae; all deuterostomes would be derived from this dimer. Dewel suggests that the ancestor of all animals with bilateral symmetry could have been metamerized, and that this metamerization could have resulted from a unique event of "colonial individuation."

Work on repeated parts thus represents an important element of continuity in the life sciences starting from the end of the 18th century; and serial homology has succeeded in redefining itself many times over within different theoretical contexts and different disciplines leading up to the present—despite the considerable transformations biology underwent in the course of this long period. These transitions have not occurred without difficulty, and there have been biologists to indicate the ambiguities that have arisen at each stage: Lankester during the transition from idealist morphology to evolutionist biology, and De Beer during the integration of homology into genetics. These problems still remain a challenge to evo-devo today.

References

Amrine, F., Zucker, F., & Wheeler, H. (Eds.). (1987). *Goethe and the sciences: A reappraisal.* Dordrecht: Reidel.

Amundson, R. (1998). Typology reconsidered: Two doctrines on the history of evolutionary biology. *Journal of the History of Biology, 13*, 153–177.

Appel, T. A. (1987). *The Cuvier-Geoffroy debate: French biology in the decades before Darwin.* Oxford, UK: Oxford University Press.

Audouin, J.-V. (1824). Recherches anatomiques sur le thorax des animaux articulés et celui des Insectes Hexapodes en particulier. *Annales des Sciences Naturelles, 1*, 97–135, 416–432.

Baer, K. E. von (1827). Über die Kiemen und Kiemengefässe in den Embryonen der Wirbelthiere. *Meckel's Archiv für Anatomie und Physiologie,* Jahrgang 1827, 556–568.

Balan, B. (1979). *L'ordre et le temps: L'anatomie comparée et l'histoire des vivants au XIXe siècle.* Paris, France: Vrin.

Bateson, W. (1894). *Materials for the study of variation treated with especial regard to discontinuity in the origin of species.* London, UK: Macmillan.

Bernard, H. M. (1900). A suggested origin of the segmented worms and the problem of metamerism. *Annals and Magazine of Natural History, 7th ser., 6,* 509–520.

Blanckaert, C. (1979). Edmond Perrier et l'étiologie du "polyzoïsme organique." *Revue de Synthèse, 3ème sér., 95–96,* 353–376.

Boyden, A. (1935). Genetics and homology. *Quarterly Review of Biology, 10,* 448–451.

Boyden, A. (1943). Homology and analogy: A century after the definitions of "homologue" and "analogue" of Richard Owen. *Quarterly Review of Biology, 18,* 228–241.

Bräuning-Oktavio, H. (1956). Vom Zwischenkieferknochen zur Idee des Typus: Goethe als Naturforscher in den Jahren 1780–1786. *Nova Acta Leopoldina. Neue Folge, 18,* 1–144.

Bräuning-Oktavio, H. (1959). *Oken und Goethe im Lichte neuer Quellen.* Weimar: Arion.

Breidbach, O., Fliedner, H.-J., & Ries, K. (Eds.). (2001). *Lorenz Oken (1779–1851), ein politischer Naturphilosoph.* Weimar: Hermann Böhlaus Nachfolger.

Breidbach, O., & Ghiselin, M. (2002). Lorenz Oken and Naturphilosophie in Jena, Paris, and London. *History and Philosophy of the Life Sciences, 24,* 219–247.

Brode, H. S. (1898). A contribution to the morphology of Dero vaga. *Journal of Morphology, 14,* 141–180.

Bojanus, L. H. (1818). Versuch einer Deutung der Knochen im Kopfe der Fische. *Isis, oder encyklopädische Zeitung, 2,* 498–510.

Bojanus, L. H. (1819). Weiterer Beytrag zur Deutung der Schädelknochen. *Isis, oder encyklopädische Zeitung, 3,* 1360–1368.

Cahn, T. (1962). *La vie et l'œuvre d'Étienne Geoffroy Saint-Hilaire.* Paris: Presses Universitaires de France.

Carus, C. G. (1828). *Von den Ur-Theilen des Knochen- und Schalengerüstes.* Leipzig: Fleischer.

Churchill, F. (1991). The rise of classical descriptive embryology. In S. Gilbert (Ed.), *A conceptual history of modern embryology* (pp. 1–29). New York, NY: Plenum.

Cuvier, G. (1835–1846). *Leçons d'anatomie comparée* (8 Vols.; 2nd ed.). Paris, France: Crochard.

Cole, F. J. (1944). *A history of comparative anatomy from Aristotle to the eighteenth century.* London, UK: Macmillan.

Coleman, W. (1976). Morphology between type concept and descent theory. *Journal of the History of Medicine, 31,* 149–175.

Crew, F. A. E. (1925). *Animal genetics: An introduction to the science of animal breeding.* Edinburgh: Oliver and Boyd.

Cuvier, G. (1812), Sur un nouveau rapprochement à établir entre les classes qui composent le règne animal. *Annales du Muséum d'Histoire naturelle, 19,* 73–84.

Darwin, C. (1859). *On the origin of species.* London, UK: Murray.

De Beer, G. (1938). Embryology and evolution. In De Beer (Ed.), *Evolution: Essays on aspects of evolutionary biology, presented to Professor E. S. Goodrich on his seventieth birthday*, (pp. 57–78). Oxford, UK: Clarendon Press.

Desmond, A. (1982). *Archetypes and ancestors: Palaeontology in Victorian London 1850–1875*. Chicago, IL: University of Chicago Press.

Dewel, R. A. (2000). Colonial origin for Eumetazoa: Major morphological transition and the origin of Bilaterian complexity. *Journal of Morphology, 243*, 35–74.

D'Hombres, E. (2012). The "Division of physiological labour": The birth, life and death of a concept. *Journal of the History of Biology, 45*, 3–41.

Di Gregorio, M. (1995). A Wolf in sheep's clothing: Carl Gegenbaur, Ernst Haeckel, the vertebral theory of the skull, and the survival of Richard Owen. *Journal of the History of Biology, 28*, 247–280.

Di Gregorio, M. (2005). *From here to eternity: Ernst Haeckel and scientific faith*. Göttingen, Germany: Vandenhoeck and Ruprecht.

Ducrotay de Blainville, H. M. (1819). Note sur les Animaux articulés. *Journal de Physique, 89*, 467–472.

Ducrotay de Blainville, H. M. (1820). Sur la concordance des anneaux du corps des Entomozoaires hexapodes adultes. *Bulletin de la Société Philomatique*, Année 1820, 33–37.

Duerden, J. E. (1923–1924). Methods of evolution. *Science Progress, 18*, 556–564.

Dugès, A. (1832). *Mémoire sur la conformité organique dans l'échelle animale*. Montpellier, France: Ricard.

Fitch, W. M. (1970). Distinguishing homologous from analogous proteins. *Systematic Zoology, 19*, 99–113.

Geoffroy Saint-Hilaire, É. (1818–1822). *Philosophie anatomique* (2 vols.). Paris, France: Baillière.

Geoffroy Saint-Hilaire, É. (1824). Composition de la tête osseuse de l'homme et des animaux: Article premier. *Annales des Sciences Naturelles, 3*, 173–192.

Geoffroy Saint-Hilaire, É. (1825). Mémoire sur la structure et les usages de l'appareil olfactif dans les poissons, suivi de considérations sur l'olfaction des animaux qui odorent dans l'air. *Annales des Sciences Naturelles, 6*, 322–352.

Geoffroy Saint-Hilaire, É. (1830). *Principes de philosophie zoologique*. Paris, France: Pichon et Didier.

Geoffroy Saint-Hilaire, I. (1847). *Vie, travaux et doctrine scientifique d'Étienne Geoffroy Saint-Hilaire*. Paris, France: Bertrand.

Goethe, J. W. (1987). *Schriften zur Morphologie* (D. Kuhn, Ed.). Frankfurt, Germany: Deutscher Klassiker Verlag.

Guédès, M. (1969). La théorie de la métamorphose en morphologie végétale: Des origines à Goethe et Batsch. *Revue d'Histoire des Sciences, 22*, 323–363.

Haeckel, E. (1866). *Generelle morphologie der organismen* (2 vols.). Berlin, Germany: Reimer.

Hall, B. K. (Ed). (1994). *Homology: The hierarchical basis of comparative biology*. San Diego, CA, and London, UK: Academic Press.

Hubbs, C. L. (1944). Concepts of homology and analogy. *American Naturalist, 78*, 289–307.

Huschke, E. (1826). Über die Umbildung des Darmcanals und der Kiemen der Froschquappen. *Isis, oder enzyklopädische Zeitung*, Jahrgang 1826, 613–627.

Huschke, E. (1828). Über die Kiemenbögen am Vogelembryo. *Isis, oder enzyklopädische Zeitung*, Jahrgang 1828, 160–164.

Huschke, E. (1832). Über die erste Entwickelung des Auges und die damit zusammen hängende Cyklopie. *Meckel's Archiv für Anatomie und Physiologie, 6*, 1–47.

Huxley, T. H. (1858). On the theory of the vertebrate skull. *Proceedings of the Royal Society of London, 9*, 381–457.

Lacalli, T. C. (1997). The nature and origin of deuterostomes: Some unresolved issues. *Invertebrate Biology, 116*, 363–370.

Lacalli, T. C. (1999). Tunicate tails, stolons, and the origin of the vertebrate trunk. *Biological Review of the Cambridge Philosophical Society, 74*, 177–198.

Lankester, E. R. (1870). On the use of the term Homology in modern Zoology, and the distinction between Homogenetic and Homoplastic agreements. *Annals and Magazine of Natural History*, 4th ser., *6*, 34–43.

Latreille, P.-A. (1820). De la formation des ailes des insectes, et de l'organisation extérieure de ces animaux, comparée, en divers points, avec celles des arachnides et des crustacés. *In P.-A. Latreille, Supplément aux Mémoires sur divers sujets d'histoire naturelle des insectes, de géographie ancienne et de chronologie* (pp. 1–21). Paris: Lanoé.

Lelorgne de Savigny, M.-J.-C. (1816). *Mémoires sur les animaux sans vertèbres* (2 vols.). Paris, France: Déterville.

Le Guyader, H. (2004). *Geoffroy Saint-Hilaire: A visionary naturalist*. Chicago, IL: University of Chicago Press.

Limoges, C. (1994). Milne-Edwards, Darwin, Durkheim and the division of labour: A case study in reciprocal conceptual exchanges between the social and the natural sciences. In I. B. Cohen (Ed.), *The Natural Sciences and the Social Sciences* (pp. 317–343). Dordrecht, Netherlands: Kluwer Academic.

Loison, L. (2010). *Qu'est-ce que le néolamarckisme? Les biologistes français et la question de l'évolution des espèces, 1870–1940*. Paris, France: Vuibert.

Maienschein, J. (1991a). The origins of Entwicklungsmechanik. In S. F. Gilbert (Ed.), *A conceptual history of modern embryology* (pp. 43–61). New York, NY: Plenum.

Maienschein, J. (1991b). *Transforming traditions in American biology, 1890–1915*. Baltimore, MD: Johns Hopkins University Press.

Minelli, A., & Fusco, G. (2013). Homology. *History, Philosophy and Theory of the Life Sciences, 1*, 289–322.

Moquin-Tandon, A. (1827). *Monographie de la famille des Hirudinées*. Paris, France: Gabon.

Nyhart, L. K. (1995). Biology takes form: Animal morphology and the German universities 1800–1900. Chicago, IL: University of Chicago Press.

Nyhart, L. K. (2011). Individuals at the center of biology: Rudolf Leuckart's Polymorphismus der Individuen and the ongoing narrative of parts and wholes. With an annotated translation. *Journal of the History of Biology, 44*, 373–443.

Oken, L. (1805). *Abriß des Systems der Biologie von Dr. Oken. Zum Behufe seiner Vorlesungen.* Göttingen: Vandenhoek & Ruprecht.

Oken, L. (1807). *Über die Bedeutung der Schädelknochen.* Jena: Göpferdt.

Oken, L. (1809–1811). *Lehrbuch der Naturphilosophie* (3 vols.). Jena: Frommann.

Oken, L. (1829). Über das Zahlengesetz in den Wirbeln des Menschen. *Isis, oder encyklopädische Zeitung, 22*, 306–312.

Owen, R. (1843). *Lectures on the comparative anatomy and physiology of the invertebrate animals, delivered at the Royal College of Surgeons, in 1843.* London, UK: Longman Brown Green and Longmans.

Owen, R. (1848). *On the Archetype and Homologies of the Vertebrate skeleton.* London, UK: Van Voorst.

Perrier, E. (1881). *Les colonies animales et la formation des organismes.* Paris, France: Librairie de l'Académie de Médecine.

Perrier, E. (1884). *La philosophie zoologique avant Darwin.* Paris, France: Félix Alcan.

Perru, O. (2000). Zoonites et unité organique: Les origines d'une lecture spécifique du vivant chez Alfred Moquin-Tandon (1804–1863) et Antoine Dugès (1797–1838). *History and Philosophy of the Life Sciences, 22*, 249–272.

Peterson, E. L. (2008). William Bateson from Balanoglossus to Materials for the Study of Variation: The transatlantic roots of discontinuity and the (un)naturalness of selection. *Journal of the History of Biology, 41*, 267–305.

Peyer, B. (1950). *Goethes Wirbeltheorie des Schädels.* Zurich: Gebr. Fretz.

Przibram, H. (1910). Die Homöosis bei Arthropoden. *Archiv für Entwicklungsmechanik der Organismen, 29*, 587–615.

Rathke, M. H. (1825a). Kiemen bey Säugthieren. *Isis, oder enzyklopädische Zeitung,* Jahrgang 1825, 747–749.

Rathke, M. H. (1825b). Kiemen bey Vögeln. *Isis, oder enzyklopädische Zeitung,* Jahrgang 1825, 1100–1101.

Reichert, K. B. (1837). Über die Visceralbogen der Wirbelthiere im Allgemeinen und deren Metamorphosen bei den Vögeln und Säugthieren. *Müllers Archiv für Anatomie, Physiologie und wissenschaftliche Medizin,* Jahrgang 1837, 120–222.

Reichert, K. B. (1838). *Vergleichende Entwickelungsgeschichte des Kopfes der nackten Amphibien nebst den Bildungsgesetzen des Wirbelthier-Kopfes im allgemeinen und seinen hauptsächlichsten Variationen durch die einzelnen Wirbelthier-Klassen.* Königsberg: Bornträger.

Richards, R. (1992). *The meaning of evolution: The morphological construction and ideological reconstruction of Darwin's theory.* Chicago, IL: University of Chicago Press.

Richards, R. (2002). *The Romantic conception of life: Science and philosophy in the age of Goethe*. Chicago, IL: University of Chicago Press.

Richards, R. (2008). *The tragic sense of life: Ernst Haeckel and the struggle over evolutionary thought*. Chicago, IL: University of Chicago Press.

Rinard, R. G. (1981). The problem of the organic individual: Ernst Haeckel and the development of the biogenetic law. *Journal of the History of Biology, 14*, 249–275.

Rupke, N. (2009). *Biology without Darwin, a revised edition*. Chicago, IL: Chicago University Press.

Russell, E. S. (1916). *Form and function: A contribution to the history of animal morphology*. London, UK: Murray.

Schmitt, S. (1999). Les apports de la génétique russe pré-stalinienne à la connaissance des mutants homéotiques de la drosophile. In C. Galperin, S. F. Gilbert, & B. Hoppe (Eds.), *Fundamental changes in cellular biology in the 20th century: Biology of development, chemistry and physics in the life sciences*, (pp. 81–87). Turnhout: Brepols.

Schmitt, S. (2000). L'œuvre de Richard Goldschmidt: Une tentative de synthèse de la génétique, de la biologie du développement et de la théorie de l'évolution autour du concept d'homéose. *Revue d'Histoire des Sciences, 53*, 381–400.

Schmitt, S. (2003). Homeosis and atavistic regeneration: The "biogenetic Law" in Entwicklungsmechanik. *History and Philosophy of the Life Sciences, 25*, 193–210.

Schmitt, S. (2004). *Histoire d'une question anatomique: La repetition des parties*. Paris, France: MNHN.

Schmitt, S. (2006). *Aux origines de la biologie moderne: L'anatomie comparée d'Aristote à la théorie de l'évolution*. Paris, France: Belin.

Schmitt, S. (2009). From physiology to classification: Comparative anatomy and Vicq d'Azyr's plan of reform for life sciences and medicine (1774–1794). *Science in Context, 22*, 145–193.

Schultz, E. (1905). Über atavistische Regeneration bei Flußkrebsen. *Archiv für Entwicklungsmechanik der Organismen, 20*, 42–47.

Sloan, P. R. (1992). Introduction. In R. Owen, *The Hunterian Lectures in Comparative Anatomy May–June, 1837* (P. R. Sloan, Ed., pp. 1–72). London, UK: Natural History Museum Publications.

Spemann, H. (1915), Zur Geschichte und Kritik des Begriffs der Homologie. In C. Chun & W. Johannsen (Ed.), *Die Kultur der Gegenwart, Allgemeine Biologie* (pp. 63–86). Leipzig, Germany: Teubner.

Spix, J. B. von (1815). *Cephalogenesis, sive capitis ossei structura, formatio et significatio per omnes animalium classes, familias, genera ac aetates digesta, atque tabulis illustrata, legesque simul psychologiae, cranioscopiae ac physiognomiae inde derivatae*. Munich: Hübschmann.

Vicq d'Azyr, F. (1778). Mémoire sur les Rapports qui se trouvent entre les usages & la structure des quatre extrémités dans l'Homme & dans les Quadrupèdes. *Mémoires* de mathématique et de physique, tirés des registres de l'Académie Royale des Sciences, 1774, 254–270.

Vogt, C. (1842). *Untersuchungen über die Entwickelungsgeschichte der Gebürtshelferkröte Alytes obstetricans*. Solothurn: Jent & Gassmann.

Wells, G. A. (1967). Goethe and the intermaxillary bone. *British Journal for the History of Science, 3*, 348–361.

Weinstein, A. (1920). Homologous genes and linear linkage in Drosophila viridis. *Proceedings of the National Academy of Sciences, 6*, 625–639.

Winsor, M. P. (1976). *Starfish, jellyfish, and the order of life: Issues in nineteenth century science*. New Haven, CT: Yale University Press.

INDEX

adaptability driver 15, 19, 142
Alberch, P. 73, 201, 211, 217, 226
Amundson, R. 20, 69, 70, 83, 171, 202, 281, 327
antimere 329
Aristotle 20, 25, 56, 201, 243, 268, 280, 317
artificial selection 146
Avalon explosion 22, 192

Baldwin effect 19, 58, 101, 113, 130, 142, 228. *See also* adaptability driver
barnacles 219
Bateson, P. 3, 8, 11, 18, 78, 97, 137
Bateson, W. 19, 43, 138, 336
behavior 4, 11, 19, 46, 59, 84, 140, 146, 202, 221, 263, 284
biased change 240, 254
Brigandt, I. 167, 212
by-product selection 74

cadherin 193
Caenorhabditis elegans 118
Carroll, S. 10, 21, 57, 70, 102, 159, 167, 192
cell-cell adhesion 193
Central Dogma 4, 58, 120, 130
chance 14, 24, 37, 55, 91, 121, 143, 239, 282

cheetah 141
Cheverud, J. M. 76, 93, 217, 289
cis-regulatory evolution 10, 169
cis-regulatory hypothesis 170, 218
constraint 8, 16, 53, 69, 73, 114, 199, 214, 223, 334
content (of evolutionary theory) 164
control theory 189
convergence 24, 49, 80, 89, 222
    of embryos 199
correlated progression 218
covariance 27, 76, 91, 123, 128, 227, 283, 287
cultural transmission 270
Cuvier, G. 39, 56, 85, 322

Danchin, E. 6, 9, 12, 17, 68, 90, 111, 288
    111, 280
Darwin, C. 8, 14, 20, 26, 38, 42, 78, 93, 117, 128, 137, 140, 151, 189, 213, 268, 280, 322
Dawkins, R. 4, 41, 59, 141, 191, 211, 266, 284
de Vries, H. 43, 304
Democritus 25, 242
Depew, D. 2, 8, 12, 14, 37, 71, 113, 165, 188
developmental form challenge (DFC) 20, 160, 167

developmental inertia 24, 224
developmental systems theory (DST) 7, 70, 265, 294
DNA 4, 26, 47, 118, 148, 201, 240, 253, 263, 282, 338
DNA damage repair (DDR) 253
Dobzhansky, T. 1, 15, 44, 54, 89, 139, 151, 282, 304
drift 44, 54, 71, 89, 115, 163, 171, 213, 252
Drosophila melanogaster 18, 114
dynamical patterning module (DPM) 22, 81, 192

ecology 1, 11, 21, 69, 84, 168, 180, 290
embryonic hourglass 22, 192, 197
empirical adequacy 159, 183
epigenetics 13, 57, 148, 149, 172
evo-devo 8, 20, 21, 24, 28, 69, 78, 99, 151, 159, 212, 227, 31. See also evolutionary developmental biology
evolutionary developmental biology 16, 24, 29, 73, 84, 159, 172, 201, 212, 340. See also evo-devo
evolvability 9, 23, 45, 51, 130, 163, 181, 211, 273, 304
explanatory schemes, alternative 69, 82, 91
exploration 131, 146, 284
  adaptive 24
  embedded 18, 120
  heritable physiological 18, 117, 122
Extended Evolutionary Synthesis (EES) 162, 180, 304
extension (of the Modern Synthesis) 8, 20, 69, 161

fecundity 159, 183, 319
Fisher, R. A. 1, 43, 71, 77, 89, 217, 282
fitness 3, 19, 27, 41, 73, 97, 118, 145, 164, 175, 202, 214, 280, 297
Ford, E. B. 46
fractionation of evolution 4, 25

Galton, F. 19, 42, 138, 281
gastrulation 193
Gayon, J. 26, 39, 70, 166
gazelle 140
gene networks 10, 216, 249
gene regulation 15, 50, 170
gene-centric 16, 196, 264, 285, 288
genetic accommodation 86, 101, 115, 147, 220
genetic assimilation 12, 17, 24, 101, 11, 146, 215
genotype-phenotype map 23, 80, 197, 216
Gerhard, J. 9, 83, 99, 248, 197, 224, 248
Gilbert, S. 9, 45, 58, 68, 283, 295
Gluckman, P. 9, 148
Godfrey-Smith, P. 106, 273, 280
Goethe, J. W. von 28, 319
Goldschmidt, R. 19, 24, 139, 210, 224, 340
Gould, S. J. 8, 16, 46, 68, 102, 138, 165, 201
gradualism 13, 22, 40, 189, 195, 224
Griffiths, P. 7, 68, 113, 131, 148, 265, 284, 294

Haeckel, E. 4, 29, 329
Hall, B. 8, 73, 82, 199, 214
Hallgrímsson, B. 214
Helanterä, H. 6, 11, 27, 137, 148, 267, 280, 285
heredity 26, 27, 53, 101, 163, 227, 228, 280, 335
  as covariance 27, 227, 287, 296
  as development 27, 294, 296
  as information 27, 289, 296
  as transmission 26, 227, 281, 296
heterochrony 17, 78, 102, 169
holozoans 192, 227
homology 80, 4, 214
  serial 28, 317
  special 318
Hox genes 19, 169, 199, 340

Huneman, P. 13, 16, 56, 68, 122
Huxley, J. 12, 16, 40, 54, 71, 91, 190
Huxley, T. H. 325
hypermutability 116, 150

idealistic evolutionary biology 22, 188, 194
idealistic morphology 28, 317
idealization 21, 179, 240
inheritance 2, 6, 13, 25, 27, 41, 71,
    163, 263
    ecological 27, 162, 301
    epigenetic 6, 11, 18, 19, 27, 56, 116, 149,
        162, 263, 269, 285, 289, 300
    extended 26, 162, 264, 273
    extragenetic (also non-genetic) 6, 27,
        72, 89, 115, 162, 263, 269, 283
    Lamarckian 18, 48, 112, 190
    mechanisms of
    Mendelian 1, 137, 190, 281
    neo-Lamarkian 112
    non-Mendelian 6, 12
    pluralism 6, 9, 266
innovation 2, 10, 23, 172, 181, 198, 213,
    241, 257, 330
intergenerational transmission 11, 27,
    269, 273

Jablonka, E. 6, 56, 68, 72, 111, 117, 149,
    197, 263, 284
Johansen, W. 4

Kant, I. 197, 245
Kauffman, S. 120, 272
Keller, E. F. 21, 45, 148, 176, 197, 249
Kirschner, M. 9, 83, 168, 212, 248, 273
Kuhn, T. H. 164, 182

lateral inhibition 194
Laubichler, M. 161, 218
learning 9, 19, 57, 121, 143, 265
Lewontin, R. C. 4, 12, 44, 69, 79, 141,
    179, 201, 221, 268, 280, 305

Love, A. 2, 10, 13, 20, 69, 88, 159

Mameli, M. 2, 268, 283, 303
Maynard Smith, J. 4, 8, 46, 75, 217,
    221, 284
Mayr, E. 6, 15, 16, 20, 44, 56, 71, 111, 190,
    213, 281, 294
Mendel, G. 190
Merlin, F. 6, 11, 13, 25, 26, 91, 111,
    122, 263
metamere 331
methylation 6, 19, 150, 263, 300
Minelli, A. 8, 23, 161, 174, 211, 317, 340
mobility 145
Moczek, A. 9, 214, 252
modularity 81, 169, 172, 214, 218, 285
Monod, J. 25, 58, 239
Morgan, T. H. 42, 54, 113, 142, 281
Müller, G. 1, 3, 8, 21, 26, 52, 57, 68, 79,
    82, 117, 159, 201, 213, 263, 281
multistable dynamics 193
mutagenicity 17, 116, 130, 139, 163
mutation 3, 9, 16, 38, 71, 82, 115, 215,
    249, 255, 300
    random 18, 25, 51, 90, 97, 122
    non-random (directed, biased) 56, 97,
        120, 149
mutational assimilation 116
mutationism 81, 190

Naturphilosophie 317, 323
Neo-Darwinism 15, 41, 111, 130, 137
Newman, S. 8, 11, 13, 19, 22, 69, 79,
    188, 214
niche construction 11, 27, 59, 86, 141, 181,
    202, 221, 264, 301
novelty 9, 25, 44, 50, 99, 112, 139, 164,
    213, 221, 249, 252, 255

Odling-Smee, J. 11, 19, 68, 89, 141, 221,
    264, 291
Oken, L. 188, 320

oscillation 189
overlap, material 27, 271
Owen, R. 80, 188, 317

Paley, W. 45, 138
phylotypic stage 23, 192
Pigliucci, M. 1, 8, 52, 68, 82, 112, 121,
    159, 201, 212, 220, 263, 304
plasticity 9, 38, 49, 58, 70, 86, 88, 99, 114,
    143, 172, 197, 220, 248, 251, 304
  genotypic 58, 101, 252
pleiotropy 43, 78, 169, 216
pluralism, inheritance. *See* inheritance
pluralism, theory 20, 84, 161, 176
Pocheville, A. 6, 9, 12, 17, 101, 111
predictability 92, 197, 225
Price equation 11, 27, 96, 267, 283, 297
Provine, W. 8, 20, 26, 43, 70, 111,
    190, 282
purpose 15, 25, 37, 197, 239, 245

ratchet, evolutionary 141
reactive genomes 250
regulatory profiles 120
repeated parts 28, 317
robustness 73, 89, 215, 240

Saccharomyces cerevisiae 98, 119.
    See also yeast
Saint Hilaire, E. G. 28, 188, 318
saltation 24, 190, 223
Schmalhausen, S. 44, 113, 197, 305
Schmitt, S. 11, 13, 28, 225, 317
Schrödinger, E. 200
sequestration hypothesis 129
SET (standard evolutionary theory)
    159, 175
sexual selection 50, 70, 140
Simpson, G. G. 170, 5, 44, 53, 113, 122,
    142, 190
Sober, E. 5, 70, 122

speciation 18, 47, 99, 137, 164
species 19, 38, 48, 70, 138, 165, 199, 214,
    222, 243, 270, 283, 317
Spencer, H. 14, 38, 113, 221
stability 9, 13, 138, 240, 253, 263, 300
stabilization 118, 228, 295
Sterelny, K. 143, 252, 265, 266
sticklebacks 48, 81
structure (of evolutionary theory)
    21, 165, 176
systematicity 159
systematics 1, 21, 160, 168, 180

teleology 47, 245
tempo (and mode) 24, 102, 221
theory presentation 21, 178, 182
timescales (evolutionary) 116, 122, 174
toolkit genes 10, 22, 59, 169, 192
trade-off 16, 72
transmission 2, 6, 26, 27, 41, 71, 118, 162,
    197, 227, 240, 263, 269, 281
  horizontal 270, 273
  vertical 267, 270
transmission bias 298, 306
type 28, 54, 84, 85, 319
  cell 196, 292

Uller, T. 6, 11, 13, 27, 137, 148, 227, 260,
    271, 280
unpredictability 92, 16

variation 8, 11, 14, 17, 23, 37, 68, 71,
    92, 99, 111, 137, 164, 201, 252,
    280, 336
  acquisition of 267, 274, 305
  blind 111, 121, 125
  environmental 220
  environmentally induced 128
  epigenetic 116, 130, 300
  genetic 74, 102, 112, 121, 212
  isotropic 8, 91

von Baer, C. E. 324
von Bertalanffy, L. 240

Waddington, C. H. 113, 121, 146, 168,
    197, 215, 340
Wake, D. 8, 80
Wallace, A. R. 137, 189
Walsh, D. 8, 10, 12, 24, 56, 70, 89, 99,
    103, 197, 239
Waters, C. K. 3, 112, 176
Weismann, A. 15, 41, 58, 111, 281

West-Eberhard, M. J. 9, 58, 70, 86, 99,
    115, 147, 197, 213, 252, 295, 304
widow bird 140
Wnt 193
Wolffia arrhizal 219
Woodward, J. 125, 247
Wright, S. 56, 71, 89

yeast 18, 98

zoonites 325